the Sacred Sphere

Thank you for Listening

EXPLORING SACRED CONCEPTS AND COSMIC CONSCIOUSNESS THROUGH UNIVERSAL SYMBOLISM

Paul D. Burley

BEAVER'S
POND
PRESS

Cover Photo: Cloud Peak Medicine Wheel by Paul D. Burley

ISBN 10: 1-59298-406-1
ISBN 13: 978-1-59298-406-0

Library of Congress Catalog Number: 2011937796

Printed in the United States of America

First Printing: 2012

16 15 14 13 12   5 4 3 2 1

Cover and interior design by James Monroe Design, LLC.

BEAVER'S
POND
PRESS

Beaver's Pond Press, Inc.
7104 Ohms Lane, Suite 101
Edina, MN 55439-2129
(952) 829-8818
www.BeaversPondPress.com

To order, visit www.BeaversPondBooks.com
or call (800) 901-3480. Reseller discounts available.

For Rose Marie and Donald

Contents

Part I: The Geometry of Myth

Part 2: A Case for Cross-Cultural Symbolism

List of Figures

Acknowledgements

Any book worth reading requires a team commitment centered on a vision. I'd like to thank you, the reader. You make our team complete.

A heartfelt thank you to everyone at Beaver's Pond Press for encouraging and supporting this project. I'm grateful to have a terrific editor in Kellie Hultgren; I accept full responsibility if any burden remains for the reader with regard to comprehending the ideas and digesting the details presented herein. Jay at James Monroe Design prepared the interior and exterior design; all credit goes to him. And thanks to Sara at Lien Public Relations for spreading the word.

I am honored by the friendship and help of Everett Poor Thunder. Hau Kola! Pilamaya yelo!

Thank you to David Tetley for his friendship and advice, and to Shane Bofto for the many times we discussed everything from quantum physics to metaphysics. I thank my father Donald for his love, and my brothers Jon and Phil for being, well, you know, brothers. Thank you to my daughters Elin and Elizabeth, my love is with you always; I've always appreciated you lending an ear when I thought I had something important to say. And thank you to my wife Nancy for urging me to put my thoughts on paper, understanding that it is what I must do. You are my inspiration.

I thank my late mother Rose Marie. She knew that the bulk of human history remains hidden from view, yet can still be heard if we listen. Her presence remains with me. She is the morning star and the evening star.

She knows the truth, as we all will know in our own time. Thank you, Mom.

The Creator and the spirits are all around us. They willingly guide us through life if we accept their invitation. They contacted me on one very special day eight years ago. They approached me, and set me on the path on which I journey today. I thank them for their guidance and assurance. I'm listening.

Paul Burley
Eveleth, Minnesota
July 2011

Introduction

One of the rocks . . . is placed at the center of the round altar; the first rock is . . . at the center of everything. . . . The second rock is placed at the west . . . the next at the north, then one for the east, one for the south, one for earth, and finally the hole is filled up with the rest of the rocks, and all these together represent everything that there is in the universe.

BLACK ELK

The Sacred Pipe

Etymologists trace the origin of the word *sacred* back to Proto-Indo-European (PIE), the ancestral language of the majority of European and Indian languages. The speakers of that language are believed to have lived in the Pontic-Caspian steppe of Eastern Europe and Central Asia between the fifth and fourth millennia BCE, and perhaps as early as the last glacial maximum over twenty thousand years ago. The PIE word *sak* meant "to sanctify" or "to make pure or holy." This is the root for the Greek word *saos* ("safe") and Latin term *sanus* ("sane, sound, whole"). Also, according to *A Latin Dictionary* by Charlton T. Lewis and Charles Short (Oxford: Clarendon Press, 1879), *sanus* meant "dedicated or conse-crated to a divinity, holy." Similarly, the Latin *sacrum* referred to deities and that which they controlled; note also the Latin *sacer* ("priest") and *sanctum* ("something located spatially apart").

The word *sphere* is derived from the Latin *sphæra* and the Greek *shaira*, meaning a ball or globe. A similar concept is *orb*, from the Latin *orbis*, which refers to a circle. Obviously, the difference between a circle and a sphere is a matter of the number of spatial dimensions under consideration. A circle is a two-dimensional (planar) representation of a line with constant curvature and forming a closed figure. In other words, a circle consists of all points in a plane that are equidistant from a given point (the center of the circle). In contrast, a sphere consists of all points in three-dimensional space that are located equidistant from a given point (the center of the sphere).

Based on etymology, then, the phrase "sacred sphere" suggests a globe-like form, whole and pure, consecrated with divine intent. Earth is often referred to as the sacred sphere in the context that it is our home, the Mother Goddess, Gaia. However, equating the sacred sphere with Earth places an unfortunate geographical limit on what I believe to be the true nature of the symbolism.

Why is the circle so prevalent as a religious symbol, across the millennia and around the world? Is there something intrinsic to a circle that has been recognized by people throughout history as an important symbolic representation of key religious concepts? Do important spiritual or religious concepts symbolized by the circle cross time and space to all people seeking to understand themselves and their role in the universe? These are questions I address in this book.

Interestingly, we can envision a sphere in two dimensions as a circle. No matter how we rotate a sphere, its orthographic form in two dimensions is a circle. This is true for a sphere and a sphere only. Placing a cube on paper, we can outline a square, but if we rotate the cube ever so slightly the two-dimensional outline becomes a rectangle or even a hexagon. We lose the symmetry of a square. The case is similar for a tetrahedron, from which we can outline on paper a triangle that is immediately reconfigured if we rotate this fundamental polyhedral structure the slightest amount.

It is this uniformity of shape that leads geometers (or geometricians, people who study geometry) to conclude that the sphere has perfect symmetry. The two-dimensional circle shares this quality. This recognition of perfect symmetry and the fact that every path therein leads one back to the starting point seems to have taken hold within human consciousness very early in our evolutionary development, even before the first *Homo sapiens* (Latin: "knowing man") walked the earth. As I demonstrate in this book, evidence for this is found in stone tools from the Paleolithic Era, over one million years ago. Even today the circle is one of the most common pictographic symbols found in the many and varied cultural traditions of the world.

In my study of the circle as a sacred symbol in ancient and indigenous cultures, I found that in virtually every case the message is intended to communicate an understanding of vital relationships between human beings, between ourselves and the world (the world often being a metaphor for the universe), and ourselves and the creator. As you will discover in this book, those relationships have been depicted symbolically in a unique spherical geometry across the world and throughout human history. Recognizing this fact was an epiphany for me.

I set out to study ancient and indigenous symbolism eight years ago. The journey began in 2003, when I encountered a previously undocumented Native American sacred site in the Rocky Mountains of Wyoming. Discovery of that site, which I later realized was detailed in tribal mythology, piqued my curiosity about the true meaning of the circle as a pervasive symbol of religious concepts. I should note that the site remains sacred and continues to be used for indigenous ritual purposes. Everyone should recognize the sanctity of any and all sacred sites encountered during our time here on Earth. Our ancestors, our brothers and sisters, and our descendants deserve nothing less.

I propose that ancient and indigenous cultures studied four-dimensional space and time, and the knowledge they received led them beyond understanding of the physical, mundane world and toward knowledge of metaphysical reality. Knowledge leads to wisdom, and wisdom can take us to the center of the world, the center of the universe, and the center of the heart. For ancient and indigenous cultures the circle symbolized life's journey to understand who we are and why we are here. The circle is the two-dimensional representation of the four dimensions of space and time, and so it is also a symbol of the sphere and our lifelong journey of discovery. The Sacred Sphere—a system symbolized by the geometry of what I call the *disdyakis dodecasphere* and its various components, most often the circle—is the sacred geometrical construct recognized by innumerable cultural traditions for thousands of years as the vehicle for receiving knowledge and gaining universal understanding.

The purpose of this book is to demonstrate that cultures from around the world, during all ages, expressed a common ideology communicated through each culture's form of geometrical symbolism. The symbols vary between cultures and fluctuate through time. Nonetheless, each symbol represents a facet of a geometry held in common by the framework of cultures across space and time. In this book I describe the fundamental geometrical structure that has been implied by these facets, the faces of sacred cultural traditions. To accomplish this goal I provide a pragmatic examination of the circle and the sphere as they symbolize concepts in religion and spirituality. Those concepts pervade human experience, suggesting that there is an innate comprehension of the nature of the universe and our place within it.

This book is the result of research in academic and theological literature and ethnographic records, including numerous mythologies and cultural traditions. However, as tribal elders and shamans know, we cannot gain knowledge from books. Books provide information. Knowledge is gained through experience, a turning away from the literature to open ourselves to the energies that fill the natural world and build relationships with other people. Most importantly, we learn by paying attention to what the universe has to say to us. When we open our hearts and our ears, we receive vital information that no book can contain. And so it is in this light that I hope you will listen to the words on these pages and continue on your own journey of discovery. I believe you will begin to see the Sacred Sphere in places you might not have expected, just as I have.

The findings presented in this book are founded on four premises.

1. Each prehistoric and historic human culture developed from a matrix of dynamic environmental parameters affecting the thoughts and actions of the individual and social group that served as the context for that culture. Those environmental parameters include chemical, physical, and biological conditions, whether readily discernable and understood by the population or not. As such, they include the geology, geography,

hydrogeology, meteorology, climate, and ecology affecting human livelihood. But they also include the human experience, our perceptions of the physical world and the cosmos, our relationships with each other, life and death, and the potential for an afterworld.

2.   Each human culture developed means of compiling the information gathered from life experiences in the form of mythologies and traditions.

3.   Mythologies are filled with both exoteric and esoteric information and purpose, typically evaluated at four levels of understanding: physical, intellectual, emotional, and spiritual (PIES). Each of those levels is important in its own right, but most are often hidden from view, buried within overt storylines filled with a cast of characters and a variety of situations. A detailed understanding of every level of meaning is difficult to achieve without considerable time and effort and assistance from those who know it well.

4.   An important aspect of each mythology and tradition is effaced in the geometry used by cultures to express important concepts associated with space, time, and relationships between both the physical and metaphysical universes.

In Part I pictographic symbols having one dimension (the point), two dimensions (lines, curves, and the circle, such as the Sacred Hoop of Native American lifeways), and three dimensions (spheres) are introduced with particular attention to the circle and the sphere. Part II presents examples of both circular and spherical symbolism from various spiritual traditions (Vedic, Sumerian, Celtic, Egyptian, Jewish, Christian, Mayan, and Lakota), as well as Freemasonry and magic.

In Part III, I introduce a specific polyhedral form, the disdyakis dodecahedron, as the archetype from which many other religious symbols may be derived, including the World Tree. Projecting the image of the disdyakis dodecahedron onto the surface of a sphere, we find nine great circles in perfect symmetry. I call this structure a *disdyakis dodecasphere*. Based on the work of R. Buckminster Fuller, I describe the physical and energy characteristics of that structure, a symmetry that reflects the basic physical and spatial characteristics of not only the five Platonic solids of sacred geometry, but also Fuller's cuboctahedron (vector equilibrium). This leads to an understanding of the fundamental nature of the disdyakis dodecasphere, its subdivisions within space, and the application of those subdivisions (such as the creation of the Sacred Hoop in two dimensions and the two fundamental shapes in space) within both sacred and mundane frameworks. I then build upon this information with descriptions of complex geometrical configurations that have long religious histories around the world. These include the Seed of Life, the Flower of Life, and the Tree of Life, as well as my discovery of the relationship between the disdyakis dodecasphere and the geometrical structure of the Otz Chiim, the Tree of Life at the esoteric heart of Kabbalah within Jewish mysticism. This previously undocumented finding may have significance regarding the source of the Otz Chiim as an evolutionary step beyond the Tree of Life as understood by ancient Egyptians.

Part IV concerns the relationship between humans, Earth, the cosmos, and time, the fourth dimension. Part V reviews the role of the shaman, the impact of hallucinogens on knowledge gained through shamanic journeys, and the meaning of the Sacred Sphere as it relates to my own journey through life.

In writing this book I have attempted to provide relevant details sufficient to support the thesis. However, space on the page is necessarily too limited to provide all of the information I would prefer to share, and I urge you to pursue further information. I have included a selected bibliography in the hope that you will seek out more on this topic. It is always difficult to communicate without some measure of subjectivity. I apologize if my personal fascination with the relationships between human cultures, engineering, and the earth sciences gets in the way of what I intend to be an objective review of the facts and an impartial presentation of the results and conclusions of my studies. My hope is that you will find this work entertaining and worth some measure of consideration. After all, it is not only about our past, but our present and future as well.

Hierology (Greek: *hieros*, "holy"; *logos*, "word" or "reason") is the study of sacred ideas or sacredness, particularly as it relates to religious or spiritual truths found in many cultural traditions or belief systems, usually encompassing sources beyond Western philosophy or religion. This book highlights the importance of a specific yet previously unrecognized structure—the Sacred Sphere—within the context of hierology. The sources of much of the information presented in the first part of this book include a breadth of historical records, not the least of which are ancient sacred texts (hierographs), sacred engravings (hieroglyphs), and cuneiform script imprinted on clay tablets dated to as early as 3000 BCE. Just as important, however, are cultural artifacts from the Paleolithic Era, the architecture of ancient and indigenous societies, and recent records including biographies, ethnographies, and photographic records of pictographic symbols from around the world. Within this mix there are questions, contemplations, assertions, and conclusions that can only come from a curious mind and an open heart.

People have recognized certain fundamental, sacred concepts for thousands of years, concepts symbolized by the circle as a common proxy for the sphere. This finding may be surprising and a thorn in the side of some in the academic community. Nonetheless, a universal and eternal understanding of sacred relationships appears bound to a unique spherical geometry, implying that there may be a cosmic consciousness within us that allows us, through the use of symbols, to communicate the universal cyclicity of birth, life, death, and rebirth to those whom will listen. This is a cord binding our past to our present and future. This is the sacred nature of the sphere.

Mitakuye Oyasin (*Lakota*: We are all related)

Part I

The Geometry of Myth

Chapter 1
The First Dimension

What the process of symbolic representation presumably does is to abstract some quality common to both referent and symbol and allow one to perceive more clearly, more imaginatively, a particular type of relationship, uncluttered by details of the referent, or reduced in magnitude to comprehensible dimensions.

RAYMOND WILLIAM FIRTH
SYMBOLS: PUBLIC AND PRIVATE

There is another theme, in which man is thought of as having come not from above but from the womb of Mother Earth. Often, in these stories, there is a great ladder or rope up which the people climb. The last people to want to get out . . . grab the rope, and snap!—it breaks. So we are separated from our source. In a sense, because of our minds, we actually are separated, and the problem is to reunite that broken cord.

JOSEPH CAMPBELL
THE POWER OF MYTH

Communication is a fundamental aspect of the universe. It is a process of transferring information from a sender at one location to a receiver at another location. It is also a one-dimensional transmission of information beginning at one moment of time and ending at another moment. Obviously, then, communication is a function of both space and time, able to occur anywhere—from microcosm to macrocosm—over any length of time—from a fraction of a second to billions of years. Knowledge of the underlying shape of space, with length defined in terms of distance between spatial points or elapsed time, leads to an understanding of the fundamental relationship between mass and energy. That relationship in turn provides a foundation for understanding the use of geometry to communicate vital information—sacred relationships—through graphic symbolism.

Listening to Stones

It changed me in an instant. One moment I was walking off my anxiety about the next morning's planned ascent to the top of a mountain peak in east-central Wyoming. A moment later my anxiousness was gone. What lay before me amidst a jumble of cobbles, boulders, and slabs of rock was unmistakable (Figure 1.1). The near-perfect circle of stones was a human creation. A small pile of rocks lay at the center of the circle. Four lines of cobbles extended between the hub and perimeter stones at roughly 90 degrees from one another, dividing the circle into quadrants. Four lines of smaller stones bisected the quadrants to form an octet of pie-shaped areas. This was a Native American medicine wheel!

I have investigated geotechnical and environmental conditions of properties across the United States and Canada for more than thirty years. Commercial, industrial, residential, farms, and ranches—I've evaluated all sorts of historic development. Most of those projects concerned the physical and chemical conditions of soil, bedrock, surface water, and groundwater. I can easily recognize surface and subsurface impacts caused by human activities rather than natural disturbances. Humans tend to leave their mark wherever they go. I've encountered prehistoric tools: stone choppers and scrapers used for butchering animals such as bison, which roamed freely across the pristine foothills of the Absaroka and Bridger mountains of Montana not so long ago.

Figure 1.1: Native American medicine wheel, Wyoming (from Burley, 2006).

Another time, while conducting an environmental investigation of a ranch alongside the upper Yellowstone River, I was crossing a gravel-covered hilltop when I found river-worn stones nearly half buried in the rocky soil. The stones formed a somewhat circular shape about three or four feet in diameter. Each stone was weathered by exposure to the sun, water, and wind for an unknown number of years. They had obviously been placed there by human hands long, long ago. I was curious about the formation, but I had work to do. It was late in the afternoon, and I needed to finish my reconnaissance of the property and then drive a hundred miles home that day. I didn't give the circle of stones much further thought.

My discovery of a Native American medicine wheel in the alpine wilderness of Wyoming, however, was different. I recognized that I was standing in a sacred place. The stone circle had settled into the soil about an inch or more, so I knew that it had been in place for many years, certainly longer than several decades. This was authentic Native American symbolism. It was the Sacred Hoop.

Why was that medicine wheel constructed there, at an elevation of over ten thousand feet, where it could be accessed during only three months of the year due to the severe alpine climate? Exactly who built it, and when? Is it still in use? I have been on a path to find the answers to those and many other questions since I first laid eyes on that circle of stones. This path leads me on a journey that will continue for the remainder of my life. The journey is not easy, and the path is not short.

I looked at the stones, and they looked back. Pay attention. Listen. The stones are speaking. Most people do not hear them, but you too can take the journey. It is a path that humans have walked for well over a million years. It is a path of culture and tradition communicated to everyone, everywhere, at any time. Listen. Listen to the stones.

What is "Life"?

The traditional definition of the term "*Life*" is the definition we all learned in school. It is "the condition that distinguishes organisms from inorganic objects and dead organisms, being manifested by growth through metabolism, reproduction, and the power of adaptation to environment through changes originating internally."[1] This definition seems quite reasonable. Most people would likely accept it. However, not everyone finds it adequate.

Professor Arnold De Loof of Katholieke Universiteit Leuven in Belgium proposes an alternative definition of life that he says is simple, plausible, and logical. He notes that Abel Schejter and Joseph Agassi consider "an adequate definition of *life* should meet the following criteria: 'Apart from its being trite and uninformational (circular, to use a traditional term), it should be neither too wide nor too narrow; it should not exclude living things and it should not include dead ones. Furthermore, it should not make biology part and parcel of chemistry and physics,'[2] meaning that there should be room for an immaterial dimension."[3] In response, De Loof states that a plausible definition of *life* must consist of "a single integrative and universally valid concept."[4] This concept is essentially the organization of life forms as compartments that send and receive information—communicate—and thereby solve problems. De Loof suggests that the content of information transferred between living matter is immaterial, but transport of the message from sender to receiver is vital if the problem is to be solved.

> Very important but usually overlooked because of being counterintuitive to some extent: any act of communication is, by definition, also a problem solving act . . . Indeed, any message, whatever its nature, is always written [in] coded form. Hence any receiver faces the problem as to how to subtract information

from the message, and how to respond to it. The more complex the level of compartmental organization is, the more acts of communication are executed at any given moment.[5]

In De Loof's opinion the concept or word *life* should be a verb (action) rather than a noun (thing). He finds that life is "nothing else than the total sum (Σ) of all acts of communication/problem-solving (C) executed by a given sender-receiver compartment (S), at all levels of compartmental organization at moment *t*."[6] In other words, life is a function of sender-receivers in time. It is in constant change and may be defined as the sum of all communications: $L (S, t) = \Sigma C (S, t)$.[7]

De Loof's definition of life addresses the very heart of the question of what differentiates living matter from nonliving matter while also encompassing issues of growth through metabolism, reproduction, and adaptation to the environment through internal changes. Each of these conditions involves communication within an organism or with other organisms, and De Loof concludes that this definition satisfies each of the criteria set by Schejter and Agassi. However, De Loof has missed an important point concerning life as a communicating entity. By his definition, life consists only of the sum of communication/problem-solving, which he defines simply as a coded message. Let's simplify further and define this "coded message" as information. We can even say that information consists solely of energy, such as sound waves in air or photons of light traveling through space, or even gamma rays emitted from the poles of the black hole at the center of the Milky Way.

Consider a sound wave emitted from a stereo speaker with an amplitude and frequency sufficient to break a pane of glass. The sending compartment (stereo speaker) effectively communicates with a receiving compartment (pane of glass) via a coded message (energy of some amplitude, wavelength, and direction) through time, resulting in the solution to a problem (breaking the glass). The sender-receiver compartments are obviously not representative of life, yet the definition of communication between parties is satisfied. Obviously there remains a problem with De Loof's definition of *life*. The issue is, of course, one of *intent*. Humans communicate with intent to transmit information from sender to receiver. The cells of an animal or plant communicate with intent by transferring chemicals and energies between each other to perform specific functions based on the biological make-up of the organism; that is what biological entities do. However, stereo speakers, based on our current knowledge of chemistry and physics, do not of themselves communicate with intent. With this understanding, De Loof's definition of *life* should be revised as shown in Figure 1.2.

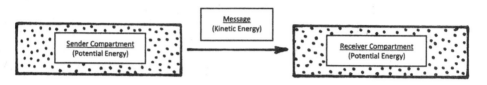

Life is communication/problem-solving with intent (i)   or   L (S, t) = C (i)

Figure 1.2: The architecture of communication with intent.

Interestingly, De Loof's definition of *life* does not require that the sender or receiver consist of matter or have a body like that of an organism, but only that it include some sort of "compartment." A compartment is a portion or division of space, like a separate room, but it can also be a particular aspect or function of something. This allows sender-receiver compartments to be associated with physical life forms as well as metaphysical entities such as the soul or spirit. The Holy Spirit is without body, yet it communicates with intent between God and man. The Holy Spirit, then, is life. Native Americans believe that every thing in existence, animate and inanimate, contains a spirit capable of communicating with intent. In this conception, stones are alive. Earth is compartmentalized (inner core, outer core, mantle, crust, water bodies, air) and communicates via energy and mass transfer (magma, plate tectonics, winds, and waves). If Earth has a spirit, then it too is life. This is the Earth Goddess, Gaia, Mother, and Grandmother. Is there any reason why a pane of glass, the sun, or a black hole cannot be life? Is it simply a matter of our lack of understanding or need for proof of spirit and intent? How much proof do we need? What is the risk of not knowing the truth?

What do the Ancients and indigenous peoples have to say about this? And are we listening?

Levels of Meaning

We can improve our understanding of ancient and indigenous cultural and religious traditions when we recognize the breadth of meaning, expressed symbolically, regarding relationships between human beings, the physical and metaphysical world, the Creator, and the cosmos. Those relationships concern matter, energy, space, and time, and therefore geometry as well. Surprisingly, archeological and anthropological evidence indicates that early humans expressed these relationships after developing an understanding of the mechanics of the universe and of the unity of matter and energy. In fact, material symbolism has been found in utilitarian and artistic forms created over one million years ago! This seems absurd. Intellectual and technical progress assumes that ancient understanding of the mechanics of the universe was naïve and ignorant of the facts. Our current understanding of the mechanics of the universe is based on scientific theories developed during the twentieth century, including the general theory of relativity and quantum theory. Even today, few people understand these theories, much less the mathematics behind modern physics.

Of course, until quite recently, humans have not explored the physical universe with technologies that could peer into the far reaches of space, nor did people express their understanding of the universe through complex mathematical analysis. Rather, their contemplations, conceptualizations, and characterizations applied qualified understanding to produce symbols based on personal experience. And yet, ancient symbolism yields strong parallels with quantitatively based theories and conclusions of modern science.

The theory of relativity commonly attributed to Albert Einstein actually consists of two theories, those of special relativity and general relativity. In his paper "On the Electrodynamics of Moving Bodies," Einstein discusses the structure of space–time formed from the three dimensions of space and the fourth dimension of time. The paper challenges classical Newtonian mechanics and proposes that space–time is curved. By 1915 Einstein developed mathematical expressions relating this curvature in space–time with mass, energy, and momentum. This was the theory of general relativity.

Quantum theory, initially developed and published by Max Planck and others soon after the theory of general relativity, addresses the wave–particle duality of matter and energy. It is most often applied at the submolecular scale (atoms, electrons, leptons, photons, and so forth) but has applications at extreme energy levels, too. Essentially, the theory derives from the observation that some changes in subatomic physics occur only in discrete amounts. The amounts are called quantities or quanta. One example of this effect is the probability density, or mathematical location, of an electron around the nucleus of an atom. This effect can be quantified as the electron's angular momentum.

Certainly the wave–particle duality of matter and energy and angular momentum at the subatomic level have parallels in physics on the macrocosmic, universal scale. The effects are evident in Einstein's equation relating energy and mass ($E = mc^2$) and our ability to measure time based on the rate of rotation or vibration of a given mass (energy). Scientists thus consider the theory of relativity and quantum mechanics to be reasonable models of the physical universe. In the last century mathematicians and physicists have suggested that there might be more to the universe than is evident by our mundane, earthly experience. Some models suggest more than three spatial dimensions. Other models suggest that our universe might be two dimensional, like a window pane, with potential for interacting with other similarly shaped universes. For most people, however, the physics of matter and energy appears constrained by the three dimensions of space coupled with the fourth dimension of time. And this appears so regardless of scale. But whether we accept theories concerning the nature of the universe that are based on higher mathematics, or merely trust in our five senses operating in the 'real' world, we can conceptualize the spatial and temporal relationships we experience in our daily lives with one fundamental idea:

Experience is the effect of interrelated geometrical energy events in space and time.

Experience results from the transmission (communication) of energy from Point A to Point B or from Time A to Time B (Figure 1.3). If life is the sum of communications transmitted with intent between sender and receiver, then we can think of experience as an effect of geometrical energy events *communicated* through space and time. In other words, energy at one point, at one location, at a particular time, moves in some manner to another location and time.

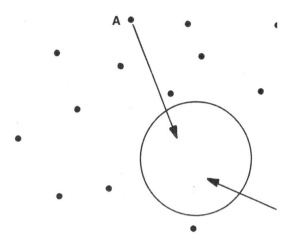

Figure 1.3: Experience results from transmission (communication) of energy from Point A to Point B or from Time A to Time B.

This energy particle or wave can be expressed by a given physical parameter (such as mass, light energy, heat energy, information, and so forth) and direction that changes as an effect of other relationships (energy events) before it arrives at its destination. This means that communication moving from one event (sender) to another (receiver) can be represented as a vector in space and time. Therefore communication can always be interpreted in terms of geometry. In our minds we form a picture from this energy transmitted between us. The message may be one of sound, light, smell, taste, or touch. It is sensed, interpreted, and symbolized. This is experience.

Raymond Firth notes that since symbols can be used to communicate complex, layered meanings (ideas) "at levels of reality not accessible through immediate experience or conceptual thought,"[8] they can be revelatory for the receiver. He suggests that meanings may not be understood by the inexperienced and therefore instruction may be necessary before the symbol is understood. Curiously, there are certain symbols that represent complex, esoteric ideas that persons completely segregated from the sending culture would not be expected to understand, and yet the symbols are understood when received. Indeed, some symbols appear to be universal and used by different cultures to represent similar if not identical ideas.

If no previous communications have occurred between those cultures, if there has been no overt interrelationship between events, then how can we explain the common understanding? In these instances, it is apparent that there is a common experience, whether physical or metaphysical, that binds people together. All humans transmit and receive information in the same ways. We interpret and symbolize information in like fashions. We have a common consciousness that helps guide our thoughts and actions.

The word *symbol* (Greek: *symbolon*) means a token, pledge, or sign by which one infers a thing; it is the putting together of something that has been divided. The word is derived from *symbellein*—to throw together.[9] A symbol can be something as mundane as a physical object, flavor, odor, sound, picture, or word. In all cases it is representative of something—an idea . The symbol represents the idea and communicates it by association. It may or may not resemble the idea in form or function, but there is usually a common knowledge between sender and receiver allowing communication to be received and understood via some convention between events. Firth states, "The essence of symbolism lies in the recognition of one thing as standing for (re-presenting) another, the relation between them normally being that of concrete to abstract, particular to general. The relation is such that the symbol by itself appears capable of generating and receiving effects otherwise reserved for the object to which it refers—and such effects are often of high emotional charge."[10]

This is certainly the case for symbols used for religious purposes. As Joseph Campbell noted, "The word 'religion' means *religio*, linking back. If we say it is the one life in both of us, then my separate life has been linked to the one life, *religio*, linked back. This has become symbolized in the images of religion, which represent that connecting link."[11] Religious symbolism acts to link us back and puts us together with what has been sundered. It helps us return to and reconnect that with which we were formerly united, and in that reunification we can experience an epiphany of the mind and the heart. This epiphany can occur at several levels.

Part II of this book considers a number of pictographic symbols and the mythologies that use them to convey meaning. Depending on the symbol, graphic symbolism can be used to communicate more than one level of meaning. It typically requires significant time and effort to understand and appreciate multiple levels of symbolic significance, and it is beneficial to keep in mind four particular levels of meaning as you read this book. The levels are:

Physical: The ideas being communicated involve specific physical realities such as geographic markers, parts of the body, or the mechanics of the physical universe; often the most transparent level of meaning.

Intellectual: Ideas are communicated through critical thought that leads to enlightenment. Conceptions are often buried at shallow depth within the overt storyline or the graphic nature of the symbol.

Emotional: The level in which the message is "charged, non-neutral" in its emotional value, often promoting bliss, joy, and unification.

Spiritual: "Projecting the mind towards the Absolute,"[12] this level of understanding links the receiver to the unknowable Creator of the universe, relating the physical to the metaphysical.

Mythology is one means of communicating knowledge. The value of mythological symbolism lies in its ability to successfully communicate these four levels of meaning—physical, intellectual, emotional, spiritual (PIES)—from sender to receiver. Communicating that knowledge via mythological symbolism from generation to generation creates tradition. The overt storyline present in myth consists of symbols expressed verbally. Recognition of this purpose of mythology leads to a greater understanding of the culture and knowledge of the people who created the mythology based on their observations and interactions with the world.

Extracting physical, intellectual, emotional, and spiritual knowledge is difficult when we are far removed from the source culture. Further difficulties arise when we are left with only the written mode of communication. The tone and inflection of voice, accentuation of particular words and sounds, and nonverbal communications such as facial expressions and physical gestures are lost in the written word. Imagine trying to reduce an array of verbally communicated cultural traditions to the written word, and then from the written word to a solitary symbol. All detail is lost. The lone symbol must encapsulate central, fundamental ideas. If it is unsuccessful in doing so, then we cannot expect to extract the associated details. Chosen carefully, the appropriate symbol will transmit a central concept, but also lead the recipient toward understanding and appreciating the details.

The Illusion of Geometry

We live in a world of three dimensions. The sun moves across the sky from east to west. The earth has two poles, north and south. And we all know that what goes up must come down. One universe, three dimensions. With basic Euclidean geometry we can use points, lines, and planes to create a myriad of geometric shapes: triangles, rectangles, circles, ovals, pyramids, cubes, spheres, and so forth. With modern technology we can model the forms and functions of far more complex, real things, like flowers, rivers, or Earth. All of these things can be studied in three dimensions. Technology makes the study of our universe and our place in it relatively simple and easy to understand: rational.

R. Buckminster Fuller had many intellectual and technical talents, not the least of which were mathematics, engineering, and architecture.[13] He is perhaps best known as the inventor of the geodesic dome, the Montreal Biosphere and Spaceship Earth at Disneyworld, Orlando, Florida, being two well-known examples. Fuller's study of polyhedra led him to identify the fundamental shape of space. He certainly understood Euclidean geometry extremely well. But he didn't accept it.

First of all, it was obvious to Fuller that all real things have substance: they fill space, or at least some part of space, no matter how small. Things also have shape, and the shape of an object is not necessarily dependent on its size. Fuller realized that size and shape in the real world are contrary to one of the basic tenets of Euclidean geometry: that points, lines, and planes have no dimensionality to them. Figure 1.4 shows a point, line, and plane as they are typically illustrated in geometry textbooks.

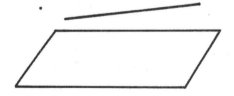

Figure 1.4: Common depictions of a point, line, and plane in Euclidean geometry.

However, points, lines, and planes are not real things. A spatial point has no thickness, no radius, no circumference, no size or shape. A line or line segment has no width and no depth; we can assign a length to a line segment, but that length is comprised of an infinite number of points extending from the beginning point to end point—and none of those points has a length, width, or depth. A plane, defined by the orientation of three points in space, has no thickness; it has an infinite length and width, but it is comprised of an infinite number of points that themselves have no thickness, size, or shape. This means the geometrical point, line, and plane are illusions. They are geometrical constructs that have no physical correspondence to real things in the universe. They are merely symbols that geometers define in a manner that allows us to study, qualify, and model the real universe.

Small Bits of Spin

An electron is one of a number of fundamental constituents in subatomic particles. The electron has the lowest mass of any stable electrically charged particle of any type. It is very small, and very resilient. However, an electron's most valuable feature is its electric charge; without it there can be no chemistry, no physics, no biology—no universe. And yet, as important as it is to the universe, the electron has no known substructure. At best we can assume it to be a point particle with a point charge and no spatial extent. This suggests that while we can see physical matter, such as a tree, we can't see the fundamental components of matter, such as electrons. Figure 1.5 illustrates this idea. If we can't see the building blocks of matter, then in fact we can't see the tree itself. Instead we see the energy effects of the subatomic particles that make up the physical universe.

An electron behaves as a wave of energy. Quantum mechanics describes the wavelike property of an electron as a complex-valued mathematical function: the wave function. Squaring this function provides the mathematical location (probability density) of an electron or other subatomic particle with respect to time. And this is the point where we jump into the realm of universal reality being stranger than fiction. Nuclear physicists contend that empty space has the ability to continually create pairs of virtual particles that quickly annihilate each other. From nothing, something. From something, nothing. Basically the idea is one of combining energy within a necessary amount of time to allow particles of matter to be created. But this borrowing of energy from space occurs within a very short timeframe. The energy required for these particles to be created can only be borrowed and returned within a span of 1.3×10^{-21} seconds. Our virtual electron is virtually nonexistent.

Figure 1.5: Instead of seeing the tree itself *(left)*, we receive waves of energy in the form of photons emitted from the tree, allowing us to visualize physical features of the tree even though it is at some distance away from us *(right)*. This cause and effect is similar to the physics of gravity. Without physical contact, energies (of unknown form) interplay between objects such as Moon and Earth, the effect being that both masses are pulled toward each other.

Theoretically, if we slice the probability density of an electron, or the wave of energy represented by the electron itself, into separate pieces, we will divide the standard unit of charge or energy of the electron into two parts. Perhaps a photon will be emitted and the electron will fall to a state of lesser energy. Photons are elementary particles of energy. They represent the fundamental unit of light. As the quantum (unit) of electromagnetic interaction, they are the basic building block of all other forms of electromagnetic radiation. Importantly, from the subatomic (microcosmic) to the universal (macrocosmic), the photon is the force carrier for the electromagnetic force. It is in constant motion, interacting with matter and energy at long distances. And like all other elementary particles, including the electron itself, the photon is governed by quantum mechanics such that it exhibits a wave–particle duality.

A photon has no mass and no electric charge[14]; it is a unit of energy that resembles a field instead of a discrete particle. However, it does carry spin or angular momentum with either right-handed or left-handed rotation. These two orientations correspond to the two potential polarization states of the photon, each of them circular.[15,16] So we can envision a photon, this fundamental quantum of light, as a spinning unit of electromagnetic energy. Whoa! Now we have something extremely interesting going on!

• Photons exhibit spinning, circular, angular momentum.

- Electrons spin, with a probability density suggesting an orbital path around the nucleus of an atom.

- Earth's atmosphere (a complex gas), in tandem with the rotation of the earth, produces weather (such as tornados and hurricanes).

- Earth's spin causes liquids to rotate, as we observe in whirlpools and lavatories.

- The moon rotates around the earth.

- The earth rotates around the sun.

- The sun rotates around the center of our galaxy.

- The center of the Milky Way is occupied by a spinning black hole.

- The Milky Way is spinning through a Local Group of more than thirty nearby galaxies.

- Our Local Group is rotating through space.

- The untold numbers of galaxies in the universe are accelerating away from the Milky Way, our Local Group, and each other.

The point here is that matter and energy, at every scale, experiences angular momentum—spin. Each particle of matter, every wave of energy, is interacting with all other matter and energy such that the entire universe moves not in straight lines, but in curves about itself. Experience, indeed, is the effect of interrelated geometrical energy events in space and time.

What about the universe as a whole? There is no scientific consensus regarding the shape of the universe, nor on whether it is rotating about some central axis. However, it is safe to say that the shape of the universe might exhibit curvature that might even be circular. What is obvious about the mechanics of the universe—from micro- to macrocosmic—is that it operates at all scales based on relationships (communication) between three elements of the universe: 1) space and time (or space-time); 2) matter and energy; and 3) physical law. In other words the dimensions of space are expressed by space-time, the substance of space consists of various forms of matter and energy, and humans characterize the structure of the universe in terms of physical laws.

Shapes of Things

Fuller recognized there were inherent relationships between space-time, matter, and energy throughout the universe. After breaking with one of the basic concepts of geometry he made it his career to discover the fundamental unit of space and to describe the qualities of space in terms of geometry and energy. He coined the term "synergetics" in his book with that title[17] to capture the depth and meaning of his work. Fundamentally, Fuller realized that:

- Geometry is the science of systems, which are defined by relationships.

- Geometry is therefore the study of relationships.

- A system is necessarily polyhedral (existing in three-dimensional space), consisting of a finite aggregate of interrelated events (in time).

- Relationships can be polyhedrally diagramed in an effort to understand the behavior of a given whole system.

All movements of mass and energy in the universe are affected by a consortium of physical effects: gravity, electromagnetic fields, and so forth. Those effects, or forces, are felt at all scales and act to change the direction of movement. Consequentially, the so-called straight line we learn about in high school geometry has no meaning, no bearing on the real universe. Mass and energy have effective lengths of travel, a length, width, and depth of field. With sufficient resolution we find that all trajectories are non-uniform as a result of geometrical relationships that are effective at all points. As Fuller said, "Physics has found no straight lines."[18] These forces, pulling and pushing between points, are commonly modeled in physics and engineering as vectors.

Interactions of matter and energy, existing within the parameters of three-dimensional space and one-dimensional time, are bound by physical laws. As we've seen, size does not matter in this regard. How about shape? Is there a fundamental shape to space? Fuller's study of the patterns of nature was a fundamental hypothesis of synergetics, his idea that "nature's structuring occurs according to the requirements of minimum energy, itself a function of the interplay between physical forces and spatial constraints."[19] He found that space does have specific properties. It has shape. This implies that space has a dimensionality to it, and the number of dimensions must exceed one. As stated by Arthur Loeb in his introduction to *Space Structures*, "Space is not a passive vacuum, but has properties that impose powerful constraints on any structure that inhabits it. These constraints are independent of specific interactive forces and hence geometrical in nature."[20] For example, the simplest geometrical form enclosing space is made up of four triangles. Three points define a plane and, of course, the corners of a triangle. Add in a fourth point not located within that plane and you create the simplest of three-dimensional forms, a pyramid with four sides, each comprised of a triangle. Nothing simpler will work in three-dimensional reality. This limitation in the fundamental form of nature is not a function of material or size, but rather the inherent nature of space. Fuller said, "*natural* is what nature permits."[21]

The Fundamental Shape of Space

Imagine our pyramid constructed from four non-coplanar points. Those points define the locations of the four three-dimensional corners (vertices) of the pyramid. There are six edges located between four vertices. Three edges define the limits of each of the four triangular faces of our pyramidal polyhedron. If we space the vertices such that each face is an equilateral triangle, then the polyhedron is called a tetrahedron, one of the five Platonic solids. We can slice the tetrahedron into four equivalent parts that each extend from the former center of the tetrahedron to a respective face. Orienting one of these parts such that it appears as a pyramid with shallow sloping edges, we notice that it has a threefold rotational symmetry: We can slice it into three pieces, each identical to the others. Surprisingly, each of these rather irregular-looking shapes can be sliced to produce two smaller parts exhibiting equivalent vertices, edges, faces, and volumes. In fact they are mirror images of the other. Fuller called this shape the *A-quanta module* (Figure 1.6), the fundamental shape of space.[22]

The universe may be observed to operate within the confines of laws and principles, but in fact, the universe could not care less how humans characterize, categorize, quantize, and qualify our perceptions of reality. And yet, in accordance with one of these observed physical laws, there appears to be a constraint to shape. The A-module can be any size, but it has only one shape allowing for only two mirror images of itself. We can

cut it into two smaller pieces, but those pieces do not express a fundamental shape of space.

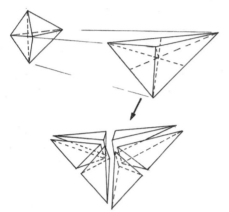

Figure 1.6: The A-module derived from the tetrahedron *(upper left)*. Each tetrahedron consists of twenty-four A-modules (based on Edmondson, 1986).

Fuller also found that another Platonic solid, the regular octahedron, can be subdivided in similar fashion into eight equivalent pyramidal parts, that can be sliced into six equivalent parts such that each one forty-eighth of the octahedron is what Fuller called the lowest common denominator (LCD). Then, by subdividing a quarter tetrahedron into six equal irregular tetrahedra (three of them mirror images of the other three), Fuller discovered what he called *B-quanta modules*. Removing the A-module from the octahedron's LCD produces this second funda-mental shape (Figure 1.). A- and B-quanta have exactly the same volume, but very different shapes. Fuller wrote, "Neither A nor B can be made from the other; they are fundamentally distinct, complementary, equi-volume modules."[23]

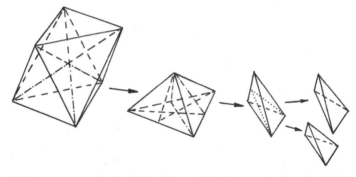

Figure 1.7: The quarter tetrahedron fits inside an eighth of an octahedron. The LCD is one forty-eighth of the octahedron. The LCD can be subdivided to form an A-module and a B-module having equivalent volumes (based on Edmondson, 1986).

Fuller had defined the fundamental shape of space, the A-module for the tetrahedron and both A-module and B-module for the octahedron. However, neither of these two modules nor any combination thereof can be used to fill the void of space. In his study of energy and form, Fuller visualized "the space-filling array of spheres in *cubic packing*"[24] and then envisioned "interconnecting the centers of all spheres—and then eliminating the spheres. Two collinear radii meeting at the tangency point between adjacent spheres form one unit vector—the length of which is equal to the sphere's diameter (Figure 1.8)."[25]

Figure 1.8: Cubic packing of spheres. The distance between centers of adjacent spheres is equivalent to one sphere diameter. The resulting matrix of unit vectors (isotropic vector matrix) consists of a space-filling network of octahedra and tetrahedra (based on Edmondson, 1986).

Fuller named the array of vectors resulting from this exercise the "isotropic vector matrix [or IVM], a space-filling network of continuously alternating octahedra and tetrahedra."[26] This cubic packing of spheres forms vertices (the spheres' centers) situated in identical locations within the matrix. Fuller then discovered that a quarter-tetrahedron and an eighth-octahedron, placed in such way that they share an equilateral-triangle face, connect geometrical centers of adjoining tetrahedrons and octahedrons in the IVM (Figure 1.9). Fuller called that structure a *Mite* (minimum space-filling tetrahedron).[27] His conclusion was that A-modules and B-modules alone cannot fill space, but the Mite is the LCD for the filling of space: "We find the Mite tetrahedron . . . to be the smallest, simplest, geometrically possible."[28]

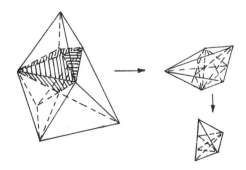

Figure 1.9: The Mite (minimum space-filling tetrahedron, *bottom right*) consists of two A-modules plus one B-module, forming the lowest common denominator and smallest geometry possible for filling space (based on Edmondson, 1986).

Universal Geometry

In architecture and engineering, we tend to think in terms of an orthogonal Cartesian coordinate system in which the three dimensions of space are oriented at 90 degrees to each other. However, nature exhibits a peculiar absence of perpendicularity. Native Americans know this, and I believe that many ancient civilizations around the world understood this to be a principle of the universe, as I will demonstrate in later chapters. Our world consists of convergence and divergence, a radiation of mass and energy exhibited in the growth patterns we see in both organic and inorganic processes. Fuller saw this manifested in the universe itself, referring to it as "Universe" (not the universe).[28] He perceived Universe as a point within some three-dimensional reality (four dimensions if we include time) made up of all energies and thoughts, a web of geometric relationships so complicated that we cannot separate the physical from the metaphysical. Therefore, our relationships with the physical nature of Universe, and our relationships with the metaphysical nature of Universe, are experience. These ideas of Fuller's fit perfectly with the definition of experience stated previously, that it is the effect of interrelated geometrical energy events in space and time.

A significant portion of Fuller's research developed through his study of energy associations and symmetry constructed from the A-module, B-module, and Mite, and those studies suggest a relationship between experience and quantum units of energy that can be represented by such symmetries. We've already seen that quantum mechanics recognizes behavior of subatomic particles in terms of the wave function, and this function relates location to time in terms of probability. Scientists conclude that relationships between elements of space-time, matter and energy, and physical law are what govern universal geometry. Recall that an electron has mass, volume, charge, and an estimated lifespan. It has no definite structure, but it exhibits a wave–particle duality. The key to understanding its structure may be the waveform or energy of this particle. For Fuller, the equivalent of the dimensionless geometrical point is an *energy event*.[30] He described every experience in the universe as an energy event, many of which may be small enough to be considered "points." So we can envision an electron simply as an energy event. The probability of an energy event being at a particular location will be related to space-time, and therefore is directly related to geometry.

We typically think of events as happening within some timeframe, and an energy event is no different. It has a magnitude and a direction of movement, and so its existence is a function of time. Size is not an issue here. A point could be infinitesimally small, or it could be an aggregate of events too far distant to be differentiated from another—perhaps a sun located on the far side of our galaxy, or a galaxy located at the far side of the universe. Regardless of size, the universe is filled with lots of matter and energy and space between those masses and energies.

Fuller concluded that this new paradigm of spatial perception and geometrical knowledge can improve our understanding of how the universe actually works. He believed that we could discover nature's inherent spatial coordinate system by observing naturally occurring phenomena, accumulating experimentally derived data, keeping an open mind about the appropriate framework for such study.

He also found that "Nature has only one department . . . *one comprehensive coordinating system.*"[31] That system functions in accordance with the "requirements of minimum energy."[32] Each and every natural system, whether a flower, a planet, or a galaxy, exhibits inherent properties of space. And so shape and volume are based on the most economical geometrical relationships, which themselves govern all structure.

Figure 1.10: The sphere provides a maximum ratio of volume to surface area (the energy interface). It has perfect symmetry.

In terms of inherently minimal energy confined by form, it is the *sphere* that provides that geometry, with its minimum ratio of surface area to volume (Figure 1.10). In other words, the spherical form provides a maximum volume with a minimum surface area (energy interface) interfacing with the surrounding environment. Recall that a vector has magnitude and direction. As a system of mass or energy, the sphere provides perfect geometrical symmetry for any given magnitude and parameter we chose—charge, spin, mass, even time itself.

If the universe is a system of mass and energy, does that mean that the shape of the universe is a three-dimensional sphere? I'll address that question in a later chapter, but at this point we can say that numerous shapes for the universe have been proposed throughout history, and mathematicians and cosmologists continue their calculations and observations, searching to identify and confirm the size and shape of the universe.

Communication Breakdown

For unknown thousands of years human beings have wondered about their purpose here on Earth, their relationship with the rest of the world and the cosmos, how this universe was created, and by whom. Surprisingly, those questions were answered thousands of years ago by people living in Africa, Asia, Australia, Europe, and North and South America. The answers can be found in many places and take a variety of forms. It took great time and effort to put the answers into forms that could be communicated to us so many thousands of years after those ancient peoples first asked the questions, paid attention to the world around them, measured and recorded it, and finally prepared their answers for us. It also takes great time and effort to decode the signs and symbols, measure and record the data, and understand the answers laid before us.

As with the Euclidean line, communications are spatially one-dimensional. Each message is transmitted from one point to another, information in the form of energy moving through time and space (Figure 1.2). Energy events can impact the message along its journey from sender to receiver. The message sent is not always the same message received. It may be interpreted differently by the receiver than was intended by the sender. Sender and receiver may not communicate in the same language, verbal or otherwise. Even the same symbol, used by both parties, might represent two totally different concepts, a problem that might go completely unrecognized by sender or receiver.

For example, Rabbi Laibl Wolf notes that "spiritual dynamics affect history."[34] In Judaism the interaction of the vertical, spiritual dimension and the horizontal, physical (human) dimension are encoded within the Torah—the books of Genesis, Exodus, Leviticus, Numbers, and Deuteronomy. But as Rabbi Laibl Wolf points out, "even these works fall far short of explaining the spiritual dynamics at play in the Cosmos."[35] It is evident that messages transmitted via symbolism, including the written word, are not always communicated effectively. Recall that the word *symbol* refers literally to the putting together of that which has been divided. If we don't know what was divided, or even that something (such as information) was divided, then how difficult will it be to reassemble all of the pertinent information and make proper sense of it?

Firth provides an anthropological perspective of symbolic forms and processes and the functions of symbolism. Human use of symbols is universal, " . . . part of the living stuff of social relationships . . . . [Ralph Waldo] Emerson wrote of the universality of the symbolic language: 'things admit of being used as symbols because nature is a symbol' (but so is culture)—'we are symbols and inhabit symbols.'"[36] Archeology has identified numerous symbols associated with the physical universe. Many of the symbols include a circle representing important traditional concepts, often universally applied and with religious overtones.

Many ancient symbols related to spiritual concepts and cosmological spheres do not yield such ready correspondence. Firth states,

> [O]n the whole in the studies of symbolism the tendency has been to emphasize the lack of "natural" links between symbol and thing symbolized—to view the symbolic attribution as a matter of cultural determination, as conventional or even "arbitrary." What is implied by such expressions is that the range of possible representations of something, particularly of an abstract quality, is so great that no exclusive choice of symbol is normally feasible by someone outside of the system. The reason why a specific symbol then appears in use, seems to depend upon some form of cultural condition; at the worst, since cultural components in the relationship of symbol to object are often hard to identify, the choice is termed inexplicable. But in stressing the conventionality, the "arbitrary" character of the relationship—and in using such a criterion in the distinction of symbol from sign or signal—it is the complexity rather than the inexplicable nature of the link that is really being considered.[37]

In other words, there are a multitude of potential meanings for any given symbol, particularly if we are unfamiliar with the source culture and traditions associated with it. Our ignorance of ancient culture and traditions can lead us to conclude that a symbol is not intuitively or rationally based, rather than recognizing the intellectual complexity of the message buried within the symbol.

Similarly, Firth notes that "if a symbol is to be an effective instrument of communication, it is essential that it should convey much the same thing to people involved—or that the range of variation in their interpretations should not inhibit the action desired."[38] A prudent method of interpreting a symbol, then, is to apply Occam's razor—expressed in Latin as lex parsimoniae, law of parsimony—using the fewest assumptions when interpreting data. The simpler and more elegant the interpretation, the better. There are two reasons for this method of interpreting ancient symbols. First, many ancient religious symbols are universal in form, suggesting that they have a common source of understanding. Second, information transmitted by the Ancients for receipt throughout the millennia was recorded in the most permanent means possible, suggesting that the message was considered vital to people in any age. The message was coded in symbolical form with an expectation that the receiver would understand the message no matter when it was received. Therefore, the message and the symbol were sent with hope that they would be universally understood.

Numbers, whether written or expressed in dimensional form, are simple and elegant symbols. Much has been made of the importance of numbers to human progress throughout the ages. Anthropologists generally believe that our prehistoric ancestors developed an ability to count things at some unknown point in their evolution, and that they did so for many centuries without numeric symbols. At first they could count whole things. Some things could be broken, so there could be $1\frac{1}{2}$ or $3\frac{1}{3}$ things. Eventually, at least several thousand years before the Common Era, humans made the leap between counting the things themselves and using symbols that stood for quantities of things. They realized there could be no things, and the concept of zero took hold. While academic tradition holds that the Greeks were the first to discover pi (3.14159 . . .), phi (1.618034 . . .), the square root of 3 (written herein as $\sqrt{3}$), and other irrational numerical values that could not be expressed in exactitude, as well as the fundamental idea of the Pythagorean theorem, it is quite evident that cultures far more ancient than classical Greece understood the importance and application of relative shape, size, and numerical equivalents to real things. Ancient civilizations were certainly conscious of the importance of numbers and geometry.

But these concepts are important only to humans. Nature does not operate on the basis of numbers or symbols. It operates on the basis of real mass and energy, what Fuller termed "requirements of minimum energy."[39] Fuller recognized that scientific principles, which are based on numerical relationships between physical parameters and which we assume govern interactions of energy events, are in reality metaphysical. Our definition of physical versus metaphysical is based on our ability to sense and relate to energy events. Ergo, Fuller's definition depends on *consciousness*.

Another ancient method of transmitting important traditional information is geometry. That geometry is found in universal mythologies is well established.[40] The symbolism and information it contains transcend time. Universal mythologies differ only in local cultural idiosyncrasies and traditions. Joseph Campbell said, "The themes are timeless, and the inflection is to the culture . . . . If you were not alert to the parallel themes, you perhaps would think they were quite different stories, but they're not."[41] In a discussion with Bill Moyers, he notes the ageless nature of the themes expressed in mythology.

> **Campbell**: Whether I'm reading Polynesian or Iroquois or Egyptian myths, the images are the same, and they are talking about the same problems. . . . It's as though the same play were taken from one place to another, and at each place the local players put on local costumes and enact the same old play.
>
> **Moyers**: And these mythic images are carried forward from generation to generation, almost unconsciously.
>
> **Campbell**: That's utterly fascinating, because they are speaking about the deep mystery of yourself and everything else. It is a mysterium, a mystery, and at the same time utterly fascinating, because it's of your own nature and being. When you start thinking about these things, about the inner mystery, inner life, the external life, there aren't too many images for you to use. You begin, on your own, to have the images that are already present in some other system of thought . . . . The myths help you read the messages.[42]

Chapter 2 provides specific examples of symbols used to express the circular nature of space, time, and our relationship with the universe, and gives additional context for understanding the information those symbols recorded for us. Part II includes numerous examples of mythological, secular, and spiritual symbolism used to transmit important information to us from as early as eight thousand years ago or more. This stream of one-dimensional communication conveys knowledge and wisdom of both ancient and primitive peoples. It contains information that was important to them, sacred knowledge considered vital enough that it was communicated in a manner that would function for millennia. The knowledge was very real for those who recorded it. It is very real today.

Chapter 2
The Second Dimension

Some such assumption of a basic "primitive" symbolic functioning of the human mind seems common to many anthropologists . . . and the presence of unconscious elements in symbolization seems to be also admitted, though evidence for such views has rarely been systematically sought. . . . But religious symbols are not simply communication media: they are held to be affectively charged, non-neutral in their emotional and intellectual value. Moreover, they can also be envisaged as possessing a spontaneous power, in themselves "projecting the mind towards the Absolute."

RAYMOND FIRTH
SYMBOLS: PUBLIC AND PRIVATE

The whole world is a circle. All of these circular images reflect the psyche, so there may be some relationship between these architectural designs and the actual structuring of our spiritual functions.

JOSEPH CAMPBELL
THE POWER OF MYTH

In chapter 1 we saw that the experiences of ancient and indigenous cultures are communicated via symbolism, which encodes vital information concerning universal and eternal relationships. The symbols express several levels of hidden meanings tied to tradition and often the circle as a metaphor for the cyclical nature of life and the physics of the cosmos. This chapter introduces numerous symbols as examples of this worldwide and timeless phenomenon, including line segments and curvilinear forms found in prehistoric tools and artwork, and application of the circle in ancient architecture. These examples are the basis for understanding geometries expressed by world mythologies in Part II.

The Symbols

Many ancient and indigenous sacred traditions from around the world are communicated to us through verbal, written, and pictorial symbolism. The circle is one of those traditional symbols. By itself or in combination with points, lines, and additional curves, it has served as a means of communicating sacred concepts. In some instances the circle is not shown, but its presence and meaning are implied by the geometry of other figures in a symbol. Specific cultures may give unique nuance to the meanings of certain symbols, but many symbols are crafted so successfully that the information they carry can be extracted and understood across cultures. Ananda Coomaraswamy concludes that "traditional symbols are the technical terms of a spiritual language that transcends all confusion of tongues and are not peculiar to any one time and place."[1] Indeed, the pervasiveness of the circle as a traditional symbol suggests that its levels of meaning are eternal and universal.

Figure 2.1 includes over eighty pictographic symbols include a circle that is either present or implied by the geometry. You might recognize some of the symbols. Each one has sacred meaning to one or more cultures, past or present. Their meanings are not provided here, but I hope that after reading this book you will begin recognizing the geometries that they all have in common—each represents a particular facet of the Sacred Sphere.

Figure 2.1: Examples of pictographic symbols with a circle or other geometrical figures such as points, lines, and curves that have sacred meaning in various cultural traditions. Each symbol represents a relationship between human beings, humans, and the world, or humans and the Creator. Many of the symbols have been used by ancient, indigenous, and modern cultures. You might recognize some of the symbols related to your own traditions.

Stone Power

Many of the symbols illustrated above communicate important concepts in mythologies and other traditions developed around the world and over the course of thousands of years. The majority of these symbols are related to Earth, Sun, Moon, and other planets and stars. Many are also associated with world geography. Some of them concern gods. However, all of them, in some manner, are intended to relay important information concerning relationships, typically between human beings and our surroundings on Earth or across the cosmos. The symbol itself is a means of communication, not itself the physical, intellectual, emotional, or spiritual concept. Rather, it represents those PIES concepts in a way that can be readily observed and understood by anyone who has a similar relationship with those ideas. In other words, there needs to be an active relationship between sender, message, and receiver if the communication is to be effective.

Modern humans have lived on Earth for more than a hundred thousand years with virtually no change in our capacity to think, reason, question, and discover. With the benefits of modern medicine and readily accessible energy, it would be far easier for a human from 100,000 BCE to survive in today's world than it would be for you or me to survive in the world a hundred thousand years ago, when humans experienced life as nature intended. Is this an indication of human progress, or have we lost something along the way? Regardless, it is reasonable to expect that symbols left by our distant ancestors can be understood today if our experience and knowledge reflects the experience and knowledge of ancient peoples. If their intention was to communicate with us via symbolism, we should be listening to what they have to say. If this was not their intention, then we are fortunate to have the opportunity to listen to them nonetheless. We just might learn something.

Circles on the Surface

It's unfortunate that our world is so full of energy. Energy brings about change, and change often means that we lose something and discover something else in its place. Given enough time, everything on the surface of Earth morphs into another thing. Most everything we make is intended for a limited lifespan. Paper, leather, steel, concrete—they all weather and erode. Sun, wind, and water are full of energy. They create and they destroy. Measured in geologic time, much of human culture is temporal, disappearing in a relative instant. But things like stone remain much longer.

The Highveld in the Republic of South Africa consists of high plateaus in the central and east-central parts of the country. To the north, the Limpopo River flows eastward for almost 400 miles (640 km) between South Africa and its neighbors, Botswana and Zimbabwe. The Orange River basin wraps around the Highveld, directing water west toward the Atlantic Ocean. The climate of much of the Highveld is relatively dry, but local variations in elevation result in development of microclimates across the region. High, grass-covered plateaus are bounded by steep scarps. Streams are undersized compared to the valleys they occupy. Water must have been plentiful at some time in the past. The Bushveld, adjoining and below the Highveld, generally receives more moisture and includes mixed dry forests. Wildlife is plentiful. Minerals are too. In fact, mining has occupied the people of the Highveld and surrounding environs for thousands of years. By far the majority of the world's reserves of platinum, manganese, and chrome are located in South Africa, as are deposits of vanadium, vermiculite, coal, diamond, zirconium . . . the list goes on.

Gold mining might have been the catalyst that first pushed the mining industry. This is the hypothesis of Michael Tellinger and Johan Heine, who have studied ancient stone ruins across South Africa. They suggest that the ruins date to at least seventy-five thousand years ago, if not older.[2] Generally comprised of stream-worn stones transported from river beds located as far as several kilometers away, most of the apparent ruins consist of curving, dry-stacked walls typically 6 to 9 feet (2 to 3 meters) in height. Many of the walls are in disarray, and lichens and mosses cover the surfaces of the stones.

None of this seems all that extraordinary until Tellinger and Heine describe the magnitude of the ruins. Several hundred structures in the region were observed by the early and mid-sixteenth century. The number of documented stone structures slowly increased as agricultural development proceeded. By 1891 that number had increased to about four thousand. Based on studies by several researchers working during the nineteenth and twentieth centuries, the stone structures were estimated to number nearly twenty thousand. Based on their own observations, Tellinger and Heine estimated in 2009 that "there are well over 100,000 of these mysterious stone ruins."[3] The walls interlace across an area of over 195,000 square miles (500,000 square kilometers). Three cities, each the size of Johannesburg or Los Angeles, are located in the area. Encountering an estimated average of 3.62 ruins per hectare (one hundredth of a square kilometer), Tellinger and Heine calculated that perhaps 3,620,000 ruins might be located in each of the ancient cities. They estimated that the number of stones at one city exceeds thirty-two billion, weighing over 1.4 trillion pounds!

The ruins appear to include dwellings, commercial and industrial facilities, roadways, agricultural terraces, and other structures associated with a large population. Most of the dry-stacked stones form curvilinear walls. There are few windows. Doorways are virtually absent. Tellinger has surveyed many of the structures, but he has only touched the tip of the iceberg. Of particular interest is what Tellinger calls Adam's Calendar, an arrangement of numerous large stones on a gently sloping grassy area above a steep escarpment. The stones appear to be oriented to mark the winter and summer solstices and the four cardinal directions. Tellinger believes that some stones represent the locations of stars in the constellation Orion and the double star system of Sirius, with the relative positioning of the stones indicative of a stellar alignment dating between 75,000 and 160,000 years ago.[4] Noting that local topography and vegetative conditions indicate that many more walls remain buried at shallow depth across the region, he estimates that erection of the structures would have required over one million people.[5]

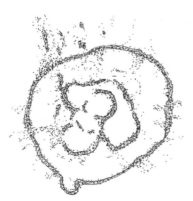

Figure 2.2: One of the many thousands of ruins in South Africa constructed of dry stack stones to create one or more circular stone walls, each enclosing an area of additional curvilinear dry stack structures with several rooms (based on Tellinger and Heine, 2009). While possibly constructed as a house, the intended purpose of the structure illustrated here remains unknown. Tellinger and Heine suggest that the numerous buildings might be tens of thousands of years old.

Of particular interest are the many thousands of ruins consisting of one or more circular stone walls, each enclosing an area of several smaller, curvilinear structures with several rooms formed from dry-stacked walls (Figure 2.2). The number of circular rooms per ruin generally ranges from three to eight, but their purpose remains unknown. Housing, factory, storage, or something else? We will have to wait for archeologists to find out. However, Tellinger and Heine's working hypothesis is that the untold number of stone structures represents the remains of cities and agricultural developments constructed to support a gold mining industry that might have necessitated use of a large slave population. They also suspect that this population might have had ties to ancient Egyptian and Sumerian civilizations.[6] Again, we will have to wait for proper studies to be conducted before these possibilities can bear weight.

During the 1960s archeologists from the University of Chicago surveyed a prominent ridge in southern Turkey. They found little of interest. In 1994 archeologist Klaus Schmidt visited the same ridge after reading notes the University of Chicago prepared of their survey. What Schmidt is unearthing on a rounded crest—Gobekli Tepe (Potbelly Hill)—of the ridge is stunning. Since the early 1900s the human transition from hunter-gatherer to farmer was understood to have occurred over several thousand years, culminating in development of cities by about 6000 BCE in Sumer, the area located between the Tigris and Euphrates Rivers in south-central Iraq. However, the architecture of Gobekli Tepe is evidence of social organization and monumental construction far earlier than expected. Five-ton megalithic (mega large, lithic of rock) limestone pillars stand 18 feet above the floor of sub-circular structures, completed with stone and clay mortar walls and founded on bedrock below the surrounding ground surface. Floor plans of three of these buildings are shown in Figure 2.3. The pillars include bas relief of various animals—mostly predatory—such as wild boars, lions, foxes, scorpions, frogs, vultures, and other birds. The purpose of the structures remains unknown, but Schmidt interprets them as temples associated with an early organized religion in which animals might have been deified.[7]

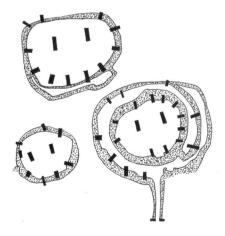

Figure 2.3: Three sub-circular structures excavated at Gobekli Tepe in southern Turkey, dated to about 9600 BCE. Constructed on bedrock with T-shaped limestone pillars (black rectangles) and dry stack stone walls, they were intentionally buried later with cobbly fill. The purpose of the structures remains unknown, although archeologists generally consider them to have been temples. The reason for burying the buildings remains a mystery.

The sacred relationship between Earth and the cosmos is expressed by the circular architectural form as well as the chiseled ornamentation found at Gobekli Tepe. The structures at Gobekli Tepe are generally devoid of subdivided space other than what appear to be few peripheral hallways or storage areas. Rather than a temple of worship with a common area separated from a sacred space such as an altar or place of offering, the temples at Gobekli Tepe have the appearance of undivided open, circular structures except for two T-shaped pillars placed off center within each structure and aligned generally in a northeast-southwest direction. The open, rounded shape is similar to a modern day arena or even the Roman coliseum where spectators observed games from peripheral seating while participants were located on the central floor area of each structure—like theater in the round. The temples are preserved because the people intentionally buried them not long (perhaps several hundred years) after their construction.

Only a fraction of the Gobekli Tepe archeological site has been excavated. Nonetheless, evidence obtained from Schmidt's work suggests the site was indeed developed as a sacred space. The temples are located on a prominent hill overlooking a broad valley. The site provided unobstructed views of both Earth and the heavens. Yet the site lacks a source of water. This fact alone shows that the development served a function of greater importance than mere secular comfort. The circular shape of the temple buildings reflects a natural, organic form. The arena-like structures would have been intended for ritual or ceremonial purposes, observation of the cosmos, and allowing a relationship to build between the human occupants and the world. Important aspects of the architecture include the massive yet ornamented stone pillars decorated with numerous carvings of animals, circles, lines, and anthropomorphic figures; the curvilinear, stepped stone walls; and floor levels situated below the surrounding ground line.

Essentially the sidewalls of the temples are like mid-sections of spheres in contact with Earth below and the cosmic dome above. Standing at floor level in one of the temples, we place ourselves symbolically at the center of the world. At night we can look up and see the moon, planets, and stars. We might perceive constellations that help us orient ourselves with the night sky. We can envision the paths of the planets, Moon and Sun crossing the sky. Those paths form a band extending across the cosmos. It is a pathway we call the ecliptic. We can also observe the Milky

Way stretching across the night sky, its bulge of stars surrounding the galactic center in the vicinity of the well-known constellations of Sagittarius and Scorpio. However, different cultures perceive different constellations. It is reasonable to assume that builders of Gobekli Tepe did not necessarily envision the same constellations as Sumerians did thousands of years later, or the Greeks who were responsible for many of the constellations we think of today. It is fortunate that the temples were buried so soon after construction, for the cosmology developed by their builders is preserved and communicated to us today, eleven millennia later.

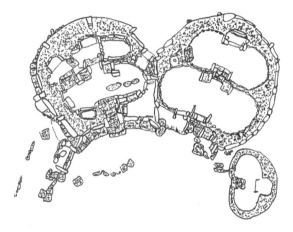

Figure 2.4: Plan view of the Mnajdra temples (c. 3200—3600 BCE) on the island of Malta (based on Hancock, 2002). Note the similarity of curvilinear forms in these stone structures and the dry stack construction in South Africa in Figure 2.2 and at Gobekli Tepe in Figure 2.3.

The shape of the temples at Gobekli Tepe and the thousands of ancient stone structures encountered in South Africa exhibit surprising similarity with stone temples located on the island of Malta in the central Mediterranean Sea, south of Italy. Figure 2.4 is a plan view of a triad of Maltese temples at Mnajdra. Pottery found on the island suggests that habitation occurred by about 5200 BCE. However, there is good reason to suspect that humans occupied the island well before that time. Evidence at Crete, located in the eastern Mediterranean, south of Greece, indicates that some of its early inhabitants were seafarers, with human habitation as early as 130,000 years ago.[8] Farming and animal husbandry were in practice by the time megalithic structures were constructed on Malta, estimated to date from about 4000 to 2500 BCE.

Those stone structures are the temples that can be seen today. Three of the largest temples are the Giants' Tower of Gigantija, perhaps the grandest and oldest (c. 3600 to 5600 BCE) of the Maltese temples; Hagar Qim, situated on a limestone mount and built about one hundred to two hundred years after Gigantija; and the temple of Mnajdra, which actually consists of three temples dating to between 3600 and 3200 BCE. Graham Hancock describes these structures.

> **Gigantija:** "[I]t faithfully and exactly reproduces what may be thought of as the 'canon' of all the Maltese megalithic temples—an outer retaining wall of cyclopean blocks . . . set out in a series of expansive, graceful curves to enclose an irregular space . . . large apsidal rooms interconnected by axial passageways . . . pairs of opposed lobes . . . the form of a clover leaf. . . . Orthodox scholarly opinion dates Gigantija to 3600—5600 BC."[9]

> **Hagar Qim:** " . . . seems to abhor straight lines, seducing the eye with patterns of curves and waves. Its flowing perimeter, flung out in a great irregular ellipse . . . Inside the temple there are the usual clusters of lobed, egg-shaped rooms arranged in pairs . . ."[10]

> **Mnajdra:** " . . . not one temple but a complex of three. Of these the easternmost, with three delicate apses disposed as a clover leaf . . . [11] All [at Mnajdra] are megalithic and all demonstrate a very high degree of architectural, engineering and mathematical competence on the part of the builders . . ."[12]

There are surprising similarities in the stone structures from the ancient cultures in South Africa, Turkey, and Malta, yet those cultures were separated by immense distances in both space and time. Hancock's descriptions of the temples at Malta could well describe Tellinger and Heine's "African Temples of the Gods." Admittedly, archeologists need to conduct additional detailed study to confirm a date (or range of dates) for the stone structures of South Africa and Turkey. Nonetheless, the architectural similarities exist. The curvilinear stone architecture is so similar, yet thousands of years separate their dates of construction. Can we resolve this enigma? Is it possible that these structures represent a common understanding of building design and construction as a metaphor for specific sacred knowledge? In other words, is vital information specified in the near-identical architecture found in these three cultural traditions?

Certainly the answer is yes with regard to the well-documented temples on Malta. However, the thousands of ruins found across South Africa and at Gobekli Tepe will require significant time and effort before their purpose can be understood. The shapes and sizes of the overall structure and internal rooms of the ruin shown in Figure 2.2 are potentially indicative of a residence sheltered from the outside world by a peripheral stone wall. Temples at Gobekli Tepe provide an unobstructed view of the sky, while the temples at Malta are oriented to take advantage of sunlight entering the sacred confines of bilaterally symmetrical rooms during specific times of the year.

Tools of the Trade

For more than forty years, James Harrod has studied prehistoric art, religion, and semiotics—cultural processes associated with communications, signs, symbols, and metaphors. Harrod promotes an interdisciplinary approach to the study of human origins. He draws on his particular interest in the prehistory of religions he seeks to identify the origins of art, symbol, and religion, anticipating that those events date to the Middle or Early Paleolithic periods—between forty five thousand and 2.5 million years ago.[13] Harrod found that human evolution proceeded over the course of four periods of time within the last three million years. In each case there occurred a spreading of "technological innovations, emergent cognitive and symbolizing capacities, and distinctive forms of representational palaeoart. . . ."[14]

Archeological and anthropological studies document an evolution of "the human capacity for spiritual-artistic representation. . . . Each stage seems to have its own unique symbol systems . . . "[15] The progression begins with creation of the first metaphor, the sense of recognizing one's self in the world. Harrod calls this "the metaphor of core essence as source of sustenance."[16] The next step was creation of symbols expressing restoration and reparation "of the core self, the reparation and mediation of social and intergroup conflict, and the first intentional sculptures in stone . . . These artists also utilized red ochre to symbolize creative energies."[17] This was followed by development of ritual art constructed upon the landscape—such as stone circles and petroglyphs—and use of "shamanic-psychological tutelary and healing spirits."[18] The progression culminates in paintings, sculptures, and engravings by Upper Paleolithic

Homo sapiens sapiens. There is the first indication of multiple levels of symbolic significance (perhaps some combination of PIES), as well as representations of "shamanic trance postures and female and male psychospiritual transformation processes . . . geometric gesture-sign sacred language to articulate these transformation processes and the basic processes of all living things."[19]

For Harrod, this apparent evolution of the cognitive capacity leading to creating symbols using various representational art forms is indicative of four significant stages in the structural development of the human brain, and a parallel evolution in the ability to create a mental picture—a model—of the world over the last two and a half million years. This progression of cognitive and creative abilities begins with mimetics, followed by conceptual-meaning, formulation of abstract ideas, and then mythic and linguistic development.[20]

In fact, the earliest record of tool use is found associated with *Homo habilis* and *Homo rudolfensis* during the Pre-Oldowan, about 2 to 2.5 million years ago.[21] The tool kit appears to include stone "flakes" obtained by striking a hammerstone against a rock from which the tool is derived, and which has been placed on an anvil stone that supports the tool-making process. The flakes were generally derived from water-worn pebbles and cobbles and commonly used for cutting. Of course, those weathered pebbles and cobbles would have had a rounded to sub-rounded shape prior to being broken to form flakes for the purpose of cutting, piercing, and so forth. However, our interest is with the hammerstone used to strike the pebbles and cobbles. With sufficient smashing of stone upon stone, hammerstones developed spheroidal forms. The classic Oldowan period is defined by artifact assemblages obtained from 1.9 -to 1.6-million-year-old sediments at Olduvai Gorge in northern Tanzania.[22] Again, flakes were obtained using a hammerstone on another stone, leaving a core of rock. The core itself becomes a tool for chopping or scraping. Classification of such cores includes such descriptive terms as discoids, polyhedrons, spheroids, and subspheroids.

Figures 2.5 through 2.8 illustrate examples of these geometries in the form of Pre-Oldowan and classic Oldowan stone tools and cores. The cores were retained by their makers, even carried from place to place as necessary for tool making at various locations. Undoubtedly these stones were recognized for their utility and differentiated from other rocks lying about. It is reasonable to assume, too, that the spherical shape of the hammerstones was recognized as an effect of the tool making process, perhaps intentionally formed as the stone was used and reused over a prolonged period of time.

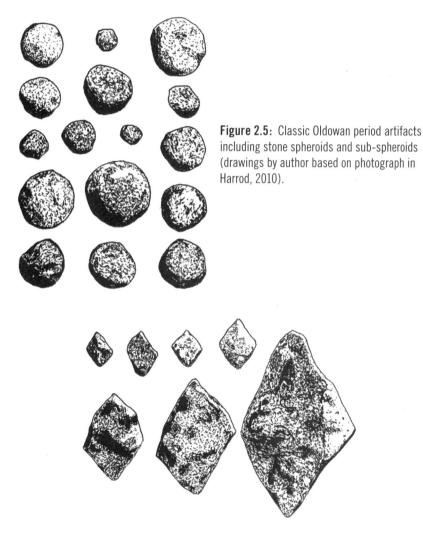

Figure 2.5: Classic Oldowan period artifacts including stone spheroids and sub-spheroids (drawings by author based on photograph in Harrod, 2010).

Figure 2.6: Classic Oldowan period artifacts including rhombohedral cores of various sizes (drawings by author based on photograph in Harrod, 2010). Note the two-fold symmetry about the vertical and horizontal axes.

Figure 2.7: Oldowan spheroidal hammerstone showing fractured face (top) and curved exterior (bottom) (drawings by author based on photograph in Harrod, 2010).

Figure 2.8: Classic Oldowan period stone artifact, an oblong faceted stone exhibiting two-fold bilateral symmetry (drawing by author based on photograph in Harrod, 2010).

Figure 2.9: Stone artifact, a chopper core exhibiting a rhomboid diamond shape and believed by archeologists to be an accidental result of flaking the core and then carried by its maker to the site where it was discovered. Note the unusual three-fold symmetry produced by flaking the stone. Harrod suggests that its shape is exemplary for understanding the mind and spirituality during the Oldowan period (drawing by author based on photograph in Harrod, 2010).

Harrod's timeline for the progression of cognitive and creative abilities suggests our ancestors during the classic Olduwan period and into the Early Acheulian period (about 1.4 to 1 million years ago) had progressed beyond mimicry of the environment toward the ability for conceptual meaning, perhaps even the formulation of abstract ideas. Those abilities indicate that the spheroidal shape of the hammerstones was an intentional by-product of tool formation. Was this a conscious intent on the part of the tool maker? Did our ancestors perceive deeper meaning to the creation and recognition of discoidal, polyhedral, and spheroidal forms? Did they recognize a value to those shapes? And did they assign this value, or were they pre-destined (encoded DNA as it were) to recognize the importance of particular shapes for understanding concepts and relationships, including those that are vital or sacred?

Harrod found that by about one million years ago, tool makers developed a "conceptual matrix with top and bottom of object either pointed/ rounded and either trimmed/natural and/or with left and right side curved either concave/convex and/or either curvilinear/rectilinear . . . shapes can be generated, including disks, spheroids, biface ('handaxe' and 'cleaver'), scraper, awl and cleaver."[23] In other words, the assemblage of tools manufactured more than one million years ago is evidence for an understanding of points, lines, planes, right angles, curves, convex and concave shapes, and bilateral symmetry. Geometrical aspects of the two-dimensional circular symbols illustrated above are evident in the three-dimensional hammerstones, tools, and debris produced from tool-making activities. I believe this is evidence of the knowledge and ability to apply these fundamental geometric concepts and constructions in other facets of the tool maker's lives.

Between one million years ago and 50,000 BCE, our Paleolithic ancestors, true *Homo sapiens sapiens*, were creating symbols in the form of rock paintings found on every continent bar Antarctica. Harrod suggests that the art of rock painting might date from a much earlier time, and that these artworks are interpreted as "geometric percepts or phosphenes . . . associated cultural interpretations of trance (altered states of consciousness)."[24] He continues,

> The "shamanic" interpretation of UP (Upper Paleolithic) art does not preclude the view, following A. Leroi-Gourhan and others, that the animals depicted in this art are symbols in a complex and sophisticated code. As in the case of the geometric signs, they are iconographic, belong to a system of juxtapositions and pairings, and have multilevels of signification.
>
> I have proposed that the animals form pairs of pairs (quaternions, fourfold, or sixfold) that are complementary and thus have the capacity to constitute a multileveled symbolic system or encyclopedia of cultural knowledge. A fourfold (or eightfold) pattern is perhaps first detectable in the European Aurignacian, e.g., Chauvet Cave, pairing bison vs. horse; mammoth vs. rhino; ibex vs. deer; and lion vs. bear . . . This system has the capacity to serve as a mnemonic device that encodes key zoological, social,

economic, and psychological information ("memes"), including . . . basic moral themes.[25]

By about 50,000 BCE, and potentially significantly earlier, humans were creating works of functional, multileveled, symbolic art with four-fold, sixfold, and eightfold rotations applied within the compositions.[26] Creation of three-dimensional tools led to conceptualized and applied two-dimensional symbolism. We have arrived at the cusp of abstract geometrical drawing and use of the circle to represent multileveled rational thought. Humans were asking the tough questions. Who are we? Where are we? Why are we here? The answers to those questions were symbolized by the circle as proxy for even deeper understanding of our place in the universe.

We can envision these four levels of abstraction at the quarter points of a circle, with each quarter connected to the center of the circle. Each level is then connected to each of the other three, and all four meet at the center.

RAYMOND FIRTH
SYMBOLS: PUBLIC AND PRIVATE

Part 2

A Case for Cross-Cultural Symbolism

Chapter 3

A Sacred Hoop on the Northern Great Plains

In the face of the rapid transformation of the earth by science and technology and the ecological crisis that has begun to unfold, leading thinkers are exploring alternative cosmologies, paradigms, and philosophies in search of models that may sustain nature rather than destroy it. Many of these thinkers have found that native cosmologies offer some of the most profound insights for the kind of sustainable relationship to place and spiritually integrated perception of nature . . . .

GREGORY CAJETE
NATIVE SCIENCE

The pattern Campbell discerns in these narratives is clear and dramatic: separation, initiation, and return. . . . Initiation in these narratives equals transformation, and the agent of this process can be either god or goddess, assume either anthropomorphic form or animal, or come in the guise of some natural phenomenon. In whatever shape, the Great Teacher's message is essentially the same wherever in mythology it is encountered. It is the inculcation of a profoundly altered vision of the self and cosmos.

FREDERICK TURNER
BEYOND GEOGRAPHY

In this chapter we consider the meaning of the Sacred Hoop of the Lakota people of the Sioux tribe in the north-central United States. We will review the migration of the Lakota onto the Great Plains subsequent to the coming of Westerners to the upper reaches of the Mississippi River. Most of the review of the migration is based on Euro-American records of Native Americans in the region, with first-person accounts dating to the early seventeenth century. That information, in tandem with ethnographic records based on interviews of Lakota holy men, and data collected at sites located across the Northern Great Plains, provides the basis for discovering what may be the largest sacred symbol ever constructed on Earth's surface. In the process we will see how the Sacred Hoop represents the lifeway of the Lakota, as it has for many other indigenous tribes of the Americas.

Native Science

The Sacred Hoop is the most representative symbol of the Native American lifeway. In its simplest form it consists of a cross dividing the interior of circle into quadrants (Figure 3.1). Simple as its geometry may be, the Sacred Hoop has deep and multiple meanings. After more than five hundred years of continuous contact with Native Americans, Western society has yet to fully understand those meanings.

In his book *Native Science: Natural Laws of Interdependence*, Gregory Cajete uses the word *science* "in terms of the most inclusive of its meanings, that is, as a story of the world and a practiced way of living it."[1] This is a foreign concept for readers raised in ethnocentric Western culture. It is certainly not a definition of the scientific method. For indigenous people, life is not lived without direct contact with or immersion in nature. It cannot be understood simply by reading books. It is not found in laboratories. Cajete drives this point home when he writes, "In native languages there is no word for 'science,'"[2] nor for philosophy, psychology, "or any other foundational way of coming to know and understand the nature of life and our relationship therein. . . . For native people, *seeking life* was the all-encompassing task."[3]

Figure 3.1: Lakota Sacred Hoop.

Let's look closer at this Native American philosophy of life. Cajete lays out in very plain terms the meaning of what should properly be called *lifeway* as it pertains to indigenous cultures, past and present. Before Europeans arrived in the Americas, Native Americans did not *practice religion*. Cajete writes that "the ultimate quest of both [the indigenous] individual and community was to 'find life,' that is, to find that place that Indian people talk about."[4] Cajete further explains:

> Native science is a metaphor for a wide range of tribal processes of perceiving, thinking, acting, and "coming to know" that have evolved through human experience with the natural world. Native science is born of a lived and storied participation with the natural landscape. To gain a sense of Native science one must participate with the natural world. To understand the foundations of native science one must become open to the roles of sensation, perception, imagination, emotion, symbols, and spirit as well as that of concept, logic, and rational empiricism.[5]

This breadth of foundation of native science is not restricted by systematic study, organization, and classification based on testable performance and predictive capabilities of Western science, of which the purpose is to gain useful, beneficial knowledge of the world. Native science is a far more holistic, philosophical approach to the study of existence and relationships between observable phenomena and the world around us. It requires a lifelong dedication to experiencing life as nature presents it, and recognizing and appreciating relationships between all forms of life, animate and inanimate. As we saw in chapter 1, experience is the effect of interrelated geometrical energy events in space and time. This effect is stored in the body and assimilated to produce knowledge. With sufficient experience and understanding of cause and effect in the world, we develop wisdom—the ability to apply experience through active participation with the world for the benefit of all creation. Native science includes experience with the metaphysical as well as the physical, expression through art and experimentation with matter and energy, and the practice of ritual and ceremony. Native science is not religion, but does require living in a sacred manner. Cajete finds:

> Much of the essence of Native science is beyond literal description. . . . [It] may be seen as an exemplification of "biophilia," or the innate instinct that all life forms share for affiliation with each other. . . . Native science encompasses such areas as astronomy, farming, plant domestication, plant medicine, animal husbandry, hunting, fishing, metallurgy, and geology—in brief, studies related to plants, animals, and natural phenomena. Yet, Native science extends to include spirituality, community, creativity, and technologies that sustain environments and support essential aspects of human life.[6]

We will discover in this and subsequent chapters that mythologies from around the world were developed to communicate traditional understandings of the human relationship with Earth and Cosmos. Those mythologies are, in fact, based on the very concepts expressed by Cajete and lived by Native Americans. Whether applied to science, religion, or lifeway, the same basic concepts appear to have been universally realized, understood, and applied across the world and throughout time. The Native American lifeway is symbolized in many ways by the Sacred Hoop.

Traditional Lakota Territory

The events that spurred the westward movement of the Sioux from as far as the southeastern United States to the Great Plains are mostly undocumented and therefore speculative. The migration appears to have been only one of many such westerly migrations soon after the arrival of Europeans in North America. Anthropologists and historians conclude that the Sioux resided in the southeast part of North America until as late as the sixteenth century, becoming a nation as they moved into the Midwest in the late 1600s.[7,8] The Lakota, one of seven nations of the Sioux, were an old nation consisting of a society with an established nomadic hunting lifestyle long before they moved onto the northern plains. Today the Sioux people include three main nations: Dakota, Lakota, and Nakota. While *Teton* was the common term used in the historical record, *Lakota* is the preferred name of the people who now occupy Sioux reservations in the area studied here, and the term I use throughout this book in reference to this proud indigenous nation.

Archeology has not supplied conclusive evidence of Sioux origins, and there are no written records of the extent of Sioux territories until the 1600s. Sioux oral tradition indicates an origin in the northern lakes region, east of the Mississippi River. Based on linguistic reconstructions, the homeland of proto-western Sioux (those nations that later occupied the western portions of what was in the 1800s considered to be their traditional territory) was in southern Wisconsin, southeastern Minnesota, northeastern Iowa, and northern Illinois.

By the mid–1600s, Sioux territory extended from what is now the forests and grasslands of central and southern Minnesota, through eastern North Dakota and South Dakota, and westward to the Missouri River. The southward and westward migration is attributed to the westward migration of the Ojibwa into land formerly occupied by the Sioux and to the relative abundance of bison in the grasslands.[9] Migration onto the grasslands occurred in two waves, the first consisting of Yanktonais (including Yanktons) and Lakotas after the middle of the seventeenth century, and the second consisting of the remaining Sioux after about 1735.[10] Father Hennepin, a Jesuit missionary exploring the upper reaches of the Mississippi River during the late seventeenth century, wrote that the Sioux of the West he encountered along the Mississippi River during the years 1679 to 1680 had walked for four moons (months) from their land to central Minnesota.[11] Assuming an average walking distance of 5 miles (8 kilometers) per day, the distance that they traveled was about 560 miles (900 kilometers), potentially well within the Powder River Basin of northeast Wyoming and southeast Montana, or farther west.

The Sioux migrated up the Minnesota River to Big Stone Lake and across the Couteau des Prairies. They traveled along Sioux Pass across the Missouri Couteau, south of the Big Bend of the Missouri River in central South Dakota, a route that can be traced and dated to the migrations of Oglalas and Brules of the Lakota nation.[12] The river crossing is depicted on a map drawn in 1825 by the Lakota Gero-Schunu-wy-ha, with a Yanktonai and Yankton camp located at the mouth of the creek and a Lakota camp situated at the west side of the river crossing. James Marshall III, a popular Lakota writer and historian, states that the Lakota crossed the Missouri River by 1680, but while the exact timeframe is unknown, that event occurred before the Lakota obtained horses.[13] Dogs served to pull loads placed on drag poles, and a usual day's travel at that time amounted to no more than six to ten miles. At the beginning of the eighteenth century, the Sioux of the West were hunters traveling by foot across the prairies between the Upper Mississippi River and the Missouri River. They encountered the Arikara tribe (not a member of the Sioux nation) in the vicinity of Fort Pierre. The Lakota appear to have followed the Cheyenne migration on to the Great Plains during the period of about 1720 to 1750.

By 1760 the Lakota were familiar with horses owned by the Arikara. Many Sioux living on the prairie had horses by the latter 1700s, and a day's travel increased to about twenty miles with horses used to haul lodge materials and other items. Smallpox epidemics during the latter 1770s resulted in heavy population losses to the tribes on the northern prairies, inviting westward expansion into the central Dakotas. With a good horse, a rider could traverse Lakota territory from the Missouri River to the Bighorn Mountains, or from the North Platte River to the Yellowstone River, in thirty to forty days.[14] Use of the horse ultimately allowed for expansion in lodge sizes, an increase in family size, and broadening of Lakota territory. Horses also brought about significant cultural changes, leading to development of a surplus economy.[15] Bison became central to the economy and lifestyle of the nomadic Sioux.

Fools Crow, a holy man of the Lakota, stated that "there was considerable change and not a little evolution on this continent; and the great Sioux migration [was] one marking an entire change in the customs and character of a tribe."[16] However, because they had no written tradition, and the details of changes in their lifestyle resulting from their migration generally remain unknown, the influence of ancient customs on the Plains culture of the Lakota remains difficult to discern. The horse intensified earlier cultural patterns as old elements melded with new elements. Cultural traditions were borrowed from other Plains tribes, such as the intertribal pipe adoption ceremony and the Sun Dance.

By 1775 the Lakota ranged from northern Nebraska through the west half of South Dakota and into North Dakota. American Horse's winter count and Standing Bear's winter count for 1775 to 1776 commemorate the Lakota discovery of the Black Hills.[17] Smallpox epidemics again devastated tribes located along the Missouri River between 1779 and 1795. The Lakota displaced other indigenous tribes such as the Crows, Kiowas, Poncas, Omahas, Arikaras, and Cheyennes by 1800. They included southern Wyoming as part of their hunting grounds during the 1830s, and the Oglala subgroup extended their hunting grounds westward into territory occupied by the tribes of the Snake River valley and northwestward into Crow country of southeastern Montana by 1840. Lakota, Cheyenne, and Arapaho continued moving west through the first half of the 1800s as a result of expanding settlement by Westerners and the decreasing bison population, displacing the Kiowa and Crow from the Black Hills. Between the 1830s and 1870s, Lakota encampments extended from alongside the Bighorn Mountains of Wyoming, north of the Yellowstone River and the Bridger Mountains of Montana, south of the Missouri River in north-central North Dakota, and the east-central region of South Dakota (east of the Missouri River). By the early 1830s the Lakotas were "fully transformed from pedestrian to mounted nomads" occupying the northern Great Plains from the Missouri River to the Bighorn Mountains and from the southern Canadian prairies to the Platte and Republican Rivers.[18]

Maps illustrating the extent of Lakota territory from the mid-seventeenth century, the time of first encounter with Europeans, through the mid-nineteenth century typically circumscribe an area including northeast Wyoming, southeast Montana, portions of North Dakota and South Dakota west and south of the Missouri River, and that portion of Nebraska located north of the North Platte and Niobrara rivers. Depictions of that area historically have varied by author. Figure 3.2 illustrates the generally understood extent of traditional Lakota territory by the mid-nineteenth century. The principal topographic features of that area are the Black Hills, sacred to the Dakota, Cheyenne, and other tribes, and traditionally considered to be the center of the Lakota world.[19]

Figure 3.2: Previously documented extent of Lakota territory during the mid-19th century. Historic depictions of that area have varied by author, but generally exhibit the area shown (from Burley, 2007).

It is important to note that bands of Lakota hunted wherever bison and other game were encountered within their territory. That territory, land associated with traditional Lakota identity as manifested by their resource consumption and lifestyle, was controlled but not owned by the Lakota. While individual tribes often occupied separate land areas, those areas were not exclusive, and the boundaries between tribal areas flexed with the temporal tribal migrations, hunting opportunities, and ceremonies and rites practiced by the Lakota. It was not until the Treaty of 1851, signed at Horse Creek near Fort Laramie, that boundaries were specified among the participating tribes, including the Lakota and others on the northern Great Plains. The treaty set in motion the limiting of tribal and subtribal movements and the loss of the ability to self-define tribal territory with regard to such parameters as relations with neighboring tribes, regional physiography, natural resource availability, and climatic conditions. The Treaty of 1868 further refined Lakota territory, setting aside South Dakota west of the Missouri River as the Great Sioux Reservation and including portions of Nebraska north of the North Platte River and portions of eastern Wyoming and eastern Montana between South Dakota and the crest of the Big Horn Mountains as unceded Indian territory.

Descriptions of the culture of Lakotas living on the Great Plains between about 1750 and 1890 are provided by the lifeways, legends, and myths told by holy men and transcribed by ethnographers and others of the twentieth century.[20] Those documents provide insight into the social structure, commonly held beliefs, and religious fabric interwoven with the daily activities, ceremonies, and rituals. They describe how the men were hunters and warriors, protecting each village and the Sioux country in general. The gifting of the sacred pipe by White Buffalo Calf Woman and the development of seven sacred rites, among other detailed legends and myths, gives context to the beliefs of the people and the geography in which they thrived, the daily life and practices of holy men, and the eventual inclusion of the Christian faith within traditional beliefs.

The multiple meanings and variations of creation and origin stories include physical, intellectual, emotional, and spiritual components of indigenous life. "A Myth of the Tetons as It Is Told in Their Winter Camps," told by Lakota holy man George Sword, describes from oral tradition the establishment of the four cardinal directions by the four sons of Tate (Wind), who was created by, and is the constant companion and messenger of, Skan (Sky).[21] Parts III through XVIII of James Walker's Literary Cycle (WLC) provide additional cultural information and context for that myth, while omitting details of kin relationships and other information that had been included in the myths by his sources.[22] WLC is a narrative adapted from the traditional myths, developed and formatted in the manner of an epic story. According to the myth as told by George Sword and as presented in WLC, the four sons of Tate leave their home to establish the four directions at the edge of the world. In turn, the sons of Tate set a direction beginning west of Tate's home and subsequently travel to the north, east, and south, returning to the point where the west direction was established and then back to Tate's home.

Circles on the Great Plains

The mythology as told in the WLC describes the northern Great Plains, an area covering virtually the entire states of North Dakota and South Dakota, as well as east and central Montana, eastern Wyoming, and northern Nebraska. From surface elevations of about 6,000 feet along the mountain fronts in east-central Wyoming to elevations of about 1,200 feet along the east side of the Couteau des Prairies in eastern South Dakota, the area is drained by the Missouri River and its major tributaries including the Yellowstone, James, and Platte Rivers. Prior to Euro-American immigration, the Great Plains were covered by semi-arid grasslands home to great herds of bison and pronghorn, pine-covered or grassy buttes, and tree-lined rivers and streams serving as oases for the indigenous population.

The Black Hills are located in the central portion of the northern Great Plains. They attain the highest elevation at Harney Peak (elevation 7,242 feet [2,300 m]), include the largest contiguously forested land surface in South Dakota, and are surrounded by rolling grasslands divided by rivers flowing eastward to the Missouri. Bear Butte, an outlier to the Black Hills about five miles northeast of Sturgis, South Dakota, is a remnant laccolith attaining an elevation of 4,422 feet (1,410 m). It is at the center of the Lakota world. Across the high plains of eastern Wyoming, the Bighorn Mountains rise to over 13,000 feet (4,150 m) along a drainage divide located about 190 miles (300 km) west of the Black Hills. Grasslands extend about 200 miles (320 km) north of the Black Hills to badlands bounding the south side of the Little Missouri and Missouri Rivers. To the east, grasslands extend to the Missouri River and beyond, crossing the Southern Missouri Couteau, the James River valley, and the eastward-facing slope of the Couteau des Prairies before encountering the mixed oak forests of the Minnesota River drainage, a distance of more than 250 miles (400 km) from Bear Butte. South of the Black Hills, grasslands roll across the White River Badlands, the Niobrara River drainage, and the sand hills of Nebraska before reaching the Platte River and the north portion of the southern Great Plains, about 200 miles (320 km) south of Bear Butte. The Lakota of the nineteenth century were well acquainted with the major landmarks of the northern Great Plains and could travel across it with little difficulty.

Native American medicine wheels constructed of cobbles and boulders on the ground surface of the northern Great Plains are encountered in a variety of shapes, sizes, and locations. John Brumley developed a general definition for medicine wheels, classifying such lithic constructs based on structural characteristics documented for sixty-seven wheels located in the north-central states and south-central Canada.[23] Brumley summarized empirical information associated with each medicine wheel within the context of eight subgroups defined therein. Direct evidence for the purpose of construction and use of medicine wheels is limited and requires an understanding of ethnographic and archeological associations with individual features. Cajete provides insight by noting that "structures and symbols of Native science, such as the medicine wheel, are used as a metaphor for native knowledge and creative participation with the natural world in both theory and practice, serving as bridges between the inner and outer realities, with those realities expressed in traditional Plains Indian culture by the medicine wheel structure."[24] Brumley concluded that to understand the purpose behind construction of each medicine wheel, research efforts should extend beyond the immediate margins of the wheel itself.[25]

According to Brumley's classifications, subgroup 6 medicine wheels are characterized by a prominent central stone cairn surrounded by a stone ring and by having two or more stone lines extending from the stone ring to the cairn.[26] Of the sixty-seven medicine wheels he evaluated, only three structures fall within the subgroup 6 classification, including the Bighorn medicine wheel located on the west flank of the Bighorn Mountains of north-central Wyoming, the Jennings site in east-central South Dakota, and the Majorville site located along the Bow River in south-central Alberta. An additional find, the Cloud Peak medicine wheel site, which I discovered during a mountaineering venture into the Cloud Peak Wilderness in 2003, also satisfies all criteria as a subgroup 6 feature and became the fourth documented lithic structure within that classification.[27]

Power Points

In a 2006 report to the Office of the Wyoming State Archeologist, I described methods and results of nonintrusive observations and measurements of the Cloud Peak medicine wheel, ancillary features, and other previously undocumented structures and natural features nearby. The medicine wheel site is uncharacteristically located near the topographically low point of an alpine valley. It is associated with ancillary features constructed for recognition of the summer solstice and situated near a mountaintop.

From onsite observations, I classified Cloud Peak medicine wheel within subgroup 6 of Brumley's classification scheme (Figure 3.3). The structure has the smallest stone ring diameter of medicine wheels within that subgroup and is the only one of the four sites in that subgroup to be located in an alpine physiographic setting above timberline and within the core of a mountain range. The orientation of the primary stone lines radiating from the center of the wheel are within about 10 degrees east of true north and 3 degrees south of east. The alignment of the primary stone lines appears to be associated with the four cardinal directions.

For the Northern Cheyenne, Crow, Sioux, and other tribes, power points are places associated with increasing spirituality. The higher the place, the closer it is to the Great Spirit. The Cloud Peak medicine wheel site includes an array of physical characteristics that serve as a sacred backdrop for rituals that could be conducted at that location. Cloud Peak is the highest mountain in the Bighorn Range, the most northeastern range in the central Rocky Mountains. It is higher than any point within a 150-mile (240 km) radius, overlooking the vast buffalo ranges of the Powder River, Yellowstone River, and Bighorn River basins. On a clear day the view from the peak extends over an area of about 36,000 square miles (105,000 sq km). The wide variety of land, water, plant, and animal features surrounding the Cloud Peak medicine wheel are key to understanding the choice of location for its construction.

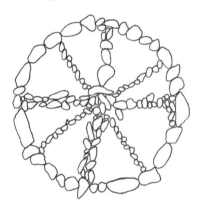

Figure 3.3: Cloud Peak Medicine Wheel in 2003. a) Oblique view photograph. b) Orthogonal sketch (from Burley, 2007).

Another important site is the Swenson medicine wheel, located in eastern Dunn County near the Little Missouri River as it enters the Missouri River in west-central North Dakota. Badlands topography extends along the south side of the Missouri River and on both sides of the

Little Missouri River across about 40 miles (64 km), extending upstream from the mouth of the river for more than 100 miles (160 km). The Killdeer Mountains, the most prominent topographic feature south of the confluence of the two rivers, are located about 36 miles (60 km) west of the Swenson medicine wheel site. The medicine wheel is situated in an upland area with moderate topographic relief, including ridges and few buttes, near the north-central edge of a prominent ridge within a larger network of ridges trending northwest-southeast for several kilometers. It is underlain by glacial till with granitic boulders, cobbles, and gravel exposed at the ground surface. The ground beyond the ridge north of the site falls northward into the badlands of the Missouri River drainage.

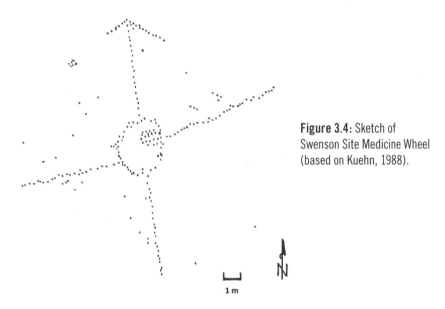

Figure 3.4: Sketch of Swenson Site Medicine Wheel (based on Kuehn, 1988).

    The Swenson medicine wheel is a subgroup 7 structure consisting of a single stone circle about 2.5 meters by 3 meters in diameter: Its four rock lines extend beyond the circle from a central rock cairn that was disturbed by the land owner prior to study of the site (see Figure 3.4).[28] Stone lines extending north and south of the stone circle form a nearly straight line oriented 4.5 degrees east of true north, with the north end of the northern rock line capped by a bent stone line suggesting an arrow point pointing northward. The stone line extending west of the stone circle, while crooked, is oriented 90 degrees west of true north. The stone line extending east of the stone circle is oriented 80 degrees east of true north, placing it in an approximate orientation between the center of the circle and sunrise during summer. The site has not been excavated, nor has it produced diagnostic artifacts. The cultural affiliation of the site has not between determined. Thus, these site observations do not provide a conclusive age, origin, or purpose of the site.[29]

Brumley summarized the physiographic and topographic setting, metric and non-metric characteristics, feature condition, and research status of another subgroup 6 structure—the Jennings medicine wheel—based on studies conducted by archeologists during the early 1980s.[30,31] The site is located in Hand County, South Dakota, on open rolling prairie about 850 feet (270 m) east of a dry lake bed. The subgroup 6 structure includes a low stone cairn, 8 feet in diameter, located in the center of a stone circle 19 feet in diameter. The cairn and circle are connected by four stone lines almost equidistantly spaced (see Figure 3.5). Limited data are available regarding the age, origin, and purpose of the site. Like the Swenson site, it has not been excavated and appears to remain undisturbed; no ancillary features or ethnographic evidence have been documented.

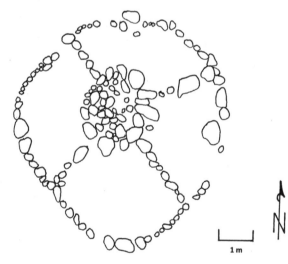

Figure 3.5: Sketch of Jennings Medicine Wheel (based on Abbott *et al*, 1982, and Rood and Overholser-Rood, 1983).

1 m

Another sacred location, Scotts Bluff, is a high promontory rising about 800 feet (240 m) above the North Platte River. The promontory is situated in the Great Plains of extreme western Nebraska, serving as a landmark for Native Americans as well as thousands of pioneers traveling west along the Oregon Trail during the mid-1800s. The great Oglala leader Crazy Horse is said to have had a vision upon Scotts Bluff.[32] The North Platte River flows southeast a little over one mile north of Scotts Bluff in a broad valley, partially wetlands, that was historically prone to flooding. The adjoining upland areas once were covered by an almost-continuous mixed- and short-grass prairie with forbs and shrubs. Scotts Spring, located in Scotts Bluff National Monument and next to the Saddle Rock Trail, was recorded in pioneer diaries as a source of clear drinking water preferable to the often turbid flow of the nearby river. Records document the historic presence of large mammals such as grizzly bears, bighorn sheep, bison, and elk in the area, and smaller mammals, reptiles, amphibians, and numerous bird species currently populate the river valley and adjoining uplands. The area also includes some of the richest fossil-bearing strata in Nebraska.

Alas, no medicine wheel has been documented in the vicinity of Scotts Bluff. The plains and bluffs of the Scotts Bluff area suffered disturbance from the passing of immigrants and livestock along the Oregon Trail during the mid-nineteenth century, and rural and urban development have continued into the present. The impacts include loss of vegetation, erosion, surface grading, and land development.

The Lakota World

Through study of the regional physiography and Lakota migration onto the northern Great Plains and review of ethnographic records, including James Walker's Literary Cycle and "A Myth of the Lakotas as It Is Told in Their Winter Camps" given by George Sword, I discovered the basis for understanding not only the source of the four cardinal directions held by the Lakota, but the setting of those directions with regard to the physiographic characteristics of traditional Lakota territory and the construction of medicine wheels defining each cardinal direction with respect to Bear Butte, the center of the Lakota world. While the timeframe of development of the mythology is not known, horses are not mentioned in either of those stories. If the myth developed prior to the introduction of the horse to the Lakota during the late 1700s, then its timeframe might be bracketed by the Lakota discovery of the Black Hills by the 1770s and the introduction of the horse before 1800.

Based on the documented extent of Lakota migration toward the west, north, and south between 1750 and 1850, I conclude that the Lakota demarcation of the limits of their territory in the four cardinal directions might not have occurred until about the 1830s. This timeframe exactly matches the age of the Cloud Peak medicine wheel site as estimated by the Office of the Wyoming State Archaeologist. Additional confirmation of these conclusions is provided in mythology. Table 1 provides descriptions of the physiographic characteristics of the mythic center of the Lakota world (Tate's lodge), the four points where the cardinal directions were founded, and descriptions of actual places that appear to correspond to those mythic locations. The suggested places are Bear Butte (center), Cloud Peak (west), the Killdeer Mountains (north), the Jennings site (east), and Scotts Bluff (south).

Comparison of Mythological Locations based on Walker (2006) and Actual Geographic Locations Based on Site Observations		
Location	Mythic Place Description	Actual Place Description
Center of World/ Bear Butte	Round lodge beyond the pines; sun shone through door at midday (south); Tate's place of honor at rear of lodge, opposite (north) of the door	Bear Butte is a sentinel on the northeast side of the Black Hills, separated from the pine-covered hills located to the southeast and exhibiting a weathered, sub-rounded topography and footprint; sacred center of the Lakota world; traditional site of Lakota ceremony and ritual including vision questing. Bear Butte Peak—Latitude 44° 28' 37," Longitude 103° 25' 41"
West Direction/ Cloud Peak, Medicine Wheel	Over a great mountain at the edge of the world and beside the trail that is around it; sky comes down at the trail; clouds at top of mountain (lodge of Wakinyan); very steep and very high; top consists of a level space; huge lodge with upright walls; no door or covering; stone placed on the trail became a huge rock where direction is fixed	Cloud Peak, highest elevation (13,167 feet) in the Big Horn mountain range of eastern Wyoming; broad, level top; peak often shrouded by cloud cover and inclement weather resulting from alpine convection; steeply eroded slopes resulting from glaciations and freeze/thaw; upper portion of peak viewed from the east appears striped vertically; large boulder situated at apex; most accessible ascent/ descent provided by narrow ridge located at west side approach with well-established climber's trail extending down slope along the Paint Rock Creek drainage and leading to the Bighorn River. Top of Cloud Peak—Latitude 44° 22' 57," Longitude 107° 10' 26"

Comparison of Mythological Locations based on Walker (2006) and Actual Geographic Locations Based on Site Observations		
Location	Mythic Place Description	Actual Place Description
North Direction/ Swenson Medicine Wheel	After traveling for one moon "We have traveled far and yet we are nowhere." Where shadows are longest; before the door of the lodge of Wazi; all stones were covered in ice, so ice placed on the trail; Wazi's lodge was beside the trail; lodge poles were icicles and its covering was snow; ice placed where direction is fixed on trail was a great bluff	Foot travel along the west side of the Bighorn Mountains, following the Bighorn River flowing northward; north of the mountain range the river continues north to its confluence with the Yellowstone River, which flows northeast to the Missouri River in northeast Montana; the Little Missouri River subparallels the Yellowstone River and Missouri River in western North Dakota, generally bounded by badlands, and empties into the Missouri River north of Dunn County; Killdeer Mountains located south of the Little Missouri River in northwest Dunn County, with tree-covered slopes and perimeter bluffs; from its confluence with the Little Missouri River, the Missouri flows east and south across central North Dakota and South Dakota. Midcontinental climate with accumulating snowfall and frozen soil conditions during winter. Killdeer Mtns (Medicine Hole)—Latitude 47° 26' 39," Longitude 102° 53' 28" Swenson Medicine Wheel, Dunn County, ND
East Direction/ Jennings Medicine Wheel	Travel beside a large lake; trail between the lake and the edge of the world was very narrow; danger of falling into the water; then came to a sandy region where no green thing grew and where there was little water; very thirsty; mirage. Grass was young and trees were budding their leaves; nearby forest; no stone found in darkness so place wand in ground; wand grown to be a huge oak tree	The right bank of the Missouri River, from the confluence with the Little Missouri River to Garrison Dam and farther south, generally is bounded by steep terrain and few perennial tributary streams; the river valley is bounded to the north and east by the Missouri Couteau extending southeast and then south across South Dakota; the couteau consists of glaciated terrain with numerous prairie pothole lakes and ponds, many streams and ponds are intermittent or ephemeral and contain alkaline water; increasingly warmer climate toward the south with significant tree cover limited to the banks of perennial streams; cobbles and boulders limited to glacial erratics; western limit of oak species east of the Missouri River. Jennings Medicine Wheel, Hand County, SD
South Direction/ Scotts Bluff	Where great trees grew close together; fruit trees. Beautiful shell placed where direction is fixed; grown to be a tipi rivaling the colors with which Anp decorated the sky; tipi shimmered in light of day	Oak, cottonwood, and other woody plants common in riparian zones of eastern South Dakota; transition to sand hills of Nebraska and the Platte River valley; bedrock exposures in western Nebraska including Scotts Bluff situated near the south bank of the North Platte river, consisting of brightly colored sandstone, a landmark during westward expansion in the 1800s. Scotts Bluff—Latitude 41° 50' 16', Longitude 103° 42' 2"

Comparison of Mythological Locations based on Walker (2006) and Actual Geographic Locations Based on Site Observations		
Location	Mythic Place Description	Actual Place Description
The World	Blue stripe around the edge; red stripe from north to south and east to west over its center to form four equal parts; twelve moons of time around the edge of the world; green disk; top of disk (north) is white; mountain at left (west) edge; colors of Anp at right (east) edge; red color at bottom (south) edge	Traditional Lakota territory generally bounded by the Bighorn River to the west; Yellowstone River to the northwest; Missouri River to the north; upper Missouri Couteau to the east; White River, Niobrara River, and sand hills to the south; North Platte River to the southwest; territory centered generally by the Black Hills and Bear Butte; territory extends into northern plains of North Dakota, the Bighorn Mountains to the west, the west side of the James River valley and the west edge of deciduous forest to the east, and colored sandstones exposures in western Nebraska to the south. Approximate length of circumnavigation is 1,535 miles (2456 km) equivalent to traveling about 4.2 miles (6.7 km) per day for one year.
Edge of the World	Trail; mountains, valleys, rivers, forests, or plains will sometimes be on one side and sometimes on the other	The perimeter of traditional Lakota territory follows major rivers, the major axis of the Bighorn Mountains, the badlands and dissected valley of the Little Missouri and Missouri rivers, crosses the Missouri river at two locations (central North Dakota and southern South Dakota), extends onto the Missouri Couteau east of the Missouri River, and across portions of the sand hills of Nebraska; including short-grass, mixed-grass and long-grass prairies, pine forests, and riparian woodlands

Indeed, if those places represent the center and cardinal directions of traditional Lakota territory by the latter 1700s, then the Cloud Peak, Swenson, and Jennings sites might provide archeological evidence for demarcating the extent of that territory from a variety of social and cultural perspectives ranging from geopolitical boundary markers to structures serving ceremonial and ritual functions, defining the limits of that territory and providing context to the traditional Lakota lifestyle and culture. The features would also have been supplemented by their under-standing of seasonal and astronomic observations. Since the myth provides reasonable descriptions of those actual places, it follows that the myth did not fully develop until after the Lakota determined the four cardinal directions at the limits of their territory.

How is this information tied to symbolism? Evidence of Lakota migration across the Great Plains, in tandem with Lakota mythology as described by James Walker, suggests that the myth of fixing the four directions by the sons of Tate was formulated during the latter eighteenth century or early nineteenth century and that the construction of medicine wheels to demarcate the actual extent of Lakota territory might have occurred at about the same time. The timeframe for development of the myth represents the latest potential period for the myth to have been fully developed, as it assumes that the myth did not develop until the Lakota were the primary occupants of the region extending from the southern Missouri Couteau to the Bighorn Mountains, and from the confluence of the Missouri and Little Missouri Rivers to Scotts Bluff. It remains possible that the myth developed prior to such occupancy by the Lakota. Nonetheless, according to the myth, the sons of Tate completed their journey in twelve months.

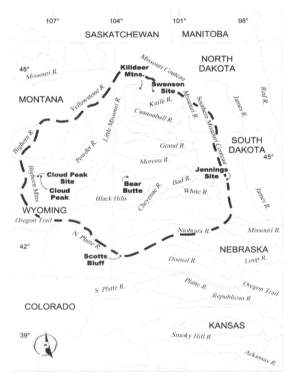

Figure 3.6: Extent of traditional Lakota territory based on historic and ethnographic information, locations of medicine wheels, and actual physiographic features (from Burley 2007).

Based on historical and ethnological information, site observations, and archeological documentation, the four cardinal directions described in Lakota mythology have parallels with natural physiographic features supplemented by medicine wheels documented at two of those locations. Those physiographic features include Cloud Peak (west), Killdeer Mountain (north), the Jennings site (east), and Scotts Bluff (south). Figure 3.6 depicts the potential extent of traditional Lakota territory based on that

evidence. Based on a 1,500-mile (2450 km) circumnavigation of this territory as depicted on Figure 3.6, the average rate of travel by the sons of
Tate along the edge of the Lakota world would have been about 4.2 miles
(6.7 km) per day. Given the myth's descriptions of the sons' slow travel
and occasional overnight stays without progress along the trail, the estimated rate of travel is a very reasonable estimate of what might have been
accomplished by Lakota traveling a similar trail during that latter 1700s.

The Sacred Hoop

Royal B. Hassrick described the Sioux as a "systematic people. They
were organizers and classifiers. As the universe was intricately patterned
into hierarchies and divisions, so was the nation. Theirs was a form of
government, a political concept that incorporated their flair for logic with
the pragmatics of successful group-living in a difficult and dangerous
world."[33] By extension, construction of medicine wheels in association
with the cardinal directions demonstrates their "inherently strong sense
of possessiveness. . . . [In] the context of their cultural meaning, there is
the suggestion that true wisdom and all it implied—living the Sioux
way—was so difficult of realization that many men would normally fail
to make the effort. If enough individuals were indisposed or unable to
accept the ideal as a real one, the very spirit, which made Sioux life operative, would have no sustenance. The Sioux way was successful only to
the extent that enough men believed in it to make it work."[34]

This certainly would have been the case for the efforts in positioning
and constructing the medicine wheels. If medicine wheels supplementing
the natural physiographic features at three of the four cardinal directions
indeed served as boundary markers on the west, north, and east edges of
Lakota territory, then those benchmarks represent one of the greatest
feats of surveying in any culture, covering an area of about two hundred
thousand square miles (over five hundred thousand sq km) with remarkably accurate astronomical and linear measurements using the unaided
eye across hundreds of kilometers and logging distances measured in
terms of days, nights, moons, and years. Such an undertaking requires
not only planning and organization, but also understanding of the
temporal and cyclical movements of the sun, moon, and stars as seen
from the earth.

However, medicine wheels also demonstrate an understanding of place and the relationship between artifact, creator, and environment. As noted previously, Cajete believes that structures and symbols of Native science, used as metaphors for indigenous knowledge and creative partic-ipation with the natural world in both theory and practice, serve as bridges between the inner and outer realities. He notes that the science of indigenous peoples is grounded on an understanding of perspective and orientation. As indigenous ceremonial cultural artifacts, medicine wheels constructed by the Lakota would have been "created with an acutely developed understanding and acknowledgement of the natural elements from which they were created . . . each component . . . carefully chosen with regard to its inherent integrity of spirit and its symbolic meaning within [Lakota] traditions."[35] Through the physiographic context of each medicine wheel, and the oneness of creative process, artifact, and creator, the wheels became artistic creations "of seeing . . . of being and of becoming."[36]

The extent of Lakota territory depicted in Figure 3.6 is based on a comparison of Lakota mythology and the actual physiography of the northern Great Plains. Those limits provide an Earth-sized representa-tion of a subgroup six medicine wheel as defined by John Brumley, with the stone ring represented by the edge of their territory (typically located along major rivers and drainage divides), two interior stone lines repre-sented by the east–west line between Cloud Peak and the Jennings site, and two additional interior stone lines represented by the north–south line between Killdeer Mountain (Medicine Hole) and Scotts Bluff. This configuration of natural features can also be symbolized by a circle—the Sacred Hoop—with Bear Butte at the center and the four features located to the west, north, east, and south, as shown in Figure 3.7.

Significantly, the Lakota and other Native American tribes use the Sacred Hoop to represent their respective nations. Thus, the limits of traditional Lakota territory may be viewed as the Sacred Hoop within which Lakota culture thrived into the mid-1800s. That hoop configura-tion would be in keeping with the importance of the circle as a metaphor for all that is natural and where power can be maintained.

William K. Powers demonstrated that "boundaries of Oglala ethnicity are synonymous with the boundaries of religious belief,"[37] further stating that "the boundary which delineates Oglala religion, then also delineates Oglala society."[38] From such strong interweaving of society, culture, religion, and an inherent sense of possessiveness, it should not be surprising that identification of natural landmarks and construction of medicine wheels would be undertaken by the Lakotas in delineating not only their territory, but their ethnicity, religious beliefs, and society as well. Such an undertaking helps to confirm the detailed knowledge the Lakota held and the importance they placed on the nature of places they inhabited. The location of medicine wheels and associated physiographic features may express for the Lakota not only an understanding of regional geography, but also the role of geography in the physical, emotional, intellectual, and spiritual components of their Native science, through what Cajete describes as "learning the language of place and the 'dialects' of its plants, animals, and natural phenomena in the context of a homeland."[39]

Figure 3.7: The Sacred Hoop envisioned by the Lakota to extend from Bear Butte to each of the four cardinal directions (from Burley, 2007). Emplacing the Sacred Hoop onto Earth's (Mother's and Grandmother's) surface was an artistic creation "of seeing . . . of being and of becoming" (Franck, 1981).

Chapter 4

The Round Red Stone of the Lakota

No distinction can be drawn between art and contemplation. The artist is first of all required to remove himself from human to celestial levels of apperception; at this level and in a state of unification, no longer having in view anything external to himself, he sees and realizes, that is to say becomes, what he is afterwards to represent. . . .

ANANDA COOMARASWAMY
THE DOOR IN THE SKY

A True History

We now turn to another Lakota myth, the bringing of the Sacred Pipe (Sacred Calf Pipe, Buffalo Calf Pipe, Calf Pipe) as described in numerous publications and ethnographic analyses.[1, 2, 3] The Lakota, like other indigenous people of the Americas, do not concur that these stories are myth. Rather, the stories describe what is held to be the true history of the Lakota people. I agree in that the stories provide context for traditional lifeways. Recall that mythologies include aspects of the physical, the intellectual, the emotional, and the spiritual. This is in complete agreement with, and understanding of, the lifeway of the Lakota people.

And it is in agreement with understanding and application of mythologies of other cultures found around the globe, as we will see in later chapters.

Black Elk was a Lakota holy man born in the 1860s. He lived through the Battle of the Little Bighorn. His account of the bringing of the Sacred Pipe to the Lakota is perhaps the most well-known version of the story, the heart of which is a detailed description of the seven religious rites of the Lakota.[4] Fools Crow states that the seven sacred rites were given to the Lakota and practiced after the Sioux had settled on the Great Plains, a timeframe that is in general accordance with the conclusion of John L. Smith, who estimated that the legend began sometime between 1785 and 1800.[5,6]

Black Elk provides an account of the Sacred Pipe as handed down by Elk Head (*Hehaka Pa*), a keeper of the Sacred Pipe, to three men, one of whom was Black Elk himself (*Hehaka Sapa*).[7] The Sacred Pipe is a symbol integrating Lakota life, ritual, and the universe.[8] Black Elk was a holy man of the Oglala, one of the seven sub-bands of the Lakota. Elk Head told Black Elk that he gave his account of the Sacred Pipe and the seven rites because "it must be handed down. For as long as it is known, and for as long as the pipe is used, their people will live; but as soon as the pipe is forgotten, the people will be without a center and they will perish."[9]

As described by Black Elk, a round rock given by White Buffalo Calf Woman (also referred to as *Wohpe*, Buffalo Calf Woman, and the Beautiful One who set moral codes and instructed the people in the mysteries of ceremony) represents Earth, Grandmother, and Mother, the place where the Lakota and all two-leggeds would live and thrive.[10] *Maka* is the Spirit of Earth, producing good or bad seasons, supplying a plentiful or scarce supply of vegetation, and providing medicines based on the invocations she hears from the people and the medicine man, based on her pleasure or displeasure at the beliefs and actions of the Lakota.[11]

Accounts of the giving of the Sacred Pipe to the Lakota typically provide a description of the initial physical appearance of the pipe; there are also some references to the appearance and condition of the pipe during the last few decades. The various renditions of the story as told by holy men of the twentieth century describe a round stone made of the same red stone as the bowl of the Sacred Pipe, the inscription of seven circles on the stone representing seven rites, the stone's purpose in representing Earth, and the use of the Sacred Pipe and stone in rituals. However, there is little definitive information regarding the stone's physical appearance, its association with the end of times as understood by the Lakota, and whether the stone remains today as part of the sacred bundle.

The Sacred Pipe, the Seven Rites, and Prophecy

Moves Walking was one of three band chiefs appointed to the high council when the tribal history began. It was Moves Walking who received the Sacred Pipe from White Buffalo Calf Woman. In Black Elk's telling of the gift of the Sacred Pipe to the Sioux, White Buffalo Calf Woman takes from a sacred bundle the pipe and "a small round stone which she placed upon the ground."[12] She says, "With this pipe you will walk upon the Earth; for the Earth is your Grandmother and Mother, and She is sacred. Every step that is taken upon Her should be as a prayer. The bowl of this pipe is of red stone; it is the Earth."[13] She touches the foot of the pipe to the round stone laying on the ground and says,

> With this pipe you will be bound to all your relatives: your Grandfather and Father, your Grandmother and Mother. This round rock, which is made of the same red stone as the bowl of the pipe, your Father Wakan Tanka has also given to you. It is the Earth, your Grandmother and Mother, and it is where you shall live and increase. This earth which He has given to you is red, and the two-leggeds who live upon the Earth are red; the Great Spirit has also given to you a red day, and a red road.[14]

This is sacred information White Buffalo Calf Woman does not want the Lakota to forget. She emphasizes that the coming of the morning sun is a holy event, as it is Wakan Tanka bringing light to the world. She tells them to keep in mind that all people are sacred and are to be treated with that understanding. She continues,

> From this time on, the holy pipe will stand upon this red Earth, and the two-leggeds will take the pipe and will send their voices to Wakan Tanka. These seven circles which you see on the stone have much meaning, for they represent the seven rites in

which the pipe will be used. The first large circle represents the first rite which I shall give you, and the other six circles represent the rites which will in time be revealed to you directly.[15]

White Buffalo Calf Woman outlines the rite of the Keeping of the Soul. She reminds the people how sacred the pipe is, and states that she is leaving but will look back on the Sioux in every age and return at the end. She then leaves, becoming in order a red and brown buffalo calf, a white buffalo, and then a black buffalo that bows to the four quarters of the universe and disappears over a hill.

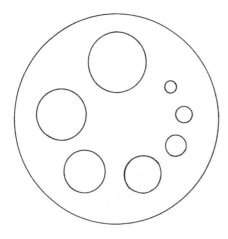

Figure 4.1: Seven circles arranged in a circle on the face of the round stone (based on a footnote illustration in Brown, 1989).

A footnote by Brown provides an illustration of seven circles arranged in a circle on the face of the round stone, the size of the seven circles progressively smaller from the "first large circle" to "the other six circles."[16] Figure 4.1 depicts the seven circles on the face of the round stone as illustrated by Brown. That arrangement of circles on the stone is not referenced by Brown to any source, including Black Elk or Brown himself. Another footnote calls out a statement by Black Elk that the pipe represents the universe and also man, and that by filling the pipe man identifies himself with his own center as well as establishing the center of the universe. Black Elk describes each of the seven rites, noting that as the holy keeper fills the pipe during each rite, he makes the ritual offering of tobacco to the six directions (west, north, east, south, up [sky], and down [Earth]), and follows with prayers. Iron Shell states that White Buffalo Calf Woman told the Sioux that they had been chosen to receive the pipe on behalf of all mankind, that the pipe was a symbol of peace to be used as such between men and between nations as a bond of good faith and the holy man smoking the pipe in communion with Wakan Tanka, the Great Mystery.[17]

Black Elk's description of the seven rites, supplemented with footnotes by Brown, includes numerous examples of the significance of and symbolic relationships between the universe, Earth, sky, rocks, the four cardinal directions, and man. Note in the following examples (three of many descriptions he provides in the book) how Black Elk mentions matter, energy, and geometry to demonstrate sacred relationships within these rites.

> Rocks . . . represent Grandmother Earth . . . they also represent the indestructible and everlasting nature of Wakan Tanka. The fire . . . used to heat the rocks . . . is as a ray from the sun, for the sun is also Wakan Tanka in a certain respect.[18]

> The great Thunderbird of the west: Wakinyan-Tanka . . . is really Wakan Tanka as the giver of Revelation (symbolized by the lightning) . . . It is . . . for the Indian the protector of the Sacred Pipe, for the pipe, like the lightning, is the axis joining heaven and earth.[19]

> The purified earth is now very carefully spread around inside the sacred central hole . . . this earth represents the universe . . . Next a cross is made by drawing a line on the ground from west to east, then one from north to south. All this is very sacred, for it establishes the four great Powers of the universe, and also the center which is the dwelling place of Wakan Tanka.[20]

In Black Elk's description of the rite Crying for a Vision, a cross is made by placing a pole in the center and four additional poles at each of the four cardinal directions. "This form has much power in it, for whenever we return to the center, we know that it is as if we are returning to Wakan Tanka, who is the center of everything . . . "[21]

> [The Sun Dance] is held each year during [June] or [July], always at the time when the moon is full, for the growing and dying of the moon reminds us of our ignorance which comes and goes; but when the moon is full it is as if the eternal light from the Great Spirit were upon the whole world.[22]

> A round rawhide circle should be made to represent the sun, and this should be painted red; but at the center there should be

a round circle of blue, for this innermost center represents Wakan Tanka as our Grandfather. The light of this sun enlightens the entire universe; and as the flames of the sun come to us in the morning, so comes the grace of Wakan Tanka, by which all creatures are enlightened. It is because of this that the four-leggeds and the wingeds always rejoice at the coming of the light. We can all see the day, and this seeing is sacred for it represents the sight of that real world which we may have through the eye of the heart. When you wear this sacred sign in the dance, you should remember that you are bringing Light into the universe, and if you concentrate on these meanings you will gain great benefit.[23]

There is much power in the circle . . . these four lines represent the Powers of the four directions. Black, you see, is the color of ignorance, and, thus, these stripes are as bonds which tie us to the earth. You should also notice that these stripes start from the earth and go up only as far as the breasts, for this is the place where the thongs fasten into the body, and these thongs are as rays of light from Wakan Tanka. Thus, when we tear ourselves away from the thongs, it is as if the spirit were liberated from our dark bodies . . . O Wakan Tanka . . . It is Your light which comes with the dawn of the day, and which passes through the heavens.[24]

In this manner the altar was made, and, as I have said before, it is very sacred, for we have here established the center of the Earth, and this center, which in reality is everywhere, is the home, the dwelling place of Wakan Tanka.[25]

The first peace, which is the most important, is that which comes within the souls of men when they realize their relationship, their oneness, with the universe and all its Powers, and when they realize that at the center of the universe dwells Wakan Tanka, and that this center is really everywhere, it is within each of us. This is the real Peace, and the others are but reflections of this.[26]

The use of this blue paint is very important and very sacred . . . for . . . the power of a thing or an act is in the understanding of its meaning. Blue is the color of the heavens, and by placing the blue upon the tobacco, which represents the earth, we have united heaven and earth, and all has been made one.[27]

[The Throwing of the Ball] as it is played today represents the course of a man's life, which should be spent in trying to get the ball, for the ball represents Wakan Tanka, or the universe . . . the sacred ball, which had been made from the hair of the buffalo and covered with tanned buffalo hide. He painted the ball red, the color of the world, and with blue paint representing the heavens, he made four dots at the four quarters; then made two blue circles running all around the ball, thus making two paths joining the four quarters. By completely encircling the red ball with the blue lines, Heaven and earth were united into one in this ball, thus making it very sacred . . . whoever catches the ball will receive a great blessing.[28]

Black Elk also describes the prophecies of Drinks Water (Drinking Cup, Wooden Cup), a Lakota holy man living west of the Black Hills during the 1800s.[29] Drinks Water foretold the coming of people from the east and south to the Lakota world, changing the land and entangling it with iron, the Lakota living in gray houses, the starving of the people, the loss of birds and four-legged animals, and the coming of great wars. But he also prophesied that Lakota culture and religion would survive. Similarly Worm, father of Crazy Horse, explained to his son that ancestors foresaw people coming from the east and destroying "our songs," but one day they would return and that instead of seven circles, there would be nine.[30] Worm said he did not know what this meant. Vinson Brown suggests that this occurred in about 1863 or 1864, when Crazy Horse would have been about twenty-four years of age.[31]

Vinson Brown and Kurt Kaltreider each describe a vision quest by Crazy Horse (His Crazy Horse) on Bear Butte, South Dakota, in about 1871. They state that Crazy Horse, having sent his cry to the seven sacred directions, asked, "O Thou Whose voice calls to me in the leaves of the pines, I have cried to you for the seven sacred circles, but tell me what is the meaning of the two circles to come?"[32,33] And in his vision, "Suddenly all the people's faces disappeared, and he saw the whole world as if it were in darkness, but there was a dawn light coming from the east and before it the daybreak star approached through the sky. It was a star with nine points, and he knew these represented the nine sacred circles, including the two new ones yet to come. Then he saw the sacred herb beginning to climb out of the earth."[34]

Circles, Round and the End Times

In Lakota culture and religion, all life forms found on Earth, as well as life forms perceived beyond Earth such as the sun, moon, and stars, are held to be *wakan*, "the dominant intangible symbol of Lakota religion, the core of Lakota belief."[35] *Wakan Tanka*, the totality of that life-force, the Great Mystery, created and embodied the universe. For the Lakota, Wakan Tanka and wakan life forms and symbols permeate human experience, their everyday life integrated with the sacred and incomprehensible. Seven Rabbits states that the Lakota did not write, instead remembering things using figures and symbols, such as using the circle to represent a camp, the sun, and the world.[36] Black Elk states that everything the Indian does is in circles, with the power of the Sacred Hoop.[37] Similarly, Lame Deer states that the circle represents togetherness of people, that the tipi was a ring in which people sat in a circle, and that all families in the village were in turn circles within a larger circle, part of the larger hoop that was the seven campfires of the Sioux, representing one nation. The nation was a part of the circular universe and made of the earth. The earth, moon, sun, stars, horizon, and rainbow are examples of circles within circles within circles, without beginning or end.[38]

In describing the rite of the Sun Dance, Black Elk states, "Since the drum is often the only instrument used in our scared rites, I should perhaps tell you here why it is especially sacred and important to us. It is because the round form of the drum represents the whole universe, and its steady strong beat is the pulse, the heart, throbbing at the center of the universe."[39]

The importance of the circle in Lakota life and beliefs is expressed in their stories. As noted in chapter 3, "A Myth of the Lakotas as It Is Told in Their Winter Camps," told by George Sword, describes the establishment of the four cardinal directions by the four sons of Tate. The four sons of Tate leave their home to establish the four directions at the edge of the world, beginning west of Tate's home and journeying subsequently to the north, east, and south, returning to the point established as west and then back to Tate's home. While the sons of Tate are on their journey, Wohpe is instructed by a witch to take a disk and paint it in a very specific way so as to divide the surface of the disk into several variously colored shapes. Part III through part XVIII of Walker's Literary Cycle (WLC) provide additional cultural information and context for that myth, noting that when the witch (Wakanka) gives the disk to Wohpe, the top (north) of the disk was white and the bottom (south) red, the left side (west) included a picture of a mountain, the colors of Amp (morning star) were on the right (east), and a lodge was pictured in the center of the rawhide disk. Wakanka states to Wohpe that the disk represents the world.[40] As I showed in chapter 3, medicine wheels at the edges of traditional Lakota territory define the boundaries of the world as illustrated on that disk, with the world centered on Bear Butte and the Black Hills of South Dakota as the *axis mundi*.

Black Elk states that the duty of the four quarters (quadrants of the earth) is to nourish and strengthen the flowering tree at the center, representing the growth of the people, and notes that the east gives the sunlight and peace, the south provides warmth, the west brings nourishing rain, and the north brings strong winds to give the people strength and endurance, all so that the tree may grow and flower.[41] This perspective allows greater understanding of humans' relationships to one another and the universe as well. Recalling his great vision, Black Elk states that when he appeared to be "standing on the highest mountain of them all, round beneath me was the whole hoop of the world," that the mountain in his vision was Harney Peak in the Black Hills, but "anywhere is the center of the world."[42] The circle was so symbolic of life for Black Elk that, sensing his inability to use the spiritual powers given to him in his great vision, he stated that he was "a pitiful old man who has done nothing, for the nation's hoop is broken and scattered. There is no center any longer, and the sacred tree is dead."[43]

Given the importance of the circle in Lakota culture and religion, it is reasonable to assume that the stone given with the Sacred Pipe would have been described as circular if it was shaped like a plate—flat with a circular periphery. However, Black Elk described the stone as being "round." Webster's Dictionary defines *round* as "that which is . . . or to be made round or rounded, as a circle, sphere, globe or cylinder."[44] As a three-dimensional object, the stone was probably more like a sphere than a plate. Lakota adjectives include *mammilla* (round), *gimlet* (round like a ball), and *gmigmiyan* (spherical, or round perhaps like a ball).[45] However, the word *mimela* is used to describe something round as like a disk and the adverb *yumimeya* for circular-wise movement.

The similarities of those words and the nuances of their meanings were potential sources of error in interpretation and transcription by ethnographers during interviews conducted with Black Elk and others. Brown simply interprets Black Elk's description of the shape of the red rock given by White Buffalo Calf Woman to the Lakota as round. However, I believe that Black Elk meant to describe a spherical object. The Lakota believe that small spherical stones have a spirit called *šicun*, an aspect of the soul lasting forever and able to be reinvested in other objects.[46] The use of the word *round* by Lakota speakers to describe a sphere or otherwise rounded three-dimensional object is supported by numerous statements by holy men referring to the shape or form of terrestrial and celestial objects. Again, as noted above, Brown interprets Black Elk as saying that birds and their nests are *round*, and Seven Rabbits recalled that the circle represented the sun and the world.

Lame Deer states that the Lakota nation was only a part of the universe, "in itself circular and made of the earth, which is *round*, of the sun, which is *round* . . . circles within circles within circles. . . . [emphasis added]."[47] Fools Crow states that his grandfather, Knife Chief, told the story of Crazy Horse returning from a vision quest on Bear Butte in the Black Hills, noting that Crazy Horse accurately described the physical shape of the world by noting that the sun rose, set, and then came up again, and therefore he knew that the world was round[48]—certainly meaning spherical, since a circular plate-like shape for the world makes no sense in this context. Fools Crow relates another story told by Knife Chief about three men who left the reservation in 1890 to travel the world and find out if the earth was round. Upon their return to the reservation, the travelers confirmed for the people that the world was indeed round. However, he goes on to say that the idea of a round world had been known to the Lakota well before the white man came to their country, noting that his great-grandfather was one of those who believed that the world was round and turned. According to his great-grandfather, the pipe was designed to be similar to the world, with a round bowl, and the universe located beyond the bowl. In addition, Fools Crow's father told him that people had debated whether stars were other worlds, that stars were round, and that they turned just as the earth.

Perhaps most telling of how Lakota holy men viewed the shape of Earth, as well as Earth's relationship to the universe, is Part XVIII of the WLC, which states, "So it is that there are four times moving in circles;

the day time, the night time, and the moon time; they circle above over the world and below in the regions under the world, but the year time moves in a circle around the world."[49] During his great vision Black Elk saw that the Sacred Hoop of the Lakota was one of many hoops that made one circle and that "anywhere is the center of the world."[50] These statements affirm that holy men knew that the world was spherical. I conclude that the description of the stone given by White Buffalo Calf Woman is a statement that the stone approximated a sphere and not the flat stone described by Brown.

In the telling of the Sacred Pipe, White Buffalo Calf Woman reminds the Sioux to remember how sacred the pipe is and says that she is leaving but will look back on the Sioux in every age and return at the end.[51] When the end will occur is not specified. As noted above, Elk Head told Black Elk that "as long as the Sacred Pipe and the seven rites are known, and as long as the pipe is used, the Lakota will live; but once the pipe is forgotten, the people will be without a center and they will perish."[52] Black Elk later stated that "the nation's hoop is broken and scattered. There is no center any longer, and the sacred tree is dead."[53] Fools Crow stated that by 1974 White Buffalo Calf Woman was in the United States accompanied by another young woman, and that this was a bad omen, suggesting that the end times were approaching. However, concluding his description of the rite The Throwing of the Ball, Black Elk said, "At this sad time today among our people, we are scrambling for the ball, and some are not even trying to catch it, which makes me cry when I think of it. But soon I know it will be caught, for the end is rapidly approaching, and then it will be returned to the center, and our people will be with it."[54]

Wheels of Stone

While between 70 and 150 wheels have been identified in Alberta, Saskatchewan, Montana, South Dakota, and Wyoming, the Bighorn medicine wheel is considered to be the type-site for medicine wheels in North America. The term *medicine wheel* was first applied to the Bighorn medicine wheel, the word *medicine* implying religious significance to Native Americans. The Bighorn medicine wheel is only reachable during the summer, when the area is free of snow for two months around the summer solstice. This certainly is the case for the Cloud Peak medicine wheel as well.

Bighorn medicine wheel (first described in a *Field & Stream* magazine article in 1895) was constructed by Plains Indians between three hundred and eight hundred years ago and has been maintained by various tribes since then. John Brumley provides several accounts attributing the wheel's construction and use variously to the Shoshone, Crow, Cheyenne, and Kiowa; he notes that a review of archeological evidence indicates no cultural stratigraphy, limited culturally diagnostic materials, and a range of explanations for construction of the structure.[55] According to the Wyoming State Historic Preservation Office (SHPO), Bighorn medicine wheel is situated within a complex of interrelated archeological and traditional use sites covering a much larger area, and those sites exhibit seven thousand years of Native American experience in an alpine landscape surrounding Medicine Mountain. The sites feature "ceremonial staging areas, medicinal and ceremonial plant gathering areas, sweat lodge sites, altars, offering locales and fasting (vision quest) enclosures."[56] The Wyoming SHPO recognizes the importance of this complex. "Ethnohistoric, ethnographic, and archeological evidence demonstrates that the (Bighorn) medicine wheel and the surrounding landscape constitute one of the most important and well preserved ancient Native American sacred site complexes in North America."[57]

The significance of Bighorn medicine wheel ranges from its use as a metaphor for the relationships between Shoshone tribes, their chief, and their gods, to a design plan for a Cheyenne medicine lodge, and as a theory of astronomical alignment for medicine wheels in general. Several accounts describe Bighorn medicine wheel as a focal point of Crow mythology and oral history, including conduct of vision quests at the site or in its immediate vicinity.[58,59] John Eddy studied Bighorn medicine wheel and suggests that the arrangements of rocks, cairns, and stone radii identify the rising and setting places of the sun at summer solstice and rising places of stars, including Aldebaran in Taurus, Rigel in Orion, and Sirius in Canis Major.[60] Based on the astrological alignments of those stars, Eddy suggested that the medicine wheel was last used for calendar purposes (marking the summer solstice and the referenced stars) between two hundred and seven hundred years ago. Astronomer Jack Robinson later found a pair of cairns at that feature marking the rising point of the star Fomalhaut with the sun twenty-eight days before summer solstice.[61]

The Royal Alberta Museum suggests that such a calendar would be used for the timing of important rituals.[60] The Sun Dance, usually held once a year during summer solstice, was the most important religious ceremony of the Plains Indians culture during the nineteenth century. Tribes who practiced sun dance ceremonies include the Arapaho, Cheyenne, Crow, Sioux, Shoshone, and Kiowa, each having occupied land at or in the vicinity of the Bighorn Mountains during prehistoric times or later. Brumley states that "Eddy's case for an astronomical alignment of the Bighorn and other medicine wheels has generated both considerable support and opposition."[62]

Cloud Peak medicine wheel is constructed of unmodified natural stones. It includes a prominent, centrally located stone cairn, a circular ring of cobbles and boulders, and eight stone lines radiating outward from the cairn to the circular ring. Reddish iron staining is apparent on stones included in the wheel's construction and numerous stones nearby. Many of the stones that make up the wheel appear embedded as much as 0.5 to 1 inch into the ground surface. The medicine wheel and ancillary features are likely not more than a few hundred years old.[63]

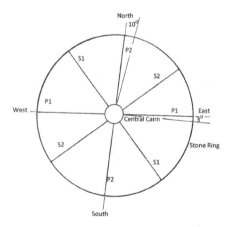

Figure 4.2: Size and orientation of primary and secondary axis of Cloud Peak Medicine Wheel (from Burley, 2007).

The diameter of the ring of Cloud Peak medicine wheel, measured from center to center of perimeter stones, is a uniform 88 inches (refer to Figure 4.2). The primary axes (P1, P2) bisect the ring through the central cairn and are constructed of cobbles and boulders resulting in a diameter of approximately 83 inches within the ring of stones. Subdividing each space between the primary axes is a secondary axis (S1, S2) constructed of gravel and cobbles 2 to 3 inches in diameter. Thus, there are eight approximately equally spaced spokes radiating from the cairn to the stone ring. The orientation of P1 is about 3 degrees south of east, and P2 is approximately 10 degrees east of true north. The two primary axes are within about 7 degrees of perpendicular, the two secondary axes are within about 5 degrees of perpendicular, and the primary axes within about 2.5 to 3 degrees of bisecting the angles between the secondary axes.

The Royal Alberta Museum states that it is very difficult to prove the astronomical hypothesis for the use of medicine wheels, noting that some astronomers in Canada and the United States "have critically evaluated the idea and have expressed severe reservations about the hypothesis."[64] They note that "simple familiarity with the night sky would likely produce an adequate estimate for timing ceremonies. Further, if great accuracy was desired, it could have been attained better by using narrow poles as foresight and back sight than by using wider rock cairns."[65] Those arguments do not consider the physiographic context of individual medicine wheels, particularly those located in areas of significant topographic relief, and specifically sites situated in alpine environments and accessible only during a few months in summer. Local climatic conditions at the Bighorn and Cloud Peak medicine wheels dictate visitation and use only when the ground is sufficiently clear of snow and ice. The rugged topography, remote locations, and general lack of resources for sustained occupancy at those features necessitates planning, organization, and the transportation of supplies to and from those locales in order to accomplish ceremonial activities to be conducted within a narrow timeframe, such as observation and acknowledgement of summer solstice.

Returning to the rituals of the Lakota, it is readily apparent that the configuration of a subgroup 6 medicine wheel, and particularly Cloud Peak medicine wheel with its central cairn, stone ring, and eight stone radii, conforms to the configuration of rocks placed to form a traditional Lakota altar as described by Black Elk:

> One of the rocks . . . is placed at the center of the round altar; the first rock is . . . at the center of everything . . . The second rock is placed at the west . . . the next at the north, then one for the east, one for the south, one for earth, and finally the hole is filled up with the rest of the rocks, and all these together represent everything that there is in the universe.[66]

Again, in use of earth [soil] and lines drawn on the ground surface:

> A pinch of the purified earth was offered above and to the ground and was then placed at the center of the sacred place.

Another pinch of earth was . . . placed at the west of the circle . . . earth was placed at the other three directions, and then spread evenly around within the circle . . . He first took up a stick, pointed it to the six directions, and then, bringing it down, he made a small circle at the center; and this we understand to be the home of Wakan Tanka. Again, after pointing the stick to the six directions . . . [he] made a mark starting from the west and leading to the edge of the circle. In the same manner he drew a line from the east to the edge of the circle, from the north to the circle, and from the south to the circle . . . everything leads into, or returns to, the center.[67]

And, in "A Myth of the Lakotas as It Is Told in Their Winter Camps,"

Wohpe is instructed to take a disk and paint it green; paint a blue stripe around the edge of the disk; paint a broad red stripe across the disk, over its center; paint another broad red stripe across the disk over its center so as to divide the disk into four equal quadrants; and paint four narrow red stripes across the disk over its center and between the red lines so as to divide the disk into equal parts [eighths].[68]

In tandem with the painted disk described in the WLC by Walker, Sword provides a description of the configuration of stones forming Cloud Peak medicine wheel (Figure 4.3), a two-dimensional depiction of the traditional Lakota territory, or world.[69] As noted in the first and second descriptions above, the configuration represents everything there is in the universe, and the center is the home of Wakan Tanka.

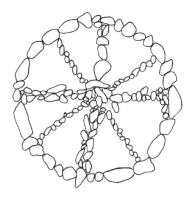

Figure 4.3: Cloud Peak medicine wheel, a two-dimensional depiction of the traditional Lakota world.

Surprisingly, when viewed from above the center of the cairn, Cloud Peak medicine wheel with a two-dimensional configuration of lines formed by the cairn, circle, and eight stone radii, matches the configuration of lines of a particular three-dimensional polygon, a structure that is also encountered in mythic symbolism recounted around the world. The highest degree of symmetry is expressed within this isometric crystal form. It approaches *sphericity,* or roundness, as the Lakota might have said. Indeed, projected onto a sphere, it yields perfect symmetry, with infinite planes of symmetry passing through its center, the presence of infinite rotational axes, and the same apparent form under any given amount of rotation.

Nine Circles

As told by Black Elk, the myth of the Sacred Pipe provides a cursory description of the round red stone and the seven rites given by White Buffalo Calf Woman, the mediator between Wakan Tanka and the people. The seven rites in which the pipe is to be used are represented by seven circles on the stone. One of the circles is referred to as the *first large circle*, representing the first rite given by Wohpe that is practiced today in the pipe ceremony. The remaining six circles represent the other six rites. Joseph Epps Brown represented the seven rites with seven circles of different sizes without evidence to support his depiction of a plate-like stone with the seven circles.

However, I believe that each of the seven circles on the stone is of equivalent size—seven *large* circles, or great circles. Lakota holy men knew, or at least perceived, that the earth was spherical. The shape of Earth was evident by the shapes of the sun, moon, stars, and those celestial life forms appearing to rotate around the world. Except for rock, the Lakota found that everything in nature was circular, rounded. The Sacred Hoop of the Lakota is one of many hoops that made one circle, circles within circles within circles: a sphere. The Sacred Hoop represents the universe, the four cardinal directions, the four associated quadrants, and the *axis mundi* extending from the center of the earth to the stars.

Wohpe's painting of the disk provides a template for subgroup 6 medicine wheels: the peripheral circle, four thick radii oriented along the four cardinal directions, four thinner radii bisecting the right angles delineated by the thicker radial lines and forming the four quadrants, and the central circle or cairn. All paths lead to the center, Wakan Tanka. Therefore Cloud Peak medicine wheel represents the universe, and the Sacred Hoop depicts the traditional Lakota territory of the late eighteenth and early nineteenth centuries—the Lakota world.

With holy men believing correctly that the shape of the earth approximated a sphere, medicine wheels are simple representations of a two-dimensional view of Earth. The four thickened radii signifying the cardinal directions, in tandem with the central cairn, or *axis mundi*, are four vertices extending orthogonally from the center of Earth through the equator, and two vertices extending from the center of Earth through the north and south poles. Five of those great circles are depicted in the Cloud Peak medicine wheel (Figure 4.4). Seven great circles on a sphere can be projected onto a plane (such as a portion of Earth's surface) to represent the four cardinal directions, the four quadrants, the *axis mundi*, and the boundary of the Lakota world, serving also as a tool for predicting the arc of the sun at summer solstice (Figure 4.5). This is the pragmatic value of the medicine wheel. As a representation of the universe, with all lines leading to the center, this is the power of the Sacred Hoop. I believe this also was the construct of seven circles on the round stone as described by Black Elk. The round stone is symbolic of Earth, with the spirit of *šicun*, lasting forever, able to be reinvested in other objects—animate and inanimate—and with the power of the Sacred Hoop including the seven sacred rites.

Figure 4.4a: Cloud Peak medicine wheel represents five great circles on a sphere (Earth).

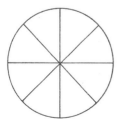

Figure 4.4): Line drawing of the geometrical structure of the medicine wheel. Straight lines depict great circles extending through the center of the circle.

What of the remaining two great circles? As stated above, there are nine great circles envisioned on the surface of the sphere. Later we will see that the intersection of each vertex becomes a node through which four great circles can be perceived extending across Earth's surface, each circle passing through two nodes at opposite ends of the surface. The result is nine great circles on the surface of a sphere. The two additional great circles mirror the two great circles that form the ellipses shown within the peripheral circle shown in Figure 4.5. They complete the framework of a sphere. However, as yet they appear to have no known cultural value to the Lakota on a medicine wheel, and no understood aspect of the Sacred Hoop. "Then I was standing on the highest mountain of them all, and round about beneath me was the whole hoop of the world," the stenographic notes taken during Black Elk's description of his vision continue, "And while I stood there I saw more than I can tell and I understood more than I saw; for I was seeing in a sacred manner the shapes of all things in the spirit, and the shape of all shapes as they must live together like one being. And I saw that the sacred hoop of my people was one of many hoops that made one circle. . . ."[70]

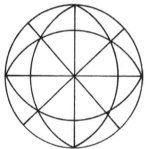

Figure 4.5: Line drawing of seven great circles about a sphere (Earth), representing the seven rites of the Lakota. Compare with Figure 4.4. Is this the true form of seven circles on the round red stone representing Earth and given to the Lakota by White Buffalo Calf Woman?

As a boy crying out for a vision, did Black Elk understand that Earth was round, a sphere, constructed of many great circles, many hoops? On a sphere the vertex (*axis mundi*) passing though any point on the surface is equivalent to any other vertex (*axis mundi*) at any other point on the surface, as they all pass through the center of the sphere, have equal lengths between the center and the surface, and are perpendicular to the surface. Is this why Black Elk said, "In this manner the altar was made, and, as I have said before, it is very sacred, for we have here established the center of the Earth, and this center, which in reality is everywhere, is the home, the dwelling place of Wakan Tanka"?[71] Is this what the round stone will represent when upon it the last two circles appear?

Before we can answer these questions in our pursuit of a worldwide connection between indigenous and ancient mythologies and sacred geometry, we need to look into the myths of ancient cultures beyond the Americas. As Cajete explains, "All the basic components of scientific thought and application are metaphorically represented in most Native stories of creation and origin. Indeed, both Native science and modern science have elements of the primal human story in common."[72] We turn now to the earliest records of civilization.

Chapter 5

Preservation of Life in Ancient Sumer

The serpent, who dies and is resurrected, shedding its skin and renewing its life, is the lord of the central tree, where time and eternity come together . . . The Garden is the serpent's place. It is an old, old story. We have Sumerian seals from as early as 3,500 BC showing the serpent and the tree and the goddess, with the goddess giving the fruit of life to a visiting male. The old mythology of the goddess is right there . . . the Goddess is the serpent. This is the symbol of the mystery of life. The male-god-oriented group rejected it. In other words, there is a historical rejection of the Mother Goddess implied in the story of the Garden of Eden . . .

JOSEPH CAMPBELL
THE POWER OF MYTH

In this chapter we peer into the beginnings of civilization and an early conception of the Sun Cross, the Tree of Life. Sumerian myths explain the origin of human beings, providing an explanation of how, why, and where man was created and questioning whether humans can attain everlasting life and, if so, how. We will decipher geometrical information provided in myth that could provide a possible early conceptualization of the spherical world. Lastly, we review information provided in two additional Sumerian myths that help us meld geometry, geography, and mythology into a context that illuminates the meaning of the sphere as a metaphor for the universe.

In the Beginning

The most commonly cited location of the cradle of civilization is the Fertile Crescent extending from the east coast of the Mediterranean to the Tigris-Euphrates rivers, between southern Turkey and the Arabian Desert (Figure 5.1). The earliest known gardens, dating from the twelfth to the fifth centuries BC, are associated with Neolithic villages located in the foothills of the Taurus and Zagros Mountains bordering northern Mesopotamia (an area that presently includes Turkey, Syria, Iraq, and Iran, and through which the Tigris–Euphrates river systems flow toward the Persian Gulf). Mesopotamia (Greek: *meso* meaning middle or between, and *potamia* meaning rivers) was the location of ancient Sumer, now referred to as Sumeria. Farming villages appeared on the north Mesopotamian steppe (*edin*) and plain by the seventh century BC and

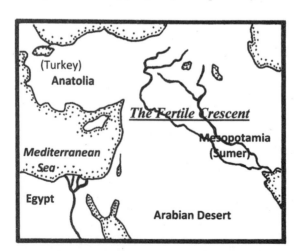

Figure 5.1: Location of the Fertile Crescent, often cited as the cradle of civilization.

extended farther south over the course of centuries. Villages with irrigated gardens supported those rural developments.[1]

Even at that early stage of civilization a shortage of labor appears to have caused concern for the urban populations. Campbell notes that Mesopotamian myths suggest man was created to relieve the gods of menial labor: "[O]ne of the chief characteristics of Levantine mythology here represented is that of man created to be God's slave or servant. In a late Sumerian myth retold in *Oriental Mythology* it is declared that men were created to relieve the gods of the onerous task of tilling their fields. Men were to do that work for them and provide them with food through sacrifice."[2]

The Akkadian Empire was located in the east portion of the Fertile Crescent, in the area we know today as Iraq. Its height of power occurred between 2300 and 2100 BCE and preceded the states of Babylonia and Assyria. The Akkadian language is related to Hebrew. By 2000 BCE the Sumerian language was replaced by Akkadian across Mesopotamia except in areas of religion and science.

The epics of both Atra-Hasis and Gilgamesh were developed by city dwellers in southern Mesopotamia before 2000 BCE. In the Akkadian epic of Atra-Hasis, dated to the eighteenth century BCE, the Akkadian gods domesticated plants and animals, developed cities, maintained irrigated city-gardens, and created man. With current archaeological evidence indicating that the city of Eridu dates from no earlier than about 6000 BCE, the myths describe what appears to be the first documented attempts at social and civic development in the southern Mesopotamian plain eight thousand years ago.

The Epic of Atra-Hasis is a creation myth. It includes poetic descriptions of the creation of humans by the gods, and the great flood that the gods caused to cleanse the world, similar in purpose as that of the biblical flood in which Noah builds his ark. The name Atra-Hasis means "extra-wise" and might be related to a name appearing on one of the Sumerian king-lists, Atrahasis, king of Shuruppak in the years before the great flood. The epic survives on three clay tablets written in cuneiform Akkadian. The first English translation of the Epic of Atra-Hasis was made by W. G. Lambert and A.R. Millard.[3] In antiquity the only title given to the Sumerian epic is provided in the colophon at the end of each cuneiform tablet, *inuma ilu awilum*, "When the gods like man."[4]

Presented from a Sumerian point of view, the Epic of Atra-Hasis describes how the gods worked and suffered before humans were created. Their work included digging and maintaining city-owned canals for the purpose of irrigation. The story introduces us to the greater and lesser gods, the seven great Anunnaki. Of note are three senior gods: Anu, ruler of heaven, Enlil, ruler on Earth, and Enki, who lives in the Apsu, an underworld of water. It is Enki who orders the Igigi (lesser gods) to build the canals. The Igigi, humiliated by this work after forty years, rebel and take their grievance to Enlil. Enlil summons Anu and Enki to unite with him against the Igigi and requests that Anu go to heaven and return with one lesser god to be put to death as a solution to the Igigi revolt. Belet-ili, the birth goddess, also is summoned. She notes the great creating skill of Enki, who must give her the clay so that she can create man. Enki prepares a purifying bath. The Igigi spit on the clay, and Enki treads upon the clay in Belet-ili's presence. Anu returns from heaven with a lesser god who is then sacrificed and his flesh and blood combined with the clay to create man, destined to continue the physical labor previously performed by the gods. This labor includes developing and maintaining the gardens at the city of Eridu, building and clearing irrigation canals, planting seed, hoeing weeds, harvesting crops, and presenting the produce to Enki at his temple.

The Eternal Search for Eternal Life

Let's look now at the Epic of Gilgamesh, the story of another mythological hero-king whose character appears to have been based on an ancient historical king, this time Gilgamesh of Uruk. Gilgamesh lived during the twenty-seventh century BCE in Sumeria, on the Euphrates River in what is now central Iraq. The Epic of Gilgamesh is another ancient Mesopotamian myth in the form of a poem or song. It is perhaps the oldest story in written form; the earliest complete version is written in cuneiform (wedge-shaped) script on twelve clay tablets that were held in the library of the seventh-century BC Assyrian King Ashurbanipal. It is well known that the Tigris-Euphrates drainage area includes the location of one of the first civilizations, and the Epic of Gilgamesh, as may be expected of mythologies in general, focuses on the adventures of a godly protagonist as metaphor explaining ancient cultural history, religion, and cosmographic ideology. Andrew George notes that this "Babylonian epic was composed in Akkadian, but its literary origins lie in five Sumerian poems of even greater antiquity. The Sumerian texts gained their final form probably as court entertainments sung for King Shulgi of Ur of the Chaldees, who reigned in the 21st century BCE."[5] Unlike latter Akkadian, the Sumerian language does not appear to have any known historic relation to other languages. Tablet 1 begins:

> The one who saw all [Sha nagba imuru] I will declare to the world,
> The one who knew all I will tell about
> [line missing]
> He saw the great Mystery, he knew the Hidden:
> He recovered the knowledge of all the times before the Flood.
> He journeyed beyond the distant, he journeyed beyond exhaustion,
> And then carved his story on stone.[6]

The epic describes Gilgamesh as two-thirds god and one-third human. His city of Uruk is already well established. Gilgamesh is the world's greatest king and strongest superhuman, but he's young and his rule over his subjects is oppressive. Newly married brides of Uruk are subject to the legal right of Gilgamesh as the lord (king) to take their virginity on their wedding night (*ius primae noctis,* law or right of the first night), while young men are subject to forced labor or other physical trials in service of the city's needs. The people petition Anu, chief god of Uruk, for help, and he creates Enkidu, a primitive (wild) man living in the *edin* surrounding Gilgamesh's lands. In their book *Hebrew Myths: The Book of Genesis*, Robert Graves and Raphael Patai note that "some elements of the Fall of Man myth in Genesis are of great antiquity . . . The Gilgamesh Epic, the earliest version of which can be dated about 2000 BC, describes how the Sumerian love-goddess Aruru [Anu] created from clay a noble savage named Enkidu."[7] Enkidu's strength is equivalent to that of many wild animals. His purpose is to distract Gilgamesh from oppressing his people, serving as counterpoint to the superhuman king.

The primary focus of our interest in the Epic of Gilgamesh is Enkidu, who becomes Gilgamesh's close companion during much of the story. Enkidu's death in the latter part of the epic causes Gilgamesh distress, and he begins a quest for immortality. But before this quest is ended, Gilgamesh will be told, "The life that you are seeking you will not find. When the heavenly gods created human beings, they kept everlasting life for themselves and gave us death."[8] We will come back to this apparently unfortunate destiny of mankind to uncover its meaning as foretold to Gilgamesh.

Tablet 1 of the epic continues with the son of a trapper checking on the rope snares his father set for the *edin*'s wild animals. Instead he discovers Enkidu running naked and tearing up the snares. The son rushes to tell his father what he saw, but Enkidu is too powerful for the hunter to confront. Gilgamesh tells the trapper that he can entrap the wild man with a naked harlot-priestess of Uruk. After Enkidu has sex with her, his animal companions will forsake him, and Enkidu will no longer tear apart the hunter's traps. The father advises his son to go into the city and take one of the temple harlot-priestesses, Shamhat, with him to the wilderness. The son leads Shamhat on a three-day journey into the wilderness to a watering hole to await the arrival of Enkidu and his companions, the gazelles, to drink at the watering hole. After three more days, the wild man and his companions arrive, and the half-naked Shamhat entices Enkidu to have sex with her. E. A. Speiser's translation of *The Epic of Gilgamesh* describes Enkidu's arrival at the watering hole:

> The creeping creatures came, their heart delighting in water.
> But as for him, Enkidu, born in the hills—
> With the gazelles he feeds on grass,
> With the wild beasts he drinks at the watering-place,
> With the creeping creatures his heart delights in water—
> The lass beheld him, the savage-man,
> The barbarous fellow from the depths of the steppe:
> "There he is, O lass, free thy breasts,
> Bare thy bosom that he may possess thy ripeness!
> Be not bashful! Welcome his ardor!
> As soon as he sees thee, he will draw near to thee.
> Lay aside thy cloth that he may rest upon thee.
> Treat him, the savage, to a woman's task! Reject him will his wild
> beasts that grew up on his steppe,
> As his love is drawn unto thee."
> The lass freed her breasts, bared her bosom,
> And he possessed her ripeness.
> She was not bashful as she welcomed his ardor . . .[9]

So Shamhat offers herself to him, and by submitting he immediately loses his strength and wildness. He can no longer run as fast as the gazelles. Enkidu's animal companions reject his companionship after he mates with Shamhat for six days and seven nights. Shamhat serves to replace the "creeping creatures" as a more suitable companion. Enkidu is no longer the wild man he once was. But at the same time he gains knowledge, understanding, and wisdom. He has "become wise like one of the heavenly gods."[10] Shamhat says:

> You are handsome, Enkidu, you are become like a god,
> Why roam the steppe with wild beasts?
> Come, let me lead you to ramparted Uruk . . .
> The place of Gilgamesh . . .
> Come away from this desolation, bereft even of shepherds.[11]

While lamenting his loss of wildness, Shamhat offers to take Enkidu into the city and introduce him to Gilgamesh, the only man worthy of Enkidu's friendship. Shamhat calls the watering hole a place of desolation and urges Enkidu to accompany her back to Uruk, leaving his animal companions.

Shamhat and Enkidu leave the watering hole and walk to Uruk, a three-day journey. Enkidu and Gilgamesh meet one evening at the marketplace in Uruk. They wrestle for a long while, but to a draw. One day they go to the cedar forest and slay the giant Humbala to remove all of the evil from the land. In so doing the King of Uruk demonstrates his strength to all the people. Upon returning to Uruk, Ishtar, the goddess of love and fertility, seeks to marry Gilgamesh, but he declines with a tirade of insults directed at her. Ishtar retaliates by sending the Bull of Heaven to kill Gilgamesh, but Enkidu leaps up, kills the bull, and issues another insult at Ishtar. Later, Enkidu becomes ill and curses the events that led him to the possibility of his own death. Gilgamesh cries alongside Enkidu's deathbed, recalling the adventures they had, and finally decides to seek how to find everlasting life.

In her introduction to the Epic of Gilgamesh, Nancy K. Sandars comments on Enkidu's curse on the harlot-priestess and the hunter who brought her to entrap him. Sanders compares Enkidu to Christianity's notion of the Fall of Adam, but in reverse:

> Enkidu, the "natural man," reared with wild animals, and as swift as a gazelle. In time Enkidu was seduced by a harlot from the city, and with the loss of innocence an irrevocable step was taken towards taming the wild man. The animals now rejected him, and he was led on by stages learning to wear clothes, eat

human food, herd sheep, and make war on the wolf and the lion, until at length he reached the great civilized city of Uruk. He does not look back again to his old free life until he lies on his death bed, when a pang of regret catches hold of him and he curses all the educators. This is the "Fall" in reverse, a felix culpa shorn of tragic development; but it is also an allegory of the stages by which mankind reaches civilization, going from savagery to pastoralism and at last to the life of the city.[12]

Gilgamesh then travels across grassy plains and scorching deserts, over mountains and across seas, to discover how to prevent his own death. Eventually he meets Urshanabi, the boatman who helped Utana-pishtim (the Sumerian equivalent and possible precursor of biblical Noah) to survive the great flood. Gilgamesh destroys sacred stone figures that Urshanabi kept in the forest, and finally Urshanabi agrees to take Gilgamesh across the deep sea to meet Utanapishtim, who may tell Gilgamesh the secret to everlasting life. Unfortunately, it turns out that it was the sacred stone figures that enabled Urshanabi "to cross the deep sea without touching the Waters of Death."[13] Nonetheless, Gilgamesh and Urshanabi cross the sea and find Utanapishtim at the boat landing, whereupon Gilgamesh pleads, "I wish to talk with you about life and death. I know that you have found everlasting life and have joined the assembly of gods. I too wish to live on earth forever. Teach me what you know, so I can live as you do!"[14] Utanapishtim replies:

> Oh, Gilgamesh, do we build a house that will last forever? Do we seal arguments forever? Do brothers divide property into equal shares forever? Does hatred persist forever? Does the river rise and flood its banks forever? Should no one experience death? Since ancient times, nothing has been permanent. The shepherd and the noble have an identical fate—death . . . When the heavenly gods gather in assembly, they decree the fate of each human being. The gods determine both life and death for every human being, but they do not reveal the day of anyone's death.[15]

Gilgamesh continues to implore him: "Tell me, how did you acquire everlasting life? How did you join the assembly of the heavenly gods?"[16] And Utanapishtim says, "Gilgamesh, I will reveal to you a secret of the gods,"[17] and he begins his tale of surviving the great flood at the city of Shuruppak, on the banks of the Euphrates River. Utanapishtim says that

Enlil, ruler of all the gods, complained that the people on Earth had become too numerous and noisy, and he decided that a great flood would be used to destroy the people's lives. Ishtar supported Enlil, but Ea (Enki, ruler of the earth and Apsu [Absu] the sweet [fresh] water on earth and below ground, god of wisdom, arts, and crafts) does not agree.

Ea devises a scheme to save the human race from the ravages of the flood. Ea tells Utanapishtim, king of Shuruppak, that the heavenly gods will create a rainstorm that will flood the earth, and that Utanapishtim should abandon his worldly possessions to preserve his life and build a large ship (ark) to be called *Preserver of Life*. Ea tells him to be sure that the dimensions of the ark are equal in width and length, and at the proper time he must bring aboard his family and possessions, craftspeople, grain, seed, animals, and birds. Utanapishtim agrees to do so, but he asks that Ea draw on the ground a plan of the ship. Within five days Utanapishtim constructs the framework of the ark. The floor space measures one field (a unit of about 3,600 square meters or 39,000 square feet), and "the length, width, and height each measured 120 cubits (about 200 feet)."[18] Utanapishtim divided "the ark so that the interior had seven floors, and [he] divided each level into nine sections."[19] On the seventh day he completes preparations and, after two-thirds of the ship was in water, he loads it with provisions, people, and both wild and tame animals, as he had been told to do.

The rainstorm continues for eight days with a powerful south wind. After the ship rocks quietly on the water for a while, Utanapishtim opens the hatch to find the world still, the waveless surface of the sea, and all humanity not on board returned to clay. The ship settles on Mount Nisir. Utanapishtim frees three birds, a dove, a swallow, and a raven. The dove and swallow return to the ship after finding no place to land and rest. But the raven circles the ark after finding that the water had receded and does not return to the ship. The flood is ended. Upon seeing that some of the humans escaped the flood, Enlil is enraged and seeks who had permitted the survival. Ea claims it was not he "who revealed the secret of the great gods. Utanapishtim, the most wise, had a dream in which he discovered how to survive [Enlil's] flood."[20] Impressed by Utanapishtim's ability to survive the flood, Enlil brings "everlasting breath" to Utanapishtim and his wife "so that, like the gods, they may continue to live for days without end . . ."[21] Utanapishtim concludes his tale:

That is how it came to pass that my wife and I became like the heavenly gods and will live for days without end. Enlil himself conferred everlasting life upon us. But Gilgamesh, king of strong-walled Uruk though you are, who will call the heavenly gods to assembly for your sake so that you can find the everlasting life you are seeking?[22]

Gilgamesh returns to Uruk, but not before Utanapishtim helps him to locate and obtain a plant, "a secret thing created by the heavenly gods. The plant that you see growing deep in the water there is like a rose. Its thorns will prick your hands when you try to pick it. However, if you can gather that plant, you will hold in your hands the gift of everlasting youth. This plant cannot make you live forever, but it will keep you young and strong all the days of your life."[23] Gilgamesh is able to collect the plant, but later he loses it along a stream bank to a serpent that "smelled the appealing fragrance of the plant. The serpent glided out of the water, slithered up the bank, took hold of the plant with its mouth, and carried it back into the water. As it returned to the water it shed its skin, emerging younger and fresher looking."[24] Upon returning with Urshanabi to Uruk, Gilgamesh remarks how impressive the brickwork of the city's walls and temples are, "one man's supreme achievement!"[25]

In the Epic of Atra-Hasis we discover the reason for the making of humankind: to serve the gods and provide the labor necessary to develop and maintain the resources of the earth for the benefit of the gods. The Epic of Gilgamesh shows that the rewards of this work are not everlasting or eternal life. Even Gilgamesh, two-thirds of a god, cannot achieve eternal life or even eternal youth. However, we also discover that the gods revealed to Utanapishtim a secret of the gods, the bringing of "everlasting breath" to man "so that, like the gods, they may continue to live for days without end."[26] But what is the secret? How can man receive everlasting breath? Exactly what did Utanapishtim do to receive such a reward from the gods? To answer this, we need to review portions of two other Babylonian myths, the story of Enki and Ninhursag, and the Incantation Song to Utu.

Source of Life

The myth of Enki and Ninhursag tells of the meeting of Enki (god of the earth and the sweet water) and Ninhursag (or Ninhursag-Ki, Earth Mother), and their falling in love:

After Time had come into being and the holy seasons for growth and rest were finally known, Dilmun, the pure clean and bright land of the living, the garden of the Great Gods and Earthly paradise, located eastward in Eden, was the place where Ninhursag-Ki, the Earth Mother, Most Exalted Lady and Supreme Queen, could be found. There she lived for a season during the Wheel of the Year, when the Earth lay deep in slumber before the onset of Spring, in the land that knew neither sickness nor death or old age, where the raven uttered no cry, where lions and wolves killed not, and unknown were the sorrows of widowhood or the wailing of the sick. And it was in Dilmun at that time that Enki, the wise god of Magic and the Sweet Waters, the Patron of Crafts and Skills, met, fell in love and [lay] with the Lady of the Stony Earth, Ninhursag-Ki.

The Earth Mother's kiss did change the carefree and sexy Sweet Waters Lord: Ninhursag had wholly captivated him through the most profound of all bonds, the thread of enchantment and passion called Love. So profound the feeling was that the God of Sweet Waters, Magic and Crafts proposed to Ninhursag, with the enthusiasm of a young lover's heart.

Ninhursag looked around the land, her stony body, and remembered the taste of the wondrous moisture of the Sweet Waters God within herself. She wondered whether the land should not feel the same loving touch without. She said then to Enki:

"I heard your heart speak, Enki dearest. But if I feel your wondrous moisture within me, I look at the earth of Dilmun, also my body, and feel its longing for the gifts that you, dear heart, for sure can bring. Thus I ask you: what is a land, what is a city that has no river quay? A city that has no ponds of sweet water?"

Taken by surprise, Enki realized that indeed he had given his whole essence to the beloved, but forgotten to look after her Earthly Body, the land. He then rose to the challenge of providing water for the land with aplomb. He replied: "For Dilmun, the land of my lady's heart, I will create long waterways, rivers and canals, whereby water will flow to quench the thirst of all beings and bring abundance to all that lives."

Enki then summoned Utu, the Sun God and Light of the Day. Together, they brought a mist from the depths of the earth and watered the whole face of the ground. Then Enki and Utu created waterways to surround the land with a never-ending source of fertile Sweet Waters, and Enki also devised basins and cisterns to store the waters for further needs. From these fertile sweet waters flow the four Great Rivers of the Ancient World, including the Tigris and the Euphrates. Thus, from that moment on, Dilmun was blessed by Enki with everlasting agricultural and trade superiority, for through its waterways and quays, fruits and grains were sold and exchanged by the people of Dilmun and beyond.

Ninhursag rejoiced in Enki's mighty prowess and said to him:

> Beloved, the powerful touch of your sweet waters, the essence of Mother Nammu that lies deep within you, transformed the land, my stony body. I feel the power of life throbbing within to be revealed without upon my surface as I give joyously birth and sustenance to the marshes and reed-beds, that from now on will shelter fish, plant, beasts and all that breathes. Thus I call myself Nintur, the lady who gives birth, the Womb of the Damp Lands by the riverbanks.[27]

So Ninhursag rejoices in Enki's mighty prowess and gives birth to daughter Ninsar, goddess of vegetation. But while Ninhursag is away to oversee the growth of the physical world, Enki falls in love with Ninsar, they make love, and Ninsar bears a daughter, Ninkurra, future goddess of mountain pastures. One day, Ninkurra encounters a well that appears where none was before. Exploring the well, Ninkurra meets Enki, god of the sweet waters. Charmed by Enki, she submits to him and soon bears a daughter, Uttu, the Spider, weaver of patterns and life desires. Although Ninhursig returns and advises Uttu to stay away from Enki, Uttu eventually submits to Enki as her mother, grandmother, and great-grandmother had before. But Ninhursig is able to wipe out Enki's seed from Uttu's womb, and Uttu buries the seed in the depths of the earth. Later, eight plants grow where the seed was buried, and Enki eats the eight plants, fruit and all. He falls ill from eating the plants, and Ninhursig is convinced to return to him and wraps herself around him. Enki complains of the pain he feels at eight locations in his body. Ninhursig absorbs his pain and then bears eight more children, four gods and four goddesses.

This story explains the unity between Earth Mother's stony body and her children, the ground and its covering of plant life, including not only vegetation, but also the weaving of patterns and life desires in nature. But life on Earth cannot survive without "the wondrous moisture of the Sweet Waters" that flow across Earth Mother's body as rivers and canals, "to quench the thirst of all beings and bring abundance to all that lives."[28] These waters include the "the four Great Rivers of the Ancient World, including the Tigris and the Euphrates."[29] According to the Bible (Genesis 2:10–14) the other two rivers are the Gihon and Pishon, which flow out of Eden. Their mouths are located near the ancient mouths of the Tigris and Euphrates rivers, which flowed into the Persian Gulf over 4,500 years ago when this myth was first developed. The outpouring of Enki's sweet waters onto and across the surface of the earth, revised in Genesis as the garden of Eden, played a significant role in man's understanding of world geography until as little as four hundred years ago.

Lastly, let's look at an Old Babylonian hymn, or incantation song (*shir-namshub*) commonly referred to as "A *shir-namshub* to Utu (Utu F)." Researchers believe that the eight known incantations were recited by priests. In the song, Inanna expresses concern to her brother Utu that she does not know about sexual intercourse prior to her marriage with the shepherd Dumuzi. While the script is somewhat fragmentary, the incantation provides important information concerning the Babylonian Tree of Life.

> Youthful Utu . . . , calf of the wild cow, calf of the wild cow, calf of the righteous son, Utu, royal brother of Inanna! He who brings thirst to streets and paths (?), Utu, he of the tavern, provided beer, youthful Utu, he of the tavern, provided beer.
>
> (Inanna speaks:) "My brother, awe-inspiring lord, let me ride with you to the mountains! Lord of heaven, awe-inspiring lord, lord, let me ride with you to the mountains; to the mountains of herbs, to the mountains of cedars, to the mountains; to the mountains of cedars, the mountains of cypresses, to the mountains; to the mountains of silver, the mountains of lapis lazuli, to the mountains; to the mountains where the gakkul plants grow, to the mountains; to the distant source of the rolling rivers, to the mountains.
>
> "My brother, come, let me . . . My brother, the midst of the sea . . . my eyes. My brother, women . . . Utu, women . . .
>
> "I am unfamiliar with womanly matters, with . . . I am unfamiliar with womanly matters, with sexual intercourse! I am

unfamiliar with womanly matters, with kissing! I am unfamiliar with sexual intercourse, I am unfamiliar with kissing!

"Whatever exists in the mountains, let us eat that. Whatever exists in the hills, let us eat that. In the mountains of herbs, in the mountains of cedars, in the mountains of cedars, the mountains of cypresses, whatever exists in the mountains, let us eat that.

"After the herbs have been eaten, after the cedars have been eaten, put your hand in my hand and then escort me to my house. Escort me to my house, to my house in Zabalam. Escort me to my mother, to my mother Ningal. Escort me to my mother-in-law, to Ninsumun. Escort me to my sister-in-law, to Jectin-ana.

"For those who venture forth single-handed, who venture forth from a man's house, for those who venture forth from a man's house, who venture forth single-handed, Utu: you are their mother, Utu, you are their father. Utu, as for the orphans, Utu, as for the widows, Utu: the orphans look to you as their father, Utu, you succor the widows as their mother. With you . . . "[30]

The incantation begins with Inanna noting Ute's familiarity with serving beer at taverns. She then requests that Utu go with her to the hills and mountains so that they may each eat cedar, cypress, herbs, and any other vegetation they find there, and so that Inanna may receive knowledge of sexual intercourse, kissing, and other "womanly duties." She then asks Utu to escort her to the house of her mother, mother-in-law, or sister-in-law after they eat, noting that Utu is looked up to by the community.

What is going on here? Why does Inanna insist on taking her brother to the mountains to eat all sorts of plants that would allow her to understand her "womanly duties"? Why is this hymn significant to our purpose here? In common transcriptions of the song, Inanna desires consumption of the cedar (Akkadian *erenu*) tree. However, Stephanie Dalley, in her notes on the Epic of Gilgamesh, states that the word *erenu* might refer to the pine instead. This suggests that Inanna was interested in eating pine or, rather, pine nuts, and likely other plants—trees, shrubs, herbs, fungi (?), and others—to obtain knowledge:

The usual translation of erenu as "cedar" is almost certainly wrong. The main grounds for a translation "pine" are: that roof-beams thus named in texts have been excavated and analyzed invariably as pine, and that the wood was obtained in antiquity not only from the Lebanon mountains, but also from the Zagros and Amanus ranges, where cedars do not grow. The Akkadian word may have covered a different and wider range of trees than the English word "pine."[31]

David Leeming boils this story down to this: "One day Inanna asks her twin brother, the sun god Utu (Shamash), son of the moon god Nanna, to go with her to earth (*kur*), where she will eat various plants and trees that will cause her to understand the mysteries of sex. Like Eve in the Garden of Eden, Inanna tastes the fruit and gains knowledge."[32] Gwendolyn Leick, in her book *A Dictionary of Ancient Near Eastern Mythology*, concurs with Leeming:

> Inanna and Utu is a mythical incident in a Sumerian hymn (BM 23631), which explains how Inanna came to be the goddess of sexual love. The goddess asks her brother Utu to help her go down to the kur where various plants and trees are growing. She wants to eat them in order to know the secrets of sexuality of which she is yet deprived: "What concerns women, (namely) man, I do not know. What concerns women: love-making I do not know." Utu seems to comply and Inanna tastes of the fruit (the same motif is also employed in Enki and Ninhursag and of course in Genesis I) which brings her knowledge.[33]

And that knowledge concerns sex. Pine nuts are rich in zinc, a mineral key to male potency, and pine nuts have been used for centuries in the creation of love potions. Pine nuts are considered an aphrodisiac in the Middle East and across southern Europe. Mythologies symbolize aphrodisiac qualities of various foods, and Mesopotamian mythologies such as "A *shir-namshub* to Utu (Utu F)" indicate that pine nuts have been used to stimulate the libido as far back as the twenty-first century BCE, and likely earlier. In particular, pine nuts from the stone pine (Pinus pinea), native to the north Mediterranean coast and eastward into the uplands of Turkey, are known to have been cultivated for over six thousand years, with natural harvesting of pine nuts extending back to prehistoric times. The Roman poet Ovid (a vegetarian), in his work *The Art of Love*, selected "the nuts that the sharp-leafed pine brings forth"[34] as an effective and powerful aphrodisiac. Arabian scholars such as Galen recommended one hundred pine nuts before going to bed.

In ancient times a distinction was made between a substance that increased fertility and one that simply increased sex drive. One of the key issues in early times was nutrition. Undernourishment potentially created a loss of libido as well as reduced fertility rates. The natural shapes of some substances symbolizing seed or semen, such as bulbs, eggs, and snails, were considered to have inherent sexual powers. Other foods were visually stimulating because they resembled genitalia. Foods considered by the ancients to be aphrodisiacs included plants like basil, carrot, salvia, gladiolus root, orchid bulbs, pistachio nuts, sage, and turnips.

Take Thy Bearings

So far in this chapter we have looked at ancient myths, poems, and songs, searching for the beginnings of civilization in ancient Babylonia. The Babylonian connection with the earliest historical conception of the Sun Cross and the Tree of Life reaches back to at least 6000 BC and likely earlier. To summarize, here are the important points drawn from each of these stories.

Epic of Atra-Hasis:

- At some time in the past, the gods worked and suffered as humans do now.

- The gods created humans from clay (earth) using sacrifice, purification, and ritual.

- Before that creation, "natural man" was believed by the gods to be physically, intellectually, emotionally, and spiritually separate from civilized humans.

- Humans are destined to continue the physical labor previously provided by the gods.

- This labor includes developing and maintaining the garden of the gods; humans are stewards of the earth/universe.

Epic of Gilgamesh:

- A counterpoint to heavy-handed leadership by a godly king is recognized as a benefit to oppressed but civilized people; duality is an important aspect of life.

- The Babylonians appear to have used mythical part-human gods of civilization (such as Gilgamesh) as metaphors for the upper echelon, the elite, of early civilized humans; civilized humans consider the wild humans to be simple clay; civilized humans are likened to fired brick.

- The wild (natural) human was coaxed and brought to civilization to serve the urbanites, perhaps as slave labor, but knowledge of love and civilized ways were also thought to lead to physical weakness and the loss of the our connection with nature. Those consequences, although lamentable, were accepted costs of the material benefits of civilization.

- Human support of civilization and protection and support of the social order, although noble acts, do not lead to eternal life; as we are born of clay, so we are destined to return to clay.

- Great achievements, even by godly kings, will not lead to everlasting life, but may only prolong the memory of our contributions to society; even those memories may not last, as material wealth of individuals, and society in general, are not everlasting.

- Everlasting life may be granted on those who listen, understand, act, and achieve in accordance with the tasks they are given in life. There are particular ideas, plans, or concepts that should be followed to achieve such ends. In the case of Utanapishtim, the concept and particulars of the plan were associated with a specified geometric construction associated with an ark built to save life from extermination. This was the secret of the gods that brought everlasting breath.

Enki and Ninhursag:

- The physical exists in unity, exemplified by rocky Earth and her covering of plant life, supported by "the wondrous moisture" of water.

- Our understanding of this relationship between minerals, water, and life has been based, at least in part, on our conceptualization of geography and geometry, expressed in this story by the waters of the four Great Rivers of the Ancient World that flowed from one source, through paradise, and watered the world.

"A shir-namshub to Utu (Utu F)":

- Knowledge (carnal or otherwise) may be obtained through experience, listening, learning, understanding, and acting to gain knowledge; knowing how to use that knowledge appropriately (wisdom) leads to the achievement of goals. For Inanna, the cedar (pine) tree represented a Tree of Knowledge.

We can see in Sumerian and Babylonian myths an early, fundamental application of metaphors to communicate an understanding of the creation of human beings (understood as *civilized*, no longer *natural*, man). Humans are placed within a geographic framework, that being the surface of the Earth—below the cosmic sky, at the lowest reaches of the atmosphere, among rocks, animals, and plants, and supplied with water from above and below. This geographic order is further defined by four additional directions—north, east, south, and west. Use of the six directions (up, down, and the four cardinal directions) in our communications continues to help us relate location and movement between ourselves and our environment. Not surprisingly, this is the context within which the Sumerians and Babylonians envisioned the gods participating in the creation of the universe, the formation of man, and the interaction of matter and life within time and space.

As such, the four rivers formed from one source and flowing outward from paradise to water the four corners of the Earth can be represented by a Sun Cross: two lines crossing at right angles at their midpoints and contained within a circle—Earth itself (Figure 5.2). The four rivers flow from the center, a sacred center, out in the four directions, representing all directions extending across our world and the universe. This is no different than the geographical and cosmological metaphors understood by Native Americans, the Sacred Hoop of the Lakota serving as one of many examples found across the Americas. And, like the Sacred Hoop, if the Sun Cross is envisioned as a two-dimensional representation of those lines extending across the surface of a sphere, then the metaphor yields an even greater assimilation with the three-dimensional world on which we find ourselves. Thus, the ideas expressed in ancient Sumerian and Babylonian mythologies may be read and understood by all cultures throughout history.

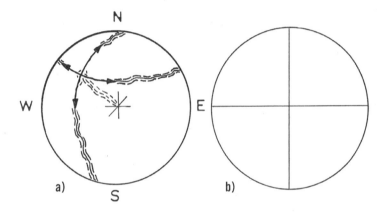

Figure 5.2: a) Four Rivers flowing outward from paradise to water the four corners of the Earth. b) The rivers represented by two perpendicular lines within a circle.

A Design for Life

As a consequence of this conceptualized third dimension for the Sun Cross, let's return to the Epic of Gilgamesh and look into the details of Utanapishtim's ark as relayed to him by Ea (Enki) during Gilgamesh's search for the key to everlasting life. Ever since the Epic of Gilgamesh was recovered in the nineteenth century and translated from cuneiform script, scholars have interpreted Ea's requirement that the ship be equal in width and length and Utanapishtim's construction of the ark with dimensions equal in length, width, and height to have resulted in an ark in the form of a cube. Indeed, at first glance this appears to make sense. Certainly the ship could not have been constructed in the form of the ocean-worthy ark generally envisioned for the ship Noah built, with an upper deck, main deck, parapets, prow, and keel sections.

However, on January 1, 2010, the *Guardian* published an article by Maev Kennedy with the headline, "Relic reveals Noah's ark was circular."[35] The headline's hook was a bit misleading. In fact, the story centered on a newly translated Babylonian clay tablet dated to about 1700 BC that describes how one man saved his family and the animals from the great flood. That man was Atra-Hasis, who we know otherwise as Utanapishtim from the Epic of Gilgamesh. But in this version of the story, Atra-Hasis constructs out of reeds a large, circular craft, not a cube, and not the "pointy-prowed craft of popular imagination" described in the article. The article states that the clay tablet was obtained in the Middle East, possibly at a bazaar, by Leonard Simmons of London, England, while he was serving in the RAF between 1945 and 1948. Leonard's son Douglas transferred the tablet to Irving Finkel, curator of the 2009 British Museum exhibition on ancient Babylon. Finkel translated the sixty lines of cuneiform script and found that the text described the shape of the ship. He stated, "In all the images ever made people assumed the ark was, in effect, an ocean-going boat, with a pointed stem and stern for riding the waves—so that is how they portrayed it. . . . But the ark didn't have to go anywhere, it just had to float, and the instructions are for a type of craft which [the Babylonians] knew very well. It's still sometimes used in Iran and Iraq today, a type of round coracle which they would have known exactly how to use to transport animals across a river or floods."[36]

In Finkel's translation, the god who decides to spare Atra-Hasis exclaims, "Wall, wall! Reed wall, reed wall! Atram-Hasis, pay heed to my advice, that you may live forever! Destroy your house, build a boat; despise possessions. And save life! Draw out the boat that you will build with a circular design; let its length and breadth be the same."[37] The god then commands use of plaited palm fiber, waterproofed with bitumen, before cabins for people and animals are constructed. According to Kennedy's article, "Finkel 'took one look at it and nearly fell off his chair' with excitement." Douglas Simmons added, "You hold it in your hand, and you instantly get a feeling that you are directly connected to a very ancient past—and it gives you a shiver down your spine."[38]

Indeed, this finding has significant implications concerning the form and functioning of Atra-Hasis's (Utanapishtim's) vessel. A circular boat would certainly be difficult to steer. Neither the Epic of Gilgamesh nor Finkel's transcription of the Mesopotamian flood story provide any information regarding any mechanisms associated with the craft, steering or otherwise. In fact, it is apparent that the occupants of the vessel were unable to see where they were, or what conditions the ark was experiencing, until the "pilot" opened a hatch after noting that the ship had rocked about for a while. With a paucity of specific design information for the vessel, let's minimize assumptions concerning the functioning and form of the craft and see if we can discern the general features of its design. We will minimize the need for assumptions in developing an hypothesis for the ark's design. The simpler, the better. We note that the specified design and as-built construction of the *Preserver of Life* (see Figure 5.3) included:

(1) dimensions equal in width and length
(2) length, width, and height of 120 cubits each
(3) floor space of one field (120 cubits x 120 cubits = 14,400 cubits^2)
(4) an interior divided into seven floors
(5) level divided into nine sections each

That's it! That is all we have to work with in constructing a vision of how the ark was composed, inside and out. But Finkel concludes that the Simmons tablet states that Atra-hasis was to "draw out the boat that [he would] build with a circular design."[39] So, we will assume that the craft includes a circular shape from fore to aft, port to starboard. This design satisfies the first design requirement. But why leave it there? Utanapishtim said that he constructed a ship equal in length, width, and height. If length and width are both 120 cubits and form a circular shape, then it is reasonable to assume that the height, having the same dimension, produced a circular shape from the keel line to the uppermost deck or roofline, port to starboard, fore and aft. Simply put, the vessel had a spherical shape! This makes sense given that the Simmons tablet states that the exterior construction materials specified for the craft are plaited palm fiber waterproofed with bitumen. These materials would lend flexibility to the structure, allowing construction of the specified circular, nay, spherical shape. The second design requirement is satisfied.

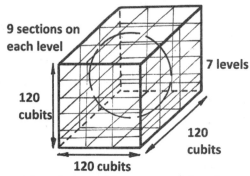

9 sections on each level

7 levels

120 cubits

120 cubits

120 cubits

Figure 5.3: Graphical depiction of specifications for Utanapishtim's ark, *Preserver of Life*.

Next, let's see if we can envision the as-built internal structure of the ship. Given a length and width of 120 cubits each for the craft, a flat floor (deck) constructed through the center point of the circular structure would have a radius (r) of 60 cubits and cover a footprint area calculated by:

$$\text{Area of a circle} = \pi r^2$$

In his book, *A History of Pi*, Petr Beckmann describes Babylonian inscriptions impressed on a cuneiform tablet and translated around 1950. The tablet "states that the ratio of the perimeter of a regular hexagon to the circumference of the circumscribed circle equals a number which in modern notation is given by $57/60 + 36/(60)^2$."[40] This form of numbering should not be unexpected, since the Babylonians used a base 60 counting system, whereas the standard numbering systems used in the modern era is base 10; hence, Babylonians used a sexagesimal system. Referring back to the ratio of the perimeter of a regular hexagon to the circumference of the circumscribed circle, that number as calculated by the Babylonians is the ratio of 6 times the radius of the circle divided by the circumference. With the definition of π equal to the circumference divided by 2 times the radius, we find that the Babylonians calculated π as:

$$3/\pi = 57/60 + 36/(60)^2$$
or
$$\pi = 3/(57/60 + 36/3600) = 3.125$$

This result is close to today's general estimate of 3.14159. However, let's continue with our analysis using the Babylonian estimate. We then find that the area of deck constructed through the center point of the Atra-Hasis's vessel would be:

$$\text{Area of circular deck} = 3.125 \text{ x } 60 \text{ cubits}^2 = 11,250 \text{ cubits}$$

This value, for a circular deck through the middle of the spherical ship, is less than the size of the floor space that according to Atra-Hasis measured one field, equivalent to 120 cubits by 120 cubits, or 14,400 square cubits, and this would appear to be a stumbling block for us when we attempt to satisfy the third design requirement. However, given the fact that the dimensions of the length and width of the ship are the same as the dimensions of one field squared (120 cubits by 120 cubits), it is readily apparent that Atra-Hasis was stating that the diameter of the ark was equal to the length and width of one field, and so the size of a circular deck constructed through the center of the ark fit exactly within the limits of one square field, and thus, the floor space measured one field. The third design requirement is, indeed, satisfied.

In the Epic of Gilgamesh, Urshanabi asks Gilgamesh to go to the forest and cut 120 poles, each 60 cubits long, and bring them back for construction of the boat that they would use to cross the Waters of Death to meet Utanapishtim. Therefore, we can certainly assume that similar poles would have been used for construction of frame and decking of Atra-Hasis's ship. Perhaps not incidentally, the 60-cubit length that Urshanabi specified would also have been the appropriate length needed to provide wooden columns, beams, and structural support within the interior of Atra-Hasis's craft, from the plaited fiber covering to the center point of the spherical ship. These wooden materials would provide the structural strength upon which to construct the decking and walls, as well as the exterior framing that supports the fibrous skin of the ark.

We have the materials available for satisfying the fourth and fifth design requirements. Note, however, that Atra-Hasis does not mention the total number and size of poles or other materials necessary for construction the ship's interior, including six decks and seven levels (including the roof of the structure). He also does not provide any information concerning the configuration of the interior compartments other than each level was divided into nine sections. So, beyond the reasonable certainty that wood would have been used for the interior construction of the vessel, we need not concern ourselves with further analysis for satisfying the fourth and fifth design requirements. In summary, we have satisfied all requirements of the specified design for the ship.

Figure 5.4 depicts a model of the Babylonian vessel used by Atra-Hasis (Utanapishtim) to survive the flood. Figure 5.5 is a plan view looking down on the craft. The exterior of the ship is covered by plaited palm fiber waterproofed with bitumen, framed with lumber so as to support the covering and withstand the stresses of the dead load, live load, and external forces applied to the craft before, during, and after the flood. As shown, the framing extends around the perimeter of the ship, using eight equally spaced arcs framed in wood. In plan view they form what appears to be an eight-legged star pattern, with sections oriented from the center (topmost point of the vessel) to the north, east, south, and west to form an orthogonal cross, and four more sections oriented toward the northwest, southwest, southwest, and northwest to form another orthogonal cross. As it turns out, each of these crosses, bounded by the circumference of the vessel, forms a Sun Cross in the plan view. To provide additional structural integrity, I show four additional arcing

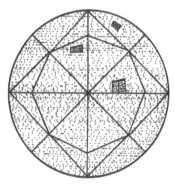

Figure 5.4: Exterior view of *Preserver of Life* based on Ea's specifications and Utanapishtim's description of the construction.

frames extending along the perimeter of the sphere, two from the north side to the south, and two from the east side to the west. Also shown are three decks illustrated by the three circles contained within the fibrous exterior covering. Sizes of these decks are based on their separate locations within the sphere, as shown in Figure 5.5.

Figure 5.5: Plan view of *Preserver of Life* showing limits and subdivision of floor levels.

Figure 5.6 is an elevation view of the vessel, looking at the side of the craft and centered from a point looking due north. Visible are four arcing frame pieces located on the south half of the ship and extending from the top center of the vessel to the bottom center point. As in Figure 5.5, note four additional arcing frames that provide structural integrity between

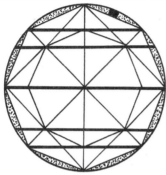

Figure 5.6: Elevation view of *Preserver of Life* showing floor levels and internal structural bracing.

the vertically oriented arcs. Five horizontal decks are illustrated: one constructed through the center (equator) of the sphere, one each at latitude 45 degrees north and latitude 45 degrees south, and two each at latitude 60 degrees north and latitude 60 degrees south. The design specifications call for a sixth deck, and this would consist of the interior space immediately above the bottom of the sphere, creating a curvilinear floor space that could be framed out for a flat floor at Atra-Hasis's discretion. With the top of the sphere representing the uppermost level of the structure, we have a total of seven levels and six decks, as described by Atra-Hasis. Also, you can see two apparent Sun Crosses resulting from the exterior framing, similar to what we saw in Figure 5.5. Indeed this would be the case if our view toward the ark was from the top, bottom, north, south, east, or west.

This is an unusually shaped watercraft. How would the vessel function once loaded and occupied during the flood? First, Utanapishtim tells us that loading of the vessel did not proceed until two thirds of the ark was submerged in the rising flood water. At least one deck was therefore located at or above the water line. Fortunately we have two decks which satisfy this requirement, one at the 45 degree north latitude, and one at the 60 degree north latitude. We can imagine the lower of these two decks being used for boarding and debarking the vessel and possible storage of materials, equipment, or passengers as necessary. The uppermost deck could be used as the living quarters for persons on the ship during the flood, allowing Atra-Hasis, his family, and other people to occupy the highest level of the ship during flood conditions. The three lower decks, then, would be used for animals, birds, and any additional materials or equipment that Atra-Hasis required.

Second, we can reasonably assume that Atra-Hasis would have organized the weight of all contents within the craft, on each deck and within each compartment of each level, so as to maximize the vessel's stability during the flood, reducing the potential for complete submersion. The vessel is described as rocking for a while at the top of a mountain; at the end of the flood, the vessel grounded on the mountain. It is possible that the top of the mountain was submerged for some time during the later stage of the flood. Much of the ark might have been submerged as it rocked to and fro on the mount. We can also expect, therefore, that the hatch Atra-Hasis opened to view the calm waters and release the birds in search of land would have been located near the uppermost part of the sphere, as I show it in figures 5.4 through 5.6.

All and all, it appears that our hypothesis for the ark's design and construction meet all applicable criteria for the ship as provided to us in the myths. This is not to say that the results of our analysis provide the only suitable vessel. However, as I will show, the design requirements for the ark and the form it takes as a result of analyzing the information provided in myth provide a most remarkable correspondence to the sacred geometry we find described around the world. This cannot be coincidence.

Trail of Mythocartology

A paper trail developed as a result of the mythical geography associated with the four Sumerian streams that entered the Garden from the Apsu of Enki. This trail followed the rivers as they flowed outward to water the world. This geographic symbolism appears in Genesis at the Garden of Eden, an apparent reformulation of the ancient Babylonian legends. In his paper "The Mythic Geography of the Northern Polar Regions: *Inventio fortunata* and Buddhist Cosmology," Chet Van Duzer examines two systems of mythology from very different cultures that relate the geography of the northern polar region in a form that he describes as "remarkably similar."[41] This similarity might not be quite so remarkable with regard to its derivation by two separate cultures. However it *is* remarkable that envisioning and describing a previously unknown arctic geography led to a common, cross-cultural geometrical understanding of our relationship with the cosmos. Van Duzer outlines his thesis:

> Sacred centers are usually located near the people in whose mythology they play a part, but there are distant spots on the earth that many different peoples recognize have a special central status: the North and South Poles. These spots are pierced by the axis of the heavens; they are the crowns of the world, about which all the stars dance, the points to which all compasses direct their needles . . . The city of Beijing is known as the "Pivot of the Four Quarters . . . the emperor's residence is the center of the world . . . the Kaaba in Mecca (for which "the center of the earth" is a common epithet among Muslims) is located directly beneath the North Star.[42]

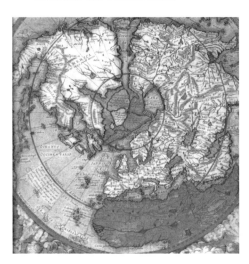

Figure 5.7: Mercator's map *Septentrionalium terrarum descriptio prepared in* 1595. Note the North Polar Region, including four islands separated by four rivers, at the center of the map.

Van Duzer describes maps of the sixteenth-century cartographer Gerardus Mercator, who developed a method for illustrating spherical forms such as Earth onto a two-dimensional plane—a map. Mercator's maps show Earth in its entirety, which would be beneficial for explorers of his day, except that to our knowledge there had been no prior or contemporary exploration and survey of the polar regions before Mercator put his cartographic skills to work. Yet Mercator's *Septentrionalium terrarum descriptio* dated 1595 displays the geography of the north polar region in surprising—if not accurate—detail. Van Duzer describes what Mercator mapped for the area north of the Arctic Circle (see Figure 5.7):

> At the center of the map, and right at the Pole, stands a huge black mountain; this mountain was made of lodestone, and was the source of the earth's magnetic field. The central mountain is surrounded by open water, and then further out by four large islands that form a ring around the Pole . . . These islands are separated by four large inward-flowing rivers, which are aligned as if to the four points of the compass—though of course there is no north, east, or west at the North Pole: every direction from this center is south.[43]

Does this description of Mercator's map seem familiar? Would it be surprising to discover that contemporaries of Mercator also depicted the region about the North Pole in similar fashion—a circle of islands separated by four rivers leading toward the center? Van Duzer lists some of the 16th century culprits of this cartographic fiction.[44]

- Orontius Finaeus' Nova et Integra Universi Orbis Descriptio, prepared in about 1519
- Abraham Orteliuss Typus Orbis Terrarum (1570)
- Fineaeus's Septentronalium regionum descriptio (1570)
- An anonymous world map published in London in George Best's True Discourse (1578)
- Cornelius Judaeis' Speculum orbis terrae of 1593
- Judaeis' Quiviriae regnum and Americae pars borealis (1593)
- Petrus Plancius' Orbis terrarum typus de integro multis in locis emendatus (1594)
- Planicus' Nova et exacta terrarum orbis tabula geographica ac hydrographica (1592)

I call this apparent lineage of cartographic symbolism based on mythological geography *mythocartography*. The source of the fictitious polar islands and watercourses is curious. A large mountain of lodestone at the North Pole (accounting for the earth's magnetism) is referenced at least as early as the thirteenth century, soon after the supposed invention of the magnetic compass. Van Duzer traces a circuitous route searching for the source of geographic information provided on the various maps, but concludes that "the source of this mythical polar geography is a lost work by an unknown author of the 14th century."[45]

Later, we will revisit Van Duzer's speculations about the source from which the *Inventio fortunata* may have derived this geography. However, for now it is worth considering his discussion of Delno West's argument that the author of the *Inventio fortunata* believed "the whirlpool at the North Pole represented an entrance to Hell, which was believed to be in the center of the earth, and also that the four inward-flowing rivers and whirlpool are the counterpart to the fountain in the Garden of Eden, whence the four great rivers branch out to water the world . . . if the rivers flow out, they must flow back in somewhere and be recycled."[46] Van Duzer finds West's theory of a connection between the northern whirlpool and the fountain in the Garden of Eden intriguing. And yet, he writes that:

> there is no evidence that Behaim, Ruysch, Mercator, and the other cartographers who followed the geography of the Inventio fortunata believed that the northern whirlpool was the counterpart to the fountain in Eden . . . I believe that the analogy between the polar geography of the Inventio fortunata and the Garden of Eden is important. . . . The persistence of the Inventio fortunata geography on maps for, say, 150 years is to some extent a testament to the esteem in which Mercator and Ortelius were held by other cartographers; it is also, I think, a testament to the great psychological and mythical power of the concept of the center.[47]

As a testament to Sumerian myths more than eight thousand years old, I could not agree more.

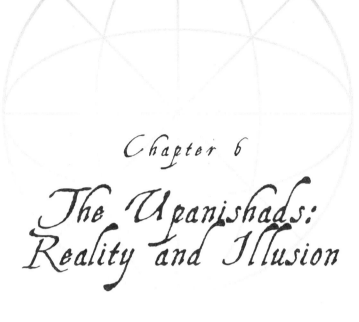

Chapter 6
The Upanishads: Reality and Illusion

In the world's false show that has known no beginning,
The soul slumbers; when it awakes,
Then there wakes in it the Eternal,
Beyond time and sleep and dreams.

MANDUKYA-KARIKA 1.16.

Perception Versus Reality

Throughout history students have gathered around their teachers for instruction on a breadth of topics including philosophy. So it is not surprising that the word *upanishad* means "to sit down near." As philosophical work, the Upanishads introduce fundamental ideas about yoga, meditation, self-realization, karma, and reincarnation. Their theme is the individual's escape from rebirth through receipt and application of knowledge of the universal underlying reality. The *Encyclopaedia Britannica* explains how this perspective came about:

Throughout the later Vedic period, the idea that the world of heaven was not the end—and that even in heaven death was

117

inevitable—had been growing. For Vedic thinkers, the fear of the impermanence of religious merit and its loss in the hereafter, as well as the fear-provoking anticipation of the transience of any form of existence after death, culminating in the much-feared repeated death (punarmrtyu), assumed the character of an obsession.[1]

Older Upanishads are associated with a particular Veda, but not so with more recent ones. The Brihadaranyaka (Great Forest Text) and the Chandogya (pertaining to the Chandoga priests) are the most important of the Upanishads. They both document traditions of sages (*rishis*) of their period, the most notable being Yajnavalkya, who pioneered newfound religious ideas. Other significant Upanishads include:

- Mandukya
- Kena/Talavakara
- Katha
- Mundaka
- Aitareya
- Taittiriya
- Prashna
- Isha
- Shvetashvatara

In the following discussion I have abbreviated each name of those works in accordance with Paul Deussen's *The Philosophy of the Upanishads*. Deussen heralds the Upanishads as the culmination of Indian doctrine of the universe.[2] Deussen, born in 1845 in Germany, was a renowned scholar of Sanskrit and Hinduism. His treatise on the Upanishads remains an influential accounting and analysis of Indian philosophy, and his expert analyses form the primary resource for much of the ancient Indian philosophy included in this chapter.

Deussen found the philosophy expressed in the Upanishads unsurpassed by subsequent religious and spiritual thought or philosophical significance through the nineteenth century. Buddhism, derived independently, nonetheless owes much to the teachings of the Upanishads. Its main fundamental thought (*nirvanum*, the removal of suffering through the removal of *trishna*, or desire) is expressed through the union with

Brahman by the removal of *kama*. And beyond Buddhism, Deussen notes that:

> The whole of religion and philosophy has its roots in the thought that (to adopt the language of Kant) the universe is only appearance and not reality (Ding an sich); that is to say, this entire external universe, with its infinite ramifications in space and time, as also the involved and intricate sum of our inner perceptions, is all merely the form under which the essential reality presents itself to a consciousness such as ours, but is not the form in which it may subsist outside of our consciousness and independent of it; that, in other words, the sum-total of external and internal experience always and only tells us how things are constituted for us, and for our intellectual capacity later.[3]

It may be that the mind, through philosophy, knows no bounds as it seeks to determine the nature of the physical universe, its circumstances, and surroundings, investigating the causal connections we perceive and the totality of the empirical reality we experience. Yet, as Deussen suggests, the explanations provided by empirical sciences seem insufficient. Additional explanation is needed to identify and understand the "real nature" and our relationship to it. Is it true that philosophy has always sought to determine a first principle of the universe, or would our vision of that principle be clouded by culture? Does the resolution of our senses prevent our experience from proving, with a more or less clear consciousness, that the empirical reality, our perception of the universe, is not the true essence of things? Kant concludes that "it is only appearance and not the thing in itself."[4] This is the focus of the Upanishads, the first records from India to advance philosophy to a clearer understanding of its calling, of the solutions humanity desires.

Reality Versus Illusion

In ancient Greek teachings, Parmenides states that this entire universe of change is merely phenomenal. Plato describes it as a world of shadows. They both search for the essential reality. The doctrine of the Upanishads pursues the same unity, which, as we shall see, also suggests a method for perceiving and understanding the real universe.

The doctrine of the Upanishads, that the physical world around us and throughout the universe, and indeed our very nature, are not the atman, what Deussen refers to as "the real 'self' of the things," but in fact *maya*, illusion.[5] The Upanishads say that the physical world and the relationships between objects have no value to themselves ("things in themselves"), but exist for the sake of, and solely for, the atman. We should not separate ourselves from the *Self* (*anyatra atmano*). "[T]his atman is the entire universe."[6]

Deussen suggests that a system can be understood as an association of thoughts collectively belonging to, and dependent on, a single center [by] an individual author. The Upanishads are the product of an unknown number of persons, and therefore a system within the Upanishads does not exist. Deussen estimates that they might have been developed between 1000 BCE and about 500 BCE, and concludes that "all these conceptions . . . gather so entirely around one common center, and are dominated so completely by the one thought of the sole reality of the atman, that they all present themselves as manifold variations upon one and the same theme."[7]

Through untold reworking and intermeshing of both fundamentally idealistic and realistic ideas concerning "requirements of the empirical consciousness,"[8] Deussen finds that development of the Upanishads climaxed the in post-Vedic period. The system of Vedanta became the universal foundation of faith and knowledge in India and has remained as such even while undergoing further development. The system of Vedanta includes four main divisions:

1. **Theology**: the doctrine of Brahman, the first principle of all things.

2. **Cosmology**: the doctrine of the evolution of the first principle in forming the universe.

3. **Psychology**: the doctrine of the entrance of Brahman, soul, into the universe that evolved from him.

4. **Eschatology and Ethics**: the doctrine of the fate of the soul upon death, and therefore the required manner of life as it should be lived.

The Indian word for symbol, *pratikam*, originally denoted the side turned toward a person, that which otherwise would be invisible. So teachers of the Vedanta will speak of symbols of Brahman. Similarly, as discussed in chapter 1, a symbol for the ancient Greek writers was understood to be a visible sign of an invisible object or circumstance.

As conceived in the Upanishads, there is no natural phenomenon that bears such ambiguity, no phenomenon derived in such proximity to the essence of things and yet so revealing of it, as that of life so manifested in the functions of the vital organs, particularly the process of breathing responsible for life itself. The significance of *prana* (breath, or vital life) and its superiority to the other vital forces, such as sight, hearing, speech, and thought, was discussed in the Brahmana period (about 800 to 600 BCE). Those discussions included identification of *prana* with Vayu, god of the wind, vital breath of the universe. Such discussions continue in the Upanishads, but they are unable to identify the first principle of the universe. Via a process of either subordination or identification, the *prana* falls into the shadow of the atman and is used infrequently as a synonym for it.

Breathe

In one of the oldest Upanishads, the *prana* appears in this serene description from Chand. 7.15: "As the spokes are inserted into the nave of the wheel, so everything is inserted into this life (*prana*). The life advances by the life (breath), the life (breath) gives the life, it becomes the life. The life is father and mother, the life is brother and sister, the life is teacher and Brahman . . . for the life only is all this." If not already apparent, *prana* is both a psychical and a cosmic principle, not just the breath of life in man, but the universal breath of all life. Let's dig deeper to understand this universal breath of life. Deussen theorizes:

> It is perhaps due to considerations of this nature [of the various organs and functions of the human organism] that as early as the hymn of the purusha, describing the transformation of the primeval man into the universe, his head becomes the heaven, his navel the atmosphere, his feet the earth, his eye the sun, his manas [mind] the moon, his mouth Indra and Agni (fire), his ears the heavenly regions, and his prana the wind. In general, precisely as we were led to recognize in prana the central organ of life . . . so that it corresponds to it in the universe, the wind, must become the vital principle of nature. . . [9]

This conceptualization of the wind—and breath—as a vital principle of the world is in total agreement with Lakota understanding, that it is Wakan Tanka, the Great Mystery, bringing life to the world via the winds that sweep across Mother and the breath of the people. The wind curves across the sky and bends the grass and trees, and as we have seen, the father Sun in its spherical form traverses the sky to warm the world. This primeval father to all creation is characterized similarly in early works of the Upanishads, in that we can discern a symbolic equivalence between Universe and Sun through metaphorical interpretation. "In Kaush. 2.7," notes Deussen:

> a ceremony is taught which by means of a worship of the rising mid-day and setting sun delivers from all sin committed by day or by night. Chand.3 19. 1 enjoins in addition the worship of the sun as Brahman; and that this representation is merely symbolic appears from what follows, where the sun is regarded not as the original creative principle, but, falling back upon representations discussed elsewhere, as the first-born of creation. With the attempts to which reference is there made to interpret these views of Brahman as the sun, and to see in the natural light a symbol merely of the spiritual light, is to be classed especially in paragraph Chand. 3.1–11, which undertakes on a larger scale to depict Brahman as the sun of the universe, and the natural sun as the phenomenal form of this Brahman. It may be regarded as a further endeavor to penetrate beyond the symbol to the substance when, in a series of passages, it is no longer the sun, but the purusha (man, spirit) in the sun, and the corresponding purusha in the eye that is described as Brahman.[10]

In the Vedanta, developed subsequent to the Upanishads, Brahman is described as being (*sat*), mind (*c'it*), and bliss (*ananda*), through a combination of three essential attributes. After much discussion with nine interlocutors, Yajnavalkya declares, "Brahman is bliss and knowledge." In a subsequent section, where Yajnavalkya arrives at the true value of six symbolical methods of representation, *satyam*, *prajna*, and *ananda* appear beside each other and three other attributes of the divine being. Taitt. 2.1 approaches the character of later formula in sublime poetry forming the apex of the development of thought:

He who knows Brahman
As truth, knowledge, infinite (*satyam jnanam anantum*)
Hidden in the cavity (of the heart) and in farthest space,
He obtains every wish
In communion with Brahman, the omniscient.

The distinction between inorganic and organic dominates the Indian view of nature. Both are derived from the atman, but there is a difference. All organisms (plants, animals, men, and gods) comprise wandering souls, the essence of the atman itself, having entered into the universe as an individual wandering soul. However inorganic bodies (the five elements of ether, wind, fire water, earth) are protected by other individual deities while also ruled by Brahman, but are not wandering souls; they are yet to be raised by Brahman to the level where the word *soul* would apply.

The Beginning of Time and Space

How did the universe evolve? Where did it come from? The Rig-Veda (praise, verse, knowledge) is a collection of sacred hymns. It is one of four Hindu texts called the Vedas, some verses of which are the oldest texts addressing religious concepts. The Rig-Veda was composed between the eighteenth and twelfth centuries BCE, and some verses might extend as far back as four thousand years or more. The verses describe the conception of the egg of the universe in the same context as the premundane *purusha* (soul). The creation myth in the Aitareya Upanishad derives from the Rig-Veda: "in the beginning atman alone was this universe; there was nothing else at all to meet the eye. He deliberated: I will create worlds." He created Earth, the atmosphere, the waters above and below, and then brought forth and gave form to the *purusha* from the waters. As he brooded over the waters, they opened "like an egg," and the Agni, Vayu, Aditya, and the other eight guardians of the universe were created and inhabit humans as the vital life forces through speech, breath, sound, vision, and other physical senses and forms of expression.

While the human frame is animated by sensory organs emanating from the *purusha*, man can exist only after the creator enters his body as individual soul through a fissure in the skull. Ait. 1.3.12 states, "After he created it, he entered into it . . . Thereupon he cleft asunder here the crown of the head, and entered through this gate." Brahman enters into all of the creatures he creates (all plants, animals, and gods) as a citizen. They become the home of Brahman as individual soul. Clearly, the *purusha*, the first principle, is a power dependent on the atman. Only the organs of man's soul are attributed to the *purusha*. The soul is ascribed to the atman.

The meaning of the Indian word for creation, *srishti*, suggests a discharge, an emission or emergence of the universe from Brahman. However, the expansion of the one atman into the manifold universe remains unknown. In Chand. 6.12 a teacher causes fruit of the Nyagrodha (banyan) tree to be brought to him and opened. The shoots of this tree grow downward and strike new roots in the earth, creating a grove from a single tree. The student finds only a very small kernel within the fruit, and the teacher tells him, "The subtle essence, which you do not observe, my dear sir, from this subtle essence in truth this great Nyagrodha tree has sprung up. Be confident, my dear sir, whatever this subtle essence is, of which this universe is a subsistence (a 'having this as its essence'), that is the real, that is the soul, that art thou, O S'vetaketu." Kath. 6.1 sheds some light on the expansion of unity into plurality:

> With its root on high, its shoots downwards,
> Stands that eternal fig-tree.

The verse illustrates the multiplicity of phenomena of the universe arising from the one Brahman as root. Unfortunately, the views expressed in the Upanishads with regard to the physical universe are inconsistent and short on detail. With a view of the world as universe, and particular regard for Indian regional geography, the horizon is limited by the Himalaya and Vindhya mountain ranges to the north and south and by the Indus and Ganges river basins to the west and east. Day begins in the ocean to the east where the sun rises, and night forms in the ocean to the west, where the sun sets. Water surrounds the earth, which has oceans, mountains, and seven islands or continents. There is a recurrence of heaven and earth perceived as two halves of the egg of the universe.

A similar view forms the basis of the cosmography in Brih. 3.3, where the identical concentric arrangement holds in the universe in correspondence with the different layers of the egg as follows: (1) the inhabited world in the middle, (2) the earth around it, and (3) the sea around the earth. The width of the world is 32 days' journey of the chariot of the sun, the width of the earth 64 days, and of the sea 128 days; thus, the diameter of the egg of the universe is 416 courses of the sun. Regarding the sphere that separates the layers of earth and heaven (the sea), "There is a space as broad as the edge of a razor or the wing of a fly, through which access is obtained to the place where the offerers of the horse-sacrifice are." Deussen suggests that this horse-sacrifice may be a reference in other verses to the "back of heaven" as being "free from suffering," where union with Brahman is obtained.[11]

In later chapters we will see that this Upanishads' view that there are three world-regions—earth, air, and heaven—to which Agni, Vayu, and Aditya correspond as rulers, is found in mythological and sacred traditions around the world. Ait. 1.1.2 states that the primeval waters extend above and below the three regions, and in Atharvasiras 6 there are nine heavens, nine atmospheres, and nine earths. There are also discrepancies in the number of heavenly regions: four are enumerated (east, west, south, and north in Chand. 4.5.2; five in Brih. 3.9.20–24; six in Brih. 4.2.4 and Chand. 7.25; and eight [four primary poles and four intermediate poles] in Maitr. 6.2, Ramap). Throughout this book we will see these and other numbers of import (such as 3, 4, 6, 8, 9, 13, 26, and 72) are found in traditions around the world. Earth, Moon, Sun, planets, and stars also play vital roles in various cultural understandings of time and space.

The Upanishads develop astronomical conceptions only slightly. Sun and Moon are principal figures, as they represent stations for the soul as it journeys to the other world. Moon holds her course, dependent on the twenty-seven constellations throughout the night. Sun traverses the same constellations during its yearly journey, according to Maitr. 6.14. Therefore, the number of constellations covered in each month of the year is 27/12, which reduces to nine quarters. The number of planets is given as nine in a late text of Ramottarat. 5, and Sun and Moon (the head and tail of the dragon) are included among them. Maitr. 1.4 describes movements affecting the cosmos as "the drying up of great seas, shattering of mountains, oscillations of the pole-star, straining of the ropes of the wind (which bind the constellations to the pole-star), sinkings of the earth, and over-throw of the gods from their place." In contrast to this startling list of cosmic events, Deussen ends his discussion of astronomical conceptions with the following verse, Kaivalya 19, in which the atman speaks of himself:

In me the universe had its origin,
In me alone the whole subsists,
In me it is lost, this Brahman,
The timeless, it is I myself!

A Forest of Trees, a Life of Illusions

It is the fate of man not to see the universe at it really exists, not to experience truth, but to live a life of illusion. Deussen finds that for ancient Indian philosophy, "the entire aggregate of experience, external and internal, always shows us merely how things appear to us, not how they are in themselves . . . and in Indian philosophy, when the Upanishads teach that this universe is not the atman, the proper 'self' of things, but a mere maya, a deception, an illusion, and that the empirical knowledge of it yields no vidya, no true knowledge, but remains entangled in avidya, in illusion."[12] Deussen argues,

> the older the texts of the Upanishads are, the more uncompromisingly and expressly do they maintain this illusory character of the world of experience [though it is] seldom expressed in absolute simplicity. . . . "This entire universe is the purusha alone, both that which was and that which endures for the future," (Rigv. X.90.2) it is implied that in the entire universe, in all past and future, the one and only purusha is the sole real. The common people however do not know this; they regard as the real not the stem, but "that which he is not, the branches that conceal him" (Atharvav. X.7.21); for that "in which gods and men are fixed like spokes of the nave," the "flower of the water" (i.e., Brahman as Hiranyagarbha) "is concealed by illusion."[13]

The Upanishads describe the heart of what the atman *is*—relationship with the spirits, and unity in the universe. Deussen writes, "thus we meet . . . at its beginning . . . a distinct entirely self-consistent idealism . . . according to which the atman, i.e., the knowing subject, is the sustainer of the universe and the sole reality; so that with the knowledge of the atman all is known."[14] The fundamental thought, the basis of the entire teaching of the Upanishads, is hidden in the conception of unity expressed in Rigv. 1.164.46: "the poets give many names to that which is one only. . . ." This verse asserts that plurality, whether in space, time, cause and effect, or "interdependence of subject and object, rests only upon words or, as was said later, is 'a mere matter of words' (Chand.6.1.4), and that only unity is in the full sense real."[15] Man can know the sole reality and sustainer of the universe, the atman, but knowledge requires a process in which the atman becomes more visible to us as we experience a transformation from waking consciousness to the "bliss of perfect consciousness." By realizing and possessing the knowledge of the atman, here and everywhere, now and eternally, we are emancipated, free of the control we ourselves created. We discover that the existence of the universe as well as our own physical existence to be an illusion (maya). Deussen summarizes The Doctrine of Emancipation in Empirical Form[16], the power of knowledge lying in the Upanishads:

1. The atman is unknowable.

2. The atman is the sole reality.

3. The intuitive knowledge of the atman is emancipation.

In these three propositions is contained the metaphysical truth of the teaching of the Upanishads. Emancipation requires that we first remove from ourselves the consciousness of plurality, and then remove all desire, which as Deussen notes, is a necessary consequence and accompaniment of that consciousness. The aim of yoga and *sannyyasa*, the two character-istic manifestations of Indian culture, is to produce these states artificially. Yoga involves withdrawing the organs from the senses, concentrating them in the direction of the inner self to free it from the universe of plurality and secure oneness with the atman. *Sannyyasa* is the casting away of home, possessions, family, and all other things that stimulate desire in oneself in favor of pursuing freedom from the physical universe and recognizing the true and supreme task of our existence here on Earth. We must recognize the subjective worth of these actions, the greatness of the sacrifice we should be making, and the degree of neces-sary self-denial (*tapas*), and self-renunciation (*nyasa*), whether or not there is benefit to others. The value we place today on sacrifice and morality is far different than it was in ancient India.

Discovering Truth

Ultimately what we discover in the Upanishads is that the soul does not migrate, that a metaphysical transformation of consciousness is not absolute philosophical truth. However, as a myth, as metaphor, this transformation holds a truth that otherwise remains incomprehensible to us, and so the metaphor serves us as a valuable substitute for real under-standing. If we could extract and understand the framework of space, time, and causality represented in the metaphor, we would realize truth.

We would discover that the cyclical return of the soul is not something to be realized in the future at some other place (such as "up" in heaven), but it is here already, in the present; this here is everywhere, and this present is eternal.[17]

"When no other (outside of self) is seen, no other is heard, no other is known, that is the infinite; when he sees, hears, or knows another that is the finite" (Chand. 7.15–24). This is evident in Ait. 1, where it is said that it is no longer the *prana-purusha* that makes its appearance as the ultimate principle, but rather the atman, and in Ait. 3, where the atman is "explained as the consciousness that comprehends all things in itself (prajna)."[18]

Note that geography plays a significant role throughout the ancient philosophy of the Upanishads. At its core, geography is little more than geometrical relationships (spatial and temporal) between physical features. Whether the immediate discussion centers on the human body or the universe, the world around us or the universe *in toto*, finite or the infinite, there is considerable attention made to specific locations and processes that are related to others—metaphors for sacred relationships.

- The universe is only appearance and not reality; with its infinite ramifications in space and time, and the intricate sum of our inner perceptions, it is merely the form in which reality presents itself to our consciousness. It is not the form in which it may exist outside of our consciousness.

- As the spokes are inserted into the nave of the wheel, so everything is inserted into this life (prana). Life is teacher and Brahman; the prana is both a psychical and a cosmical principal, not just the breath of life.

- The wind (prana, cosmical and psychical) is celebrated as the thread of the universe (satram) which holds together all beings.

- The organisms Brahman creates are citadels into which he enters as a citizen; they become the home of Brahman as individual soul; only the organs of man's soul are attributed to the purusha; the soul is ascribed to the atman.

- The multiplicity of the phenomena of the universe arises from the one Brahman as root.

- "There is a space as broad as the edge of a razor or the wing of a fly, through which access is obtained to the place where the offerers of the horse-sacrifice are." Deussen suggests that this horse-sacrifice may be a reference in other verses to the "back of heaven" as being "free from suffering,"[19] union with Brahman is obtained. Taitt. Ar. 10.1.52.

- With the world as universe, the horizon is located by the Himalaya and Vindhya mountain ranges to the north and south, and the Indus and Ganges river basins and mouths to the west and east; day begins in the ocean to the east, and night forms in the ocean to the west; water surrounds the earth, which has oceans, mountains, and seven islands or continents; heaven and earth perceived as two halves of the egg of the universe.

- There are three world-regions—earth, air, and heaven—but primeval waters may extend above and below the three; discrepancies exist in the Upanishads regarding the number of heavenly regions: there might be four (east, west, south, and north), five, six, or eight (four poles and four intermediate poles); or nine heavens, nine atmospheres, and nine earths.

- Sun and Moon are principal figures, representing stations for the soul as it journeys to the other world; the moon holds her course, dependent on the twenty-seven constellations of the night; the sun traverses the same constellations during its yearly journey; the constellations are covered in nine celestial quarters during each of the twelve months; the number of planets is given as nine, with the sun and moon (the head and tail of the dragon) reckoned with them.

- In the transformation of primeval man into the universe, his head becomes the heaven, his navel the atmosphere, his feet the earth, his eye the sun, his manas the moon, his mouth Indra and Agni (fire), his ears the heavenly regions, and his prana the wind, the central organ of life.

- Gods and men are fixed like spokes of the nave.

- The entire aggregate of experience, external and internal, always shows us merely how things appear to us, not how they are in themselves.

Several basic principles result from this philosophy:

- Science cannot, and will not, answer the fundamental questions we have about our relationship with the cosmos: who are we, why are we here, how did the universe begin, how will it end, is there a creator, and if so, how should we relate to that creator? This failure of science is a result of the physical world since relationships between objects having no value to themselves.

- Rather, the universe exists purely for Self; it is everything, everywhere and eternal; what we perceive the universe to be (having the qualities of separateness in being, time, and space), is not reality; it is illusion.

- This Self, or oneness, can be realized through transformation from waking consciousness to the "bliss of perfect consciousness"; through this transformation we are freed to realize our oneness with the universe, free from suffering.

- Prana, the Great Mystery, the Cosmic Breath, and other similar personifications of the life source of the universe as expressed in all religious traditions, is the matrix holding the universe together; through it our individual souls are connected to all others, including the atman.

- The world, this universe, contains all physical and metaphysical qualities and quantities, that in reality are one in space and time.

Visualizing Truth

This oneness of the universe is a difficult, if not nearly impossible, fact to accept and visualize. By our very nature we separate, discriminate, categorize, interpret, and find favor or disfavor in the reality we perceive. Mandalas provide a medium for understanding the psychical and cosmic oneness in the universe (Figure 6.1). They do this even though they present the universe in graphic form with a variety of separate images for us to interpret.

Specifically, a mandala as defined in Romio Shrestha's book *Celestial Gallery* is:

a symbolic, graphic representation of a tantric deity's realm of existence and, more generally, a symbol of innate harmony and perfection of being. The Tibetan dkyil 'khor means center and periphery. The mandala represents the transformation of seeming complexity and chaos into a pattern and a natural hierarchy through the experience of meditation. The outer world, one's body and state of mind, and the encompassing totality can all be seen as a mandala. The constructed form of a mandala has as its basic structure a palace with a center and four gates in the cardinal directions.[20]

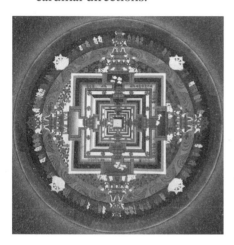

Figure 6.1: Mandalas depict an understanding of the psychical and cosmic Oneness in the universe. Note the four-fold symmetry. The center point and uppermost roof of a temple is shown at the mandala's center, indicating the third dimension in the painting, and therefore two additional directions—up and down—which reflects the form and symmetry of the Lakota Sacred Hoop.

Mandalas are visual representations of *mantras*, sacred sounds. What's more, the mandala is a two-dimensional art form that includes a third dimension that must be envisioned by the beholder, a discernment typically achieved through meditation while viewing the mandala. Perfecting the ability to envision this third dimension comes through learning, understanding, and wisdom.

Figure 6.2: *Medicine Buddha Mandala* painted by Romio Shrestha (image courtesy Romio Shrestha). The painting was the centerpiece of an exhibit at the American Museum of Natural History in New York during 2011. A circle envelops the blue Medicine Buddha at center. Again, note the four-fold symmetry of the temple.

The life goal of each one of us, although it might take thousands of lifetimes to achieve, is to become an awakened and enlightened human being, reaching the peak of evolution through development of compassion and wisdom until they reach unity and perfection. This is the definition of Buddha. In compassion we are "selfless kindness," wishing to "alleviate the suffering of others," while "spontaneously expressing indivisibility of emptiness and luminosity."[21] Wisdom is "[t]he most important mental power in Buddhism—as only wisdom can bring a being's liberation from suffering. It is the perfect experiential knowledge of the nature of reality, which is sheer relativity free of any intrinsically established persons or things. Wisdom can be personified as the mother of all Buddhas, a goddess, or the female Buddhas among the Tantric archetype-deities . . ."[22]

Mandalas painted by Romio Shrestha are visions of emancipation, expressing our individual potential for achieving psychical and cosmological understanding and enlightenment (Figure 6.2). Shrestha is a director of a school of artist-craftsmen in the Kathmandu valley of Tibet and a master of the artistic traditions of both Nepal and Tibet. Drawn "from the ancient wisdom of Buddhist and Hindu traditions . . . his paintings express the possibility of total spiritual freedom."[23] While Shrestha's contemporary paintings are unrestrained by religious and artistic convention, his mandalas retain the compositional symmetry that the traditional form has always expressed. This typically includes a central circular illustration exhibiting a fourfold rotational symmetry with secondary two-fold or threefold symmetry between the cardinal directions. These symmetries also support cubic forms within and outside of the primary circular structure that include Buddhas, palaces, lotus flowers, and other traditional motifs. Similar illustrations surround the central form, providing a universal context in support of the primary scheme. Because Shrestha does not limit his mandalas to traditional forms and symmetries, they do not necessarily comply with the traditional illustrative expressions of the Tibetan Buddhist religion. The popularity of his paintings may be, at least in part, attributable to the more inclusive universality of the geometric configurations included in his compositions.

Another interesting configuration of fourfold symmetry is found in statuary associated with the deity Shiva (Sanskrit: "Auspicious One"), who is worshipped by Hindus across India, Nepal, and Sri Lanka. Hindu

Shaivite sects consider Shiva paramount to all other deities. Scholars believe that the modern attributes of Shiva are an amalgamation of ideas from religious traditions of numerous regional sects over time. This deity derives from Rudra, mentioned only a few times in the Rig-Veda, and an alternate name for Shiva that remains in use. Rig-Veda 2.33 refers to Rudra as the Father of the Maruts, gods associated with storms. But Rudra's stature among Hindu deities developed further with the attribution of important aspects such as fertility. The Rig-Veda includes sacred Hindu hymns, some of which are addressed to Rudra and in fact refer to him as Shiva. In time, Rudra became Shiva, who is part of the Hindu trinity along with Vishnu and Brahma.

Recall that some verses of the Rig-Veda might extend as far back as four thousand years or more. The scripture describes atman's conception of the egg of the universe. As the atman brooded over the waters he had created over and under the earth, the waters (*purusha*, soul) opened like an egg, and eight guardians of the universe came forth. The Rig-Veda also uses the term *Shiva* as an epithet for many other deities, including Indra and Agni. This identification of Shiva with numerous other gods helped to ensure Shiva's paramount stature with regard to the physical, intellectual, emotional, and spiritual levels of Hindu mythology.

The relationship between Rudra and Agni described in the Vedas is particularly important for understanding Shiva's role in the universe. The relationship is so intimate that in an early text called the Nirukta the names of the two deities are equivalent: "Agni is called Rudra also."[24] Recall Deussen's theory that the verses describe "the transformation of the primeval man into the universe."[25] It is therefore apparent that together Rudra and Agni symbolize storms and fire. Adding to this understanding of the increasing association of Shiva with fertility, it becomes easier to appreciate the complexity of relationships among these deities. Stella Kramrisch remarks, "The fire myth of Rudra-Shiva plays on the whole gamut of fire, valuing all its potentialities and phases, from conflagration to illumination."[26] There is a sensuous, emotional, fiery subtext to Shiva.

He is also complex, as we might expect from one who symbolizes fiery emotions and fertility. His nature appears contradictory—both cosmic destroyer and restorer, an ascetic, the gatherer of souls, and an avenger of anyone who may cross him. Lord Shiva's chosen mode of transportation is a bull, Nandi; as it turns out, Agni is referred to as a bull in the Vedas, which mentions Agni's horns. By riding the bull Shiva represents the preserver of the eternal principle of the universe, that of dharma; Shiva's complexity is further exemplified by various attributes that artisans commonly include in paintings, engravings, and statuary. They include:

- a third eye symbolizing higher consciousness, with which he reduced Desire (Kama) into ash

- agni (fire) carried in his hand, an additional sign of Shiva's ability to destroy

- an ornamental crescent moon on the side of his head, symbolizing the cyclical nature of time, love, and fertility

- ash smeared across his body, symbolic of his asceticism

- long, curly and matted hair

- a blue throat, the result of drinking poison from the churning ocean, or possibly strangulation

- the Ganges River flowing from his hair, the nectar of immortality

- a tiger skin as his seat—ascetic Shiva holds down the tiger, which represents lust

- a garland of skulls representing the cycle of life, death, and rebirth;

- five snakes entwined around him, symbolizing reincarnation, eternity, and wisdom

- a deer held in his upper left hand, symbolizing his ability to contemplate, to concentrate on individual tasks rather than let the mind wander, just as the deer runs and jumps

- a trident in his lower right hand representing sovereignty and the use of mind and matter, and the qualities creation, preservation, and destruction in nature

- a small, hourglass-shaped drum representing the act of creation in one hand and a nearby hand forming in the gesture (mudra) symbolizing fearlessness

- a seat on the bull Nandi, denoting Shiva as preserver of the universal principle of dharma, order and conformity

- Mount Kailash, Shiva's worldly home, conceptualized as lingam and symbolizing the center of the world and, therefore, universe. It is the central pillar of the worldwide mandala; it is the source of four rivers, the Brahmaputra, Indus, Karnali and Sutlej, flowing toward the four corners of the world.

Shiva's consort and wife is Parvati. He is a loving husband, reverent and generous toward her. Shiva is also a dancer. He is Nataraja—the cosmic dancer, Shiva and Parvati held within one body, androgynous. And it is this form that artists manifest Shiva in his glory (Figure 6.3). Surrounded by a circle of flame (Agni), the Nataraja symbolizes the source of all movement in the universe. Nataraja is intent on freeing us from maya, illusion. He dances for the reincarnation of the cosmos, an apocalypse followed by creation, and the transformation of mankind toward power and enlightenment. He does this with five synchronous gestures and accoutrements:

1. Embodiment indicated by his right foot on the ground

2. Protection and preservation signified by the mudra gesture expressing "do not fear"

3. Destruction resulting from fire held in his hand

4. Creation represented by the drum

5. Grace symbolized by lifting his left foot upward and outward

Figure 6.3: Statue of Shiva as Nataraja. Note the ring of fire about Nataraja, and his stance which calls us to consider the symbolism in three dimensions—the ring of fire is not a circle, but a sphere.

Nataraja's right foot is not actually on the ground; it is firmly planted upon the body of Apasmara, a demon symbolizing ignorance and confusion, particularly with regard to acceptance of maya. This lack of knowledge and understanding is represented by the cobra that Apasmara holds.

Nataraja dances at the end of each cosmic age. He transforms the world by removing maya. The transformation necessitates destruction, but Shiva makes clear that it is not wanton removal of all life, but rather an act of cleansing and creation. With his long locks of matted hair aloft behind him, he renounces the comforts of life and yet ensures the eternal sanctity of the world through his acts of destruction and creation. He is surrounded by rings of water and cosmic flame, the universal consciousness. The flame is Mahakala, the great blackness, the Lord of Time, one of the protectors of dharma. This fire, too, is Shiva.

This strange dichotomy found in Shiva—the dark and terrible destroyer and the fearless protector and creator—seems baffling. Exactly who is he? What does he represent? The answers come when we remember that Hinduism can be characterized as a system of thought and action appropriate for attaining emancipation. We previously saw how the power of knowledge was contained in the metaphysical truth of the Upanishads. Emancipation requires that we remove from ourselves the consciousness of plurality and also all desire. These are two vital aspects of Shiva. As unity he is removed from the consciousness of plurality, and as an ascetic he has no needs or wants other than to protect and preserve the universe. He is liberated and self-preserving. In short, he is the world, the Milky Way, and the universe. Most often Shiva is found deep in meditation, but he may dance upon the demon of ignorance and destroy in a terrible and avenging manner when it is necessary to preserve atman, the individual soul of the universe. And so, he is the atman. He represents unity. He is eternal. And eternity is cyclical: life, death, and rebirth.

There is great speculation that Shiva represents our galaxy and specifically the black hole that inhabits the center of the Milky Way. As the source of all matter and energy in the relative immediacy of the galaxy, we can perceive this black hole to be the creator and destroyer of the world and the cosmos. Billions of stars, planets, nebulae, and motes of interstellar dust are created in and rotate out from this center like a cosmic pinwheel. The rotating black hole expels a column of electromagnetic energy perpendicular to the general plane of the galaxy, raining it across galactic space. The geometry is familiar. It is that of the electron, the earth, the sun, and our solar system.

We have seen that liberation can be achieved by a variety of means such as yoga and sannyyasa. The austerity of *tapas* ("consuming of heat" or "striving") is one of the most highly regarded means to liberation. In fact, it is said that at one time Shiva's *tapas* consumed so much heat that he transformed into a pillar of fire—a fiery *lingam* that could destroy the world. No other deity could attenuate Shiva's burning energy. But a *yoni* ("divine passage") miraculously appeared, catching the lingam of Shiva, containing its heat, and averting destruction of the cosmos. This was cosmic liberation in the extreme. The Shiva Linga is a common aspect of his worship. But Shiva's third eye is another attribute which exudes light and heat, metaphors for higher consciousness. All life may be destroyed by the fire in Shiva's third eye. And so the power of Shiva to create and destroy is acknowledged, and his grace is petitioned.

The text known as Chidambara Mummani Kovai cries,

O my Lord, Thy hand holding the sacred drum has made and ordered the heavens and earth and other worlds and innumerable souls. Thy lifted hand protects both the conscious and unconscious order of thy creation. All these worlds are transformed by Thy hand bearing fire. Thy sacred foot, planted on the ground, gives an abode to the tired soul struggling in the toils of causality. It is Thy lifted foot that grants eternal bliss to those that approach Thee. These five actions are indeed Thy Handiwork.

KUMARAGURUPARAR [27]

But these actions are one.

Chapter 7

European Maces, Whirlpools, and Gristmills

In Neolithic Europe and Asia Minor (ancient Anatolia)—the era between 7000 BCE and 3000 BCE—religion focused on the wheel of life and its cyclical turning . . . the focus of religion encompassed birth, nurturing, growth, death, and regeneration, as well as crop cultivation and the raising of animals. The people of this era pondered untamed natural forces . . .

MARIJA GIMBUTAS^
THE LIVING GODDESS

The five actions of Shiva as he dances in the role of Nataraja are, in fact, one action—the universe is unity—the cycle of life, death, and rebirth is an eternal constant. This theme of the cyclical nature of the universe is common to almost all sacred traditions. Human beings across the world came to this realization many thousands of years ago as they observed the physical and metaphysical workings of the cosmos from here on Earth. So, the fact that these ideas are common to Hinduism originating in southeast Asia and ancient Celtic religion across central and northern Europe should not be surprising, even less so given that most of the languages of Europe, the Iranian plateau (including ancient Mesopotamia), South Asia, and Central Asia are of the Indo-European family, with a common recorded history going back more than six thousand years. The purpose of this chapter is to outline several important cultural and religious traditions associated with Europe prior to the coming of Christianity.

Some Background

One of the various cultures that developed from proto-Indo-Europeans, the Celts occupied much of Europe including the British Isles by about 2000 BC. Unfortunately, they did not leave a written record of their history. Instead, much of what is known of the people who inhabited north and central Europe during the Bronze and Iron Ages comes from the records of other cultures, primarily the Romans. We also have stone and metal artifacts of Celtic origin. While primary-source details are few, and many of the historical records present a Roman spin on the history of Europe north of the Alps, it is possible to find some measured understanding and appreciation for Celtic traditions in the little information we have.

The British Isles were one of the last strongholds of the Celts as the Roman Empire expanded across Europe. The island now occupied by Northern Ireland and the Republic of Ireland never came under Roman control. However, surviving documents from the Greek colony of Marseilles quote earlier records that reference the British Isles as early as 2,600 years ago. Marseilles, located on the coast of what is now France, was an early Mediterranean port for Greek seafarers. It also served as a way station for trade routes northward into Gaul, which was also occupied by Celts. But human habitation in and around Marseilles goes much further back, as much as thirty thousand years, based on cave paintings, some of which are now under the waters of the Mediterranean Sea. Archeologists have excavated brick-lined structures dating to about 6000 BC at Marseilles, indicating the value placed on trade and settlement in that area over the millennia. Humans occupied Great Britain by about 12,000 BC, when sea levels were lower and the island was still connected to the continent. Ireland was colonized within four thousand years after that. So, while the earliest written information we have concerning the British Isles dates to about 600 BCE, it would not be unreasonable to assume that communications between Celts on the continent and the isles, and with seafaring peoples originating from other areas, existed much earlier.

Irish Celtic Mythology

In the myth of The Ages of the World, the Túatha Dé Danaan are portrayed as talented and learned gods.[1] They were the people of Danu, the Mother Goddess. Their skills included arts, crafts, and magic learned while they lived on the islands of northern Greece before immigrating to Ireland. They held divine power from their wisdom, magic, and four talismans. The greatest of the Túatha Dé Danaan were Daghda the Good (also called the Daghda, god of fertility), Lugh, or Lug, of the Long Arms (god of the sun), and Nuada of the Silver Arm (king of the Túatha Dé Danaan).

The four treasures of the Túatha Dé Danaan were Daghda's bronze Cauldron of Plenty; the Stone of Plenty; the Spear of Nuada that always hit its intended target, bringing death to all it wounded; and the Sword of Lugh, flashing and roaring with flames and bringing victory to those who wielded it. Lugh's sword was ever thirsty for blood and always sought appeasement of its hunger. Lugh had to keep the sword in a container of poppy-leaf juice; the narcotic put the sword to sleep until it was needed. The Cauldron of Plenty and the Stone of Plenty could provide sustenance for all life on Earth. In addition to these four treasures, Daghda the Good owned a great club capable of both bringing death and restoring life. This eight-pronged war club was so heavy it required eight powerful men to carry it. The Daghda pulled it into battle on a cart; he would then lift it to smite nine men in a single motion.

Daghda also had a harp. King Nuada once commanded Lugh to play Daghda's harp for him and the other gods. They all cried as Lugh played a sad melody. Then they all laughed when he played a cheerful melody. Lugh also played a melody that put the gods to sleep until the next day. Obviously the harp, a source of vibrational energy, was of great significance in ancient mythological Ireland. And for the Celts of Ireland, their island was the world.

Scholars conclude that Irish and Scottish monks living between the twelfth and seventeenth centuries recorded Celtic myths entailing a tradition that was more authentic than other ancient myths documented from the British Isles.[2] If we consider the myth of The Ages of the World as a true accounting of ancient Irish history, as Irish tradition holds to be the case, then there are several fascinating coincidences between Hindu and Celtic mythologies. Ideas expressed within each tradition are common to both cultures. We previously reviewed some of the mythology contained within the Upanishads and Rig-Veda, so let's now look at pertinent information from the Celtic ages of the world.

Celtic myth tells us that we live in the sixth age of the world, with gods remaining here on Earth with mortals. These deities include the god of fertility, the god of the sun, a king, and numerous other masculine and feminine gods. Daghda the Good deals out both death and life with his club (Figure 7.1). The playing of his harp can create sorrow or joy, and also cause sleep. In the great battle at the end of the fifth world, sun-god Lugh enters the fight with the other gods and exclaims, "Be of good courage! It is far better to face death in battle than to live as a slave and pay taxes to the conqueror!"[3] The myth, of course, is addressing the ability of the gods to create, destroy, and restore life. The god of the sun tells us not to live our lives as slaves or face death with fear. Life is to be lived! The gods protect us! Certainly death can be sudden and terrible, but the gods can also create—they have created the world (universe)—and restore life as it is intended to be. Taken as a whole, in unity, are these gods not the Celtic equivalent of Shiva as Nataraja?

Figure 7.1: Daghda the Good depicted on the Gunderstrup Cauldron. The cauldron is housed at the National Museum of Denmark. Notice the attendant holding a wheel at Daghda's side, and the various mythical animals surrounding him.

A Unity of Myth

In the history of the world as represented in Irish mythology, there is a trail of development in which peoples have come and gone. Myth states that the world already existed by the time people first populated the island. Ireland has been the site of peace and war, technological development, plagues, the habitations of aristocrats and common people, a place of gods and human beings. Great cycles have been at work here, and they continue in the present. This, too, is the cyclical nature of eternity acknowledged in the Upanishads. It is preservation of the *atman*. The fog surrounding Ireland at the beginning of each age is *maya*, illusion.

The point here is that there is an understanding held in common between cultures separated by space and time. This understanding is so common to world mythologies that we must conclude that it contains vital information. Ancient cultures held in common a fundamental acceptance of the nature of the universe as a cycle of creation, destruction, and re-creation. Each cultural tradition has its own interpretation of re-creation as a cleansing, rebirth, resurrection, or other metaphor for protecting and supporting the order of the universe. Each culture or religious tradition develops its own interpretation of this recycling of the universe, but all ancient traditions recognize that it is fundamentally so. As such, the traditions of the Upanishads assign us the task of removing the illusion, discovering the atman, and recognizing that each one of us is an important component (Fuller's equivalent of an energy event) within the cycles of space and time. If you think this is heady stuff, remember that these ideas appear to have been universally held to be truth by our ancient ancestors, wherever and whenever they lived.

One ancient Celtic god demonstrates the importance of the cyclicity of universal processes. The god Cernunnos (Figure 7.2) was held in common by Celts across the European continent. Little information available about Cernunnos can be directly sourced to the Celts, again due to their lack of written records. However, a limited number of Celtic engravings and carvings have been found that illustrate certain characteristics helpful to our understanding of the importance of this god. Considered in tandem with documents provided by the Romans, our picture of Cernunnos resolves sufficiently to provide us understanding of his importance within the Celtic pantheon.

Figure 7.2: Detail of Cernunnos on the Gunderstrup Cauldron at the National Museum of Denmark. Note Cernunnos'antlers. He holds a torc and a snake, and is surrounded by animals.

The common form of Cernunnos—the horned one—is a bearded man with two stag's antlers, a symbol of Celtic royalty, emanating out of and above his head. He is usually shown in a sitting position holding a snake in each hand, or holding a snake in one hand while holding a neck ring (torc) in the other. In some depictions Cernunnos holds a bag from which coins are falling, while in other scenes he is surrounded by animals. He has been interpreted to be a hunter, a fertility god, and the primary deity associated with the reawakening of nature in the spring and cycling of the season into summer. He is equated with the Daghda (The Good God) of Irish mythology. Also, Cernunnos has been compared to Shiva.[4] Virupaksha Shiva is referred to as the master of animals, and the sitting position of Cernunnos is reminiscent of the yogic lotus position often associated with Buddhist and Hindu traditions. In fact, more than three hundred figures of Celtic deities shown in the lotus position have been found across Europe.[5] Few depictions of Cernunnos have been found, yet Figure 7.2 shows him seated accordingly.

In her essay "Who is Cernunnos?" Alexa Duir proposes that, based on the various attributes prescribed to this horned deity of the Celts, Cernunnos "was directly associated with divinity, wealth and animals, and potentially indirectly associated with regeneration, healing, fertility and death."[6] With the dearth of direct evidence we have concerning the attributes and importance of Cernunnos, there is no doubt it is difficult to identify attributes peculiar to the horned god. However, it is evident from the numerous pre-Christian, polytheistic religious traditions found around the world that humans assigned specific anthropomorphic aspects to each of their deities, and also that the respective pantheon was overseen by one particular god whose attributes were believed greater than those of any other. Man's attempts to separate, classify, and categorize the attributes of the universe seem to have almost invariably led to his ultimate conclusion that each aspect of the universe is connected to all others. The universe exists as unity. Duality is illusion. If one aspect is not connected, then the universe is out of balance. A correction is not only needed, but necessary.

Duir notes that the name Cernunnos is provided on no more than four inscriptions: the Pilier des nautes (Pillar of the Boatmen) from Paris, which includes a carving and the earliest record of the name, derived from the Gaulish *carnon* or *cernon* which means "antler" or "horn"; two metal plaques from Seinsel-Rëlent at Luxembourg include the phrase *Deo Ceruninco*, "to the God Cerunincos"; and a Hellenistic inscription from Montagnac in Hérault, Languedoc-Roussilion, France, which includes the name *Karnonos*. Duir says that the Parisian pillar was constructed by a guild of Gaulish boatmen living among Celts and controlling trade along the Seine River. A torc hangs from each of the figure's two carved horns. The lower portion of the relief is missing, but Duir envisions the figure sitting in a cross-legged position. She also lists several features common to the majority of Celtic images of Cernunnos: the horns; torcs (typically adorning the necks of Celtic gods); a purse or cornucopia; his three heads or faces; the holding of a ram-headed snake; nearby animals (most notably stags); and his seated, usually cross-legged position. The horn aspect may be a misinterpretation, as Duir states that Cernunnos appears to wear ram's horns in one particular rendition but is otherwise depicted with what appear to be deer antlers. She correctly surmises the difference between antlers and horns, as the former, once fully formed, are comprised of dead bone and shed on an annual basis, while the latter consist of a core of living bone with a proteinaceous covering and not shed.

Duir suggests that the antlers are indicative of "the seasonal nature of the god,"[7] but their true meaning to the ancient Celts remains a puzzle.

The thought that antlers symbolize virility is countered by images of antlered goddesses. Duir mentions the theory that Cernunnos was Lord of the Hunt within a wilderness setting and then counters this thought by observing the cross-legged position and "arms raised in a Buddhic style, as seen on the Gunderstrup Cauldron,"[8] which may contradict the supposed wild nature of this deity. She also notes that other figures depict Cernunnos seated on a bench, suggestive of a more peaceful presence.

And what of the ram-headed snake? Duir lists the most common associations with the serpent as "fertility, death, the under-world . . . regeneration . . . healing."[9] The ram often symbolizes the planet Mercury, conflict, and strong virility. Duir notes that this association between Cernunnos and Mercury has "a less direct association of triplicity by [Cernunnos'] iconography being found, on several occasions, associated with triple-headed figures. Mercury is found alongside Cernunnos on the Reims relief."[10] She continues,

> The Pillar of the Boatman links him with sailors and commerce and, again, one recalls the association with Mercury in the Reims relief. As mentioned, Mercury is associated with healing and holds his caduceus of entwined snakes; he is also usually identified with the Greek Hermes, who, amongst other things, was a psychopomp, who escorted the dead to the under-world, as well as being a divine keeper of herds.[11]

Duir completes her outline of Cernunnos by noting depictions of animals—stags, boars, rats, hares, dogs, dolphins, and lions—situated around this god.

> As mentioned, this gives rise to the commonly held attribution of the god as Lord of the Hunt and, since hunting involves death, a connection with the underworld. The image of the Gunderstrup Cauldron is often compared to that of Shiva Pash-upati, the Yogic "Lord of Beasts," as shown on at least one well-known image, the Marshall Harappan seal. In this, the horned Pashupati is surrounded by animals and has his legs crossed. The resemblance is striking and may have influenced the design of the Cernunnos plate of Gunderstrup, which may have its origins in Romania or Thrace, which stood between Greece and the east.[12]

Associating Cernunnos with the underworld is a reminder of gods from other times, in other traditions. Duir contemplates this idea:

> If there is a connection with the underworld, does this raise a possible connection with the Celtic god Dispater? . . . Of the Gallic gods [Caesar wrote], "They worship chiefly the god Mercury. After him they worship Apollo and Mars, Iuppiter and Minerva. About these they hold much the same beliefs as other nations. Apollo heals diseases . . . All the Gauls assert that they are descended from Dispater, their progenitor." . . . Dispater . . . is obviously a reference to a god of the dead, and to wealth, which comes from the earth . . . this name also appears on the Pillar of the Boatmen. . . . The identity of Dispater remains elusive, and some people more readily identify him with the Irish gods Donn or the Daghda.[13]

Duir concludes, "on the basis of what we have evidence for, that Cernunnos was directly associated with divinity, wealth and animals, and potentially indirectly associated with regeneration, healing, fertility and death. We have little to explain the cross-legged pose so characteristic of many images, although it may relate to either a common Celtic position of a hunter, or to something more akin to Buddhic calm."[14] We have seen the symbolism before. The images, few as they are, express much more than the cycle of life, the life of the hunter, and the calming effect of meditation. In fact, the purpose of Duir's essay is to argue against Margaret Murray's idea that the Celts worshipped a horned god as early as the Paleolithic. According to Murray, "in spite of his Latinized name, [Cernunnos] was found in all parts of Gaul. It was only when Rome started on her career of conquest that any written record was made of the gods of Western Europe, and those records prove that a horned deity, whom the Romans called Cernunnos, was one of the greatest gods, perhaps even the supreme deity, of Gaul. Cernunnos is recorded in writing and in sculpture in the south of Gaul."[15] In contrast, Duir proposes that "the [Celtic] concept of one goddess and one god of whom other goddesses or gods are but aspects"[16] is a figment of twentieth-century academic minds, popularized by Murray.

Based on the evidence, Duir finds that various attributes of Cernunnos conflict with Murray's Paleolithic source for this deity. I believe Murray discovers in Cernunnos the tendencies of a deity with far more spatial and temporal (holistic) significance. Let's look again at the accoutrements of Cernunnos. Our knowledge of ancient symbolism associated with the myths and traditions of Sumeria, Babylon, India, Egypt, Judea, and the Americas can help us discover the symbolism of this Celtic god.

- **Antlers**: an annual growth on the male skull of most deer species, where tissue protrudes beneath velvet-covered flesh and calcifies. Once the velvet is shed, the antlers are used by the stags during the fight for and protection of does during the mating season. Antlers are shed annually. Female reindeer and caribou grow antlers. Antlers symbolize seasonality, the cyclicity of time, re-growth, renewal.

- **Snakes**: We have already seen snakes symbolizing renewal and the cyclicity of time, like the antler. The ram-headed snake is a symbol of Mercury and Mars—war, protection, healing.

- **Three heads/faces**: The tripartite nature of deities is found in the Hindu conceptions of Vishnu, Brahma, and Shiva; the ancient Egyptian sun as Atum, Re, and Khephra; the Christian God, Christ, and the Holy Ghost; and many other deities from ancient and modern traditions. While depicting different aspects of the universe, they are, nonetheless, one.

- **Torcs**: a circular adornment worn around the neck by royalty, women, and warriors; often made from precious metals; depicted on antlers or in the hand of Cernunnos.

- **Seated position**: a common position of Cernunnos, Buddha, Shiva, and numerous other deities in depictions from around the world; also used for portraits of royalty. A position of rest and contemplation.

- **Coins**: Spilling coins from a bag he holds, Cernunnos appears not to be miserly in protecting his wealth, but rather spreads it, offers it, gives it beyond himself; a sign of beneficence.

These attributes are common to many deities found around the world. We can with certainty conclude that Cernunnos was directly associated with more than just "divinity, wealth and animals, and potentially indirectly associated with regeneration, healing, fertility and death."[17] To say that Cernunnos represents the Lord of the Hunt is to fall short of the greater implication of the symbolism. After all, deer do not hunt. They're prey. They are the hunted, sought by the hunter, the seeker. The hunter has something to gain from the deer. And what is to be gained?

Clearly, the symbolism identifies Cernunnos as a protector and a healer, a defender of the cyclicity of life, of nature, of the universe. He is perceived as royalty, overseeing this Earth, providing the sustenance we need—all the wealth of the world. He gives life and he takes life for the benefit of all. And again, we have seen a parallel to Cernunnos before. Why, it is Shiva, just as MacKenzie concluded over eighty years ago! And it is the sun god of the Maya and the Aztec! And Wakan Tanka! In contrast to Duir's conclusions, I do not find Cernunnos' identity problematic. Nor does the "idea that Celtic gods were perceived as 'separate individuals' prior to the third century CE and Neo-Platonic philosophy"[18] hold much merit. And, of course, there remains no evidence of people of Paleolithic Europe not worshipping a horned god. In fact, they might have, or they might not. The concern is trivial. Rather, it is the Celtic conception of Cernunnos as creator, destroyer, and preserver of life that should be held in regard.

Early in this chapter I noted that scholars have equated Cernunnos with Daghda the God of Irish mythology. Daghda's name may be derived from *dago devas*, although some texts suggest it comes from *dag dae* "good hand" (skillful hand), or *daeg dia* "god of fire."[19] *Dago devas* seems to be the preferred source of his name based on current information but, of course, each derivation appears to provide a suitable explanation for the name of a god as important as Daghda, known in separate traditions as Cernunnos, Shiva, and so forth. Other names given to the Daghda provide additional clues to his importance in Irish Celtic mythology: Eochu Ollathair ("Horse Great-Father," indicative of a god of fertility and the elder of other deities) and Ruadh Rofhessa ("Red One Great in Knowledge"—the color signifying sacrifice and courage). Mary Jones's *Celtic Encyclopedia* states, "Now, looking at this information, we can see several things. First, by his kingship and his titles 'the good god,' 'horse great-father' and 'red one great in knowledge,' we can see that he is a leader of the gods . . . The Daghda is more like the father-figure and druid of the gods (which he is explicitly called in several texts)."[20]

As we've seen in the Celtic myth of The Ages of the World, one of the four treasures of the Túatha Dé Danaan was the Daghda's bronze cauldron containing sustenance enough for all people of Ireland. It is evident that with the Cauldron of Plenty, as well as his harp and giant club, the Daghda was seen as capable of providing more resources than necessary for the good of the people, that he had the ability to sway other gods, and he could wield his power to produce terrible destruction with a single blow.

The Daghda is a god of fire, worshipped as an earth-god wielding great power. He rains fire upon enemies of the world with wisdom and inspiration. As such, he is god of and for the world, the swift and mighty Regenerator of the World who observes the world from behind his brown or milky gray cloak. He is pure, but can also be resolute and harsh, strong and tough as an oak tree—an Indo-European metaphor associated with thunder and lightning. He uses his terrible club as a weapon upon the earth as he creates and regenerates for the betterment of the world and the rebirth of humanity.

Now we are travelling in very familiar territory. There is no need to belabor the obvious deduction from the scholarly transcription of Celtic mythology: The Daghda is the world-god, a protector and provider. He is caring. He is the creator and destroyer. His means of destruction is swift and terrible, but his purpose is regeneration. We can conclude that the Daghda, like Cernunnos, is a Celtic parallel for Shiva. Even the color of his cloak suggests that we may look skyward to find him. These similarities demonstrate the unified understanding and long-held acceptance of the qualities of the deity humans believe to be the creator, protector, destroyer, and regenerator of the world. But there is one additional piece of evidence that supports this common tradition. It is a geometric construct used worldwide as a symbol relating life to death, sacrifice to rebirth, protector to protected, and Earth to Sun. It is the sign of the beneficent giver and taker of life, and protector of the universe. On the obverse of a Celtic silver coin from Hampshire is an engraving of an antlered head—Cernunnos. Between his antlers is an eight-rayed sun wheel. Green considers the "association between Cernunnos and sun symbolism . . . curious and unique."[21] Of course, considering the few artifacts we have with the figure of Cernunnos, we shouldn't be surprised by this unusual association between this deity and the sun. Based on universal characteristics attributed to the creator and regenerator, it is difficult to understand why this association would be "curious." In fact, it is quite appropriate.

A Second Look at Mythocartography

Looking farther north, to Scandinavia and Finland, other mysteries related to geography and its ties to the cosmos lay buried in mythology. In chapter 4 we found that cartographers located a fictitious mountain of lodestone at the North Pole by the thirteenth century, but discovering the source of the suspected polar islands and the rivers that course through them involves a tortuous journey along a trail of medieval records that may include the lost *Inventio fortunata*. Recall that Van Duzer concluded, "Thus the source of this mythical polar geography is a lost work by an unknown author of the 14th century."[22]

However, Delno West argued that the author of the *Inventio fortunata* believed that "the whirlpool at the North Pole represented an entrance to Hell, which was believed to be in the center of the earth, and also that the four inward-flowing rivers and whirlpool are the counterpart to the fountain in the Garden of Eden, whence the four great rivers branch out to water the world (see Genesis 2:10–14) if the rivers flow out, they must flow back in somewhere and be recycled."[23] The Jesuit polymath Athanasius Kircher stated that, "a whirlpool at the North Pole sucking in the waters of all the oceans, also asserts that the waters emerge again, not in Eden, but at the South Pole."[24] Kircher's early chart of global ocean circulation included in his *Mundus subterraneus* followed the geography of the *Inventio fortunate*. Van Duzer concluded that, "The persistence of the *Inventio fortunata* geography on maps for, say, 150 years is . . . a testament to the great psychological and mythical power of the concept of the center."[25]

What center is the author of the *Inventio fortunata* referring to? According to Van Duzer, ancient traditions held that "the North Pole was the true center of the earth, and the author of the *Inventio fortunata* gave an account of the geography that was so mythologically satisfying, that it continued to be believed or at least repeated well past the time when scholars and explorers knew that the account was false."[26] As we have seen in previous chapters, and as Van Duzer outlines in his essay, many cultures have aligned sacred centers to the four cardinal directions with respect to the axial spin of Earth or the directions of the solar equinoxes and solstices. However, neither of these sets of directions is useful where the center is defined as either the north pole or south pole. At the north pole, for example, the sun does not rise or set during the solstices, and the directions east and west have no meaning, since all directions from the pole lead southward. So, how can we locate the poles? What geographic feature would we find at each?

In Norse mythology, the answer is a whirlpool. Van Duzer lists sources for a polar geography including "a great northern whirlpool in Norse legends of the world's well, 'Hvergelmer,' which causes the tides by pushing and pulling water through its subterranean channels."[27] In fact, Mercator cites the *Topographia hibernica,* written by archdeacon of Brecon, Gerald of Wales (1146–1220), as his source for describing this polar whirlpool. Van Duzer also provides the following quote from the eighth-century Lombard author Paulus Warnefridi:

> And not far from the shore which we before spoke of, on the west, where the ocean extends without bounds, is that very deep abyss of waters which we commonly call the ocean's navel. It is said twice a day to suck the waves into itself, and to spew them out again; as is proved to happen along all these coasts, where the waves rush in and go back again with fearful rapidity. . . . By the whirlpool of which we have spoken it is asserted that ships are often drawn in with such rapidity that they seem to resemble the flight of arrows through the air; and sometimes they are lost in the gulf with a very frightful destruction. Often just as they are about to go under, they are brought back again by a sudden shock of the waves, and they are sent out again thence with the same rapidity with which they were drawn in.[28]

We now have a strange, mythical, geographic dichotomy. Is the north pole the location of a mountain, or a whirlpool? In the former case our sight is directed upward, skyward. In the latter we are drawn down into Earth. Which is the correct geography? Perhaps it's both. Van Duzer outlines the Hindu/Buddhist conception of Earth from the *catur-dvipa vasumati,* a four-continent Earth. Our interest here is the relationship between geography and geometry as conceived in Bhraamic, Hindu, and Buddhist beliefs. The important findings are that:

- The sacred center of Earth is located in the far north, at Mt. Meru, believed to be the *axis mundi* about which the heavens revolve.

- Four mountainous continents are located around Mt. Meru, forming the image of a lotus flower. The four land masses are aligned to the four cardinal directions.

- The ultimate source of the world's rivers is Lake Anotatta, located near Mt. Meru.

- Mountains surround the lake, and each of four rivers flows through a high pass shaped like the head of a lion, elephant, ox, or horse, and toward a respective cardinal direction.

- The pole star is located above Mt. Meru, situated at the center of Earth's surface, envisioned to take the form of a disc or slightly cup-shape.

Van Duzer notes that transferring this flat terrestrial geography to a cosmological globe necessitated several adjustments to account for the spherical form of the globe, including placing Mt. Meru at the North Pole rather than simply at the center of the world; the surrounding continents were then situated immediately south of, but still surrounding, the mountain.[29] Figure 7.3 shows illustrations of models and devices used by ancient Chinese astronomers to study and interpret the skies. Based on Van Duzer's analysis, parallels between the polar geography in *Inventio fortunata* and Brahmanic Hindu and Buddhist mythology are now obvious.

Figure 7.3: Drawings of cosmological models and devices used by ancient Chinese astronomers to study and interpret the skies.

Firth states, "The essence of symbolism lies in the recognition of one thing as standing for (re-presenting) another, the relation between them normally being that of concrete to abstract, particular to general. The relation is such that the symbol by itself appears capable of generating and receiving effects otherwise reserved for the object to which it refers—and such effects are often of high emotional charge."[30] Recall, too, that mythology often consists of several layers of meaning that are generally hidden within the overt storyline. And so it is surprising to have Van Duzer conclude that, while the similarities between the two geographies are impressive, ultimately "the one is a secular or geographical mythology, the other divine . . ."[31] We should recognize that each is a construct resulting from two cultures, two traditions, but which are intended to symbolize the same concepts of the axis mundi and a cosmological geography. This may be implied when Van Duzer states that he is "inclined rather to see the fact that two so similar mythographies of the northern polar regions should arise and persist in two so different cultures as a testament to both the creativity of these two cultures, and to the degree to which these mythographies match our innate transcultural conception of what a sacred center should be."[32]

Norse Mythogeography and the Number Nine

As a final example of the relationship between universally perceived cosmic geometry and the mythological sacred center of the world, let's turn to Norse mythology, formed in proto-Norse (Scandinavian Indo-European) Nordic traditions prior to the introduction of Christianity in northern Europe. Much of the symbolism of the mythic tradition has fallen into folklore, but ancient understandings of the relationship between Earth and the cosmos can be mined from the prose and poetry provided to us by medieval historians such as the Icelander Snorri Sturluson (collector of the Prose Edda), the Dane Saxo Grammaticus, and others. For our purposes, the focus is on the mythic geometry of the world and the gods representing the physical aspects perceived on Earth and in the cosmos.

The culture of the Norse is formed by the traditions of a seafaring people inhabiting a subarctic, mountainous landscape. Their mythic vision of the universe consists of nine regions or worlds, each connected to Yggdrasill, the World Tree. The etymology of the name remains a point of conjecture, but the symbolism is universal. Similarly, the configuration of nine worlds is not understood, but they are tied to Yggdrasill nonetheless. Based on the contents of the Prose Edda, Poetic Edda, and other medieval records, each world is occupied by, and named for, particular beings or elements.

- Midgard, the biosphere of Earth, occupied by humans

- Asaheimr, the abode of Norse Gods

- Vanaheimr, region of Vanir

- Jotunheimr, land of Frost Giants

- Alfheimr, world of Light Elves

- Svartalfaheimr, land of Dark Elves

- Muspellsheimr, world of fire

- Niflheimr, world of ice and mist

- Helheimr, abode of the dead

Yggdrasill is said to extend three roots, one each into Asaheimr, Jotunheimr, and Helheimr. Those three worlds provide water to nourish Yggdrasill, with the root at Jotunheimr tapping the spring of wisdom and understanding.[33] This configuration of three roots extending between four locations may indicate a fundamental geometry that we will investigate later.

As Norse myth explains, in the beginning there are two worlds, those of Muspellsheimr and Niflheimr, fire and ice.[34] The combination of warm air and icy cold creates the giant Ymir and a cow, the ancestors of all giants and gods. Earth was formed from the body of Ymir, killed by the gods Odin, Vili, and Ve. Those gods created two humans from a pair of tree trunks and provided Midgard for them to live upon. In time came the ancient giant Fornjot, who was the father of Logi (fire giant), Kári (wind giant), and Ægir (giant of the sea). Ægir's power is expressed by the oceans. His wife, Rán, is the goddess of the sea. Ægir became known for hosting feasts for the gods. Food and drink are made plentiful for all who attend, and it is at such an affair that Ægir notices that Rán holds a net she uses to catch men who fare upon the sea. (In Norse tales, the sea is at times referred to as "Rán's Road," suggestive of the importance of

the ocean to Norse life.) In Brodeur's translation of the second part of the
Prose Edda, the Skáldskaparmál, Snorri states:

> Gymir's wet-cold Spa-Wife
> Wiles the Bear of twisted
> Cables
> Oft into Ægir's wide jaws,
> Where the angry billow
> breaketh.
> And the Sea-Peak's Sleipnir
> slitteth
> The stormy breast rain-driven,
> The wave, with red stain running
> Out of white Rán's mouth.[35]

Ægir and Rán can be a dangerous duo on the open sea, and their nine
daughters perhaps no less so. These daughters, "nine skerry-brides,"[36]
are commonly interpreted to represent ocean waves. Their names are
translated as the Pitching One, Bloody-Hair, Riser, Frothing Wave,
Welling Wave, Billow, Foam-Fleck, Cool Wave, and That Through
Which One Can See the Heavens (understood to refer to water being
transparent).[37] This last name, Himinglæva, "stirs up the roar of the sea
against the brave."[38] Seafaring was not easy, "[w]hen hard gusts from the
white mountain range teased apart and wove together the storm-happy
daughters of Ægir, bred on frost."[39]

Another Norse god is Heimdall, the watchman of the gods and the
Norse version of the ancient Indo-European god of fire.[40] His name may
derive from *himinn* ("sky, heavens") and þöll ("pine-tree" or "river") and
his home, Himinbiörg ("heavenly" or "sky mountain").[41] Heimdall is
said to have nine mothers, all sisters. Some scholars conclude that those
nine women are indeed the nine daughters of Ægir and Rán.[42] Heimdall
stands guard at the Bifrost Bridge—the rainbow-extending between
Midgard and Asgard. Ever alert, he is said to hear grass grow. He will
blow his Gjallarhorn, a calling horn, when the foretold Æsir-Vanir battle
begins at the start of Ragnarök (doom of the gods). The Æsir live in
Asgard and include the principals of the Norse pantheon: Odin, Thor,
Baldr, and others. The Vanir are gods of wisdom, prognostication, and
fertility, and live in Vanirheimr. Many of the gods, including Odin, Thor,

and Heimdall, will die during the war. Heimdall will be one of the last to be killed. The world will end, but it will be regenerated.

Life and death on the seas, a daily consideration in the lives of the Norse, is highly dependent on one's ability to determine bearings, set a course, and achieve the end of the journey in a reasonable timeframe without sighting land for navigation. In other words, ancient seafarers depended on the sun, moon, and stars to guide their course. The mythic vision of the universe, including Earth itself, consists of nine regions, including the sky and the cosmic dome in its apparently eternal rotation above. These worlds are connected to Yggdrasill. The worlds beyond Earth are the abodes of gods, giants, elves, fire and ice, and the dead. Yggdrasill is fed by these worlds, including the nourishing water of wisdom and understanding.

Earth was formed soon after warm air and icy cold melded together. The giant (also possibly "ancient destroyer") Fornjot lived to the east (in what is now Sweden or Finland) and perhaps represented the rising Sun. Fornjot was the father of fire and wind, and he ruled the sea from which the sun rose and set each day. The sea is a symbol for the cosmos. The power of space is given to Ægir. The journey, whether on land or at sea, can be captivating, as the skywatcher's attention is drawn not only to the breadth of space, but also to the constant change observed over time—whether the changes are related to the microcosm of the weather or the macrocosm of the cosmos. Suitably attentive seafarers survived their journeys. The consequences of inattention were dire. The nine daughters of Ægir represent the ever-changing scenery of the sky. Their names seem reflective of physical features of the sky as well as waves: Bloody-Hair (red sky, Mars), Frothing Wave and Billow (cloud formations), Cool Wave (cold front), and That Through Which One Can See the Heavens (the air itself). This last name certainly is appropriate for the one who "stirs up the roar of the sea against the brave"[43] assisted by that deity's uncle Kári, the wind giant.

As for Heimdall, he represents Earth's rotational axis perceived as a rainbow, pine tree, or river, extending outward to the heavenly mountain where Earth connects to the cosmic sphere. His mothers, nine female giants located at the edge of Earth, are the nine daughters of Ægir. Heimdall is alert for the arrival of Ragnarök, when the gods will battle and many will die, the world will end and be reborn. Heimdall will sound the beginning of the end when he blows his horn, but he will die before the war is over, struck by something called Man's Measure. What is this? There is a hint of what this means when we recall that one of Heimdall's nicknames means "bent stick." Surely it means more than the belief that a set of ram's horns were on his helmet, and might this nickname help explain how Earth's rotational axis could be perceived as curved like a rainbow or meandering river?

The true meaning relates to Heimdall's home at the end of the heavenly mountain at the far end of the polar axis, the pine tree, on the cosmic sphere. This point in the heavens is the polestar, currently Polaris. However, as we've seen, precession causes Earth's axis to circumnavigate a circular area over the course of about 26,000 years. At some time in the past, and at a time in the future, the axis will move away from Polaris such that it points to open space before closing in on another polestar. Thus, the pine tree is envisioned to *bend* and become reoriented over the millennia. A rainbow is similarly curved, and so is a river's meander.

This aspect of precession is known to Western science, but it was also understood by ancient cultures around the world for thousands of years. The implications of this slow change in tilt of the Earth's axis are buried in world mythology and documented in great detail by de Santillana and von Dechend and others. There is little need to address this finding much further here. However, Norse myth does provide several clues as to how precession was understood to operate, which can lead us to a further understanding of the geometrical construct of the nine worlds and the apparent movement of the cosmos about Earth. Surprisingly, this construct fits well with the geometrical model that we have been building since we began looking at world mythologies in chapter 3. Importantly, we need to keep in mind our idea that the simpler the model, the more likely it is to be representative of the idea expressed in mythology.

We begin at the center: The name given to Earth—Midgard—is a reasonable choice for a centrally located world, particularly given that this is the abode of mankind (Figure 7.4). We are told that in the beginning there were two worlds—Muspellsheimr and Niflheimr, fire and ice—and the coming together of fire and ice created life, which led to the creation of Earth. It seems reasonable, then, to place the worlds of fire and ice at opposite ends of the earth. Next, we know that Heimdall lives at the heavenly mountain at the end of the polar axis, or Polaris. He stands guard at the rainbow and is in communication with the principal gods at Asaheim, so we place Asaheim next to the "pine tree," within the circular area defined by precession of the polar axis. At the opposite end of this axis, extending outward from Earth's south pole, we place the world of the dead, Helheim, certainly in keeping with the trend we've seen in other world mythologies. We then have four remaining worlds: Vanaheim, Jotunheim, Alfheim, and Svartalfaheim. As with the other four worlds, we place these four regions with equivalent spacing around Earth. The myths give us no specific priority to their locations.

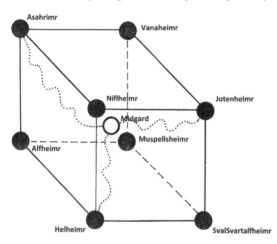

Figure 7.4: A model of the Nine Worlds of Norse mythology. The model forms a cube centered at Midgard—Earth. The cube, and cubic symmetry, represents the sphere of the universe.

Now, by connecting these nine worlds with lines, we can begin to recognize the geometry resulting from our placements. Midgard is at the center, with each of the other eight worlds located an equal distance from the center. If we envision Asaheim at the top, then Helheim is at the bottom, and the six other worlds are located in pattern about midway between Asaheim and Helheim. We can also see that those eight worlds are located at points on a sphere with the center defined by Midgard. Furthermore, the eight peripheral worlds define the locations of the eight corners of a cube. We applied our simpler-is-better scheme to locate the nine worlds of Norse mythology and discovered that their relative locations can be defined by cubic symmetry, or what we may conclude to be perfect symmetry as it regards the sphere.

Two additional points must be made concerning the geometry implied by Norse myth. The issues are related to the nine skerry-brides and two of those women in particular. De Santillana and von Dechend detailed their conclusion that Earth's rotational axis is symbolized in numerous mythologies from around the world as a huge mill, drill, or churn that rotates and causes the celestial dome to be seen as revolving around Earth.[44] This has been shown to be the case in the mythologies of ancient Egypt, Greece, India, Central America, and other cultures. In Finland the axis was described as a *Sampo* (Sanskrit: *skambha*, meaning "pillar" or "pole") that grinds, leading de Santillana and von Dechend to conclude that it is a mill. They evaluate various translations of chapter 16 in Snorri's Skaldskaparmal, of which they believe the following "appears to be the most carefully translated" concerning the mill:

"T'is said," sang Snaebjörn, "that far out, off yonder ness, the Nine Maids of the Island Mill stir amain the host-cruel skerry-quern—they who in ages past ground Hamlet's meal. The good chieftain furrows the hull's lair with his ship's peaked prow. Here the sea is called Amlodhi's Mill."[45]

Figure 7.5: *Rán and the Wave Girls* (1831). Rán is instructing her nine daughters on the movement of waves upon the world (from Alkuna: Nordische und nordslawische Mythologie).

Scholars interpret the Nine Maids in this passage to be the nine daughters of Ægis. Figure 7.5 shows Rán instructing her nine daughters treading upon the sea. Skerry is a Norse term for a small, seabound island, and the maids are seen as pounding the island with wave upon wave, transforming it into grains of sand. But in the case of the kenning 'skerry-quern, the nine maids are perceived as working to rotate the mill. In other words, they turn the axis of the world, causing the cosmic dome to rotate about the polestar from "beyond the skirts of the earth" or "off yonder promontory."[46] So, as it appears, the nine maids are turning the mill—Heimdall's pine tree—from above. We can envision the turning of the cosmic dome by those nine skerry maidens as a process in which each maiden exerts a force in a unique direction along one of nine great circles beyond Earth, as depicted in Figure 7.6, a spherical geometry we have seen in other sacred traditions.

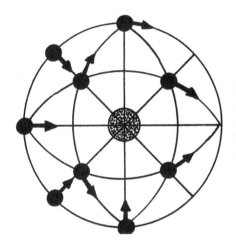

Figure 7.6: The nine skerry maidens depicted by blackened circles, each exerting a unique force along one of nine great circles about Earth. The geometry is common to many sacred traditions.

In the Norse myth, King Frodhi, an alias for one of the knowledgeable gods of Vanir, owns the mill that cannot be turned by mortal means. He recruits two sister giants, Fenja and Menja, to operate the mill, and they do so until Frodhi's greed forces them to work both day and night (Figure 7.7). They make a plan to be rid of Frodhi. A sea-king arrives and kills Frodhi but then takes ownership of the mill and orders the sisters to continue working the mill from on board a ship. They grind out salt, but the mill collapses, the vessel sinks, and a whirlpool forms in the ocean where the millstone had been turning. The deeper meaning of this story comes about by turning this scene upside down. A god of Vanir, apparently thinking quite highly of himself, orders Fenja and Menja to turn the mill, rotating Earth's axis. But the sisters are worked too hard, and the god is killed by the sea-king—a possible alias for Ægis, who we know to represent the ruler of the cosmos; thus, the salt that Fenja and Menja grind may be the stars. Soon the mill shakes and goes off-kilter, and a whirlpool forms; this is the circular path of the polar axis through precession. "And from that time there has been a whirlpool in the sea where the water falls through the hole in the mill-stone. It was then that the sea became salt."[47]

Figure 7.7: Two sister giants, Fenja and Menja, turn the mill that causes Earth's axis to rotate (drawing by Gunnar Forssell from Fredrik Sander's Poetic Edda, 1893).

We know that mythology contains several layers of meaning. So, who or what do Fenja and Menja represent? If these two giants who turn the mill cause the celestial dome to appear rotating about the Earth, and if the mill continues rotating throughout the precession of the equinoxes, then Fenja and Menja may represent the two equinoctial points where the celestial equator and the ecliptic intersect at opposite directions from Earth.

The Sami and the Sioux

The question of why traditional mythologies from around the world are so similar remains unanswered. A singular cause is likely insufficient to explain all of the similarities. However, before ending our discussion of the symbolism expressed in mythology and ideas concerning universal sacred geometry, there is one example worth mentioning as it brings us full circle from our review of symbolism, beginning with the Lakota and ending with the ancient Norse mythology. Ben Baird describes "incredible parallelism of mythic themes"[48] between the indigenous Sami (Saami) of northern Sweden, Norway, Finland, and extreme northwest Russia, and the Dakota, or Sioux, Indians of North America.

- The Dakota and Sami each recognized the sacredness of nature. "The physical world was viewed as the manifestation of this animating force and therefore the Sioux, like the Sami, viewed all natural objects as indistinguishable from this sacred spirit. Both the Sami and the Sioux also express this existential truth through the symbol of the world pillar or world tree."[49]

- Ancient spiritualities of both cultures contain the same fundamental ideas and use virtually identical symbols to convey those ideas. Both spiritual traditions metaphorically describe the sun as the father of all and the earth as the mother of all.

For the Sioux and the Sami, all nature is viewed as being alive; this is radically different from our modern perspective, emphasizing a sacred connection and interdependence of all things—including their prey and their enemies.

- The shaman, or *noaidi*, was an important feature of Sami culture and spiritual tradition. The shaman was able to fulfill many practical purposes with his special talent and function as a leader in the community. Although the shaman had no formal authority, he traditionally held a dignified position and was well respected. By listening closely to the drum (called a *meavrresgarri*), attending to its "speech" or watching the particular pointers (*arpa*) while drumming with his hammer, it is said that a shaman could predict future events. The drumming of the shaman also served practical purposes for reindeer herding, finding lost objects, and hunting. Most important, however, the shaman served as a spiritual guide or priest, a mediator between this world and the spiritual realm, and a healer of illnesses. According to the Sami, illness was caused by a person's soul becoming lost or invasion of a person's body by a hostile object. The shaman would either retrieve the lost soul while in trance or expel the foreign object by invoking the aid of spirits or powers. Although the Sioux do not specify a cause for all illness, the shaman or medicine man employs a generally similar technique of entering into a trancelike state and calling for the assistance of natural powers. Black Elk recalls how he was "drumming as I cried to the Spirit of the World, and while I was doing this I could feel the power coming through me from my feet up, and I knew that I could help the sick boy."[50]

Another similarity of Sioux and Sami traditions is recognition of a cyclical conception of time. We'll take a look at the geometry of time in chapter 19. For now, however, it is important to understand that ancient and indigenous peoples related time to the diurnal cycle of the sun, monthly waxing and waning of the moon, and change in seasons throughout the year. Baird finds that "[t]his cyclical movement of time also applies to the lives of humans and other living beings as all things come into being, live, and then come to an inevitable end. However, there is also the notion that the cycle of time does not ultimately exist, and that the flux of the temporal world rests on the eternal source of being which is infinite and always 'now.'"[51]

Beyond the similarities of culture and tradition found in the Sami and Dakota, the impact human progress has made on these two indigenous peoples is remarkably similar and raises the question of whether such progress is ultimately a benefit or cost to the human condition. The impact of Western civilization on Native Americans over the last five hundred years—in the name of religion, economics, and Eurocentric idealism—is well documented, and Baird notes these same impacts on the Sami.

However, hundreds of years of subjugation under the boot of Christian hegemony have not totally crushed the ancient Sami spiritual worldview. Some remain critical of the form of spirituality brought by Christian missionaries and skeptical of missionaries' ability even to understand their own spiritual symbols: "you speak of eternal life . . . without knowing . . . what eternal is . . . what life is . . . and even you contain . . . infinity . . . the universe . . . strength, power . . . undiscovered . . . unused."[52]

History is written not by the winners of wars, nor by the losers, but rather by the survivors. Human survival means the continuation of culture, tradition, and human understanding of our place in the universe. All cultures and traditions add value to our understanding. We all lose when any culture and tradition around the world is destroyed. We lose more than just vital information. We have cut the cord that binds us together.

So we are separated from our source. In a sense, because of our minds, we actually are separated, and the problem is to reunite that broken cord.

JOSEPH CAMPBELL
The Power of Myth

Chapter 8
Pillars of Ancient Egypt

The expressive value of what is called a collective symbol may lie in fact in a set of variations on a common theme rather than in any uniform conceptualization.

RAYMOND FIRTH
SYMBOLS: PUBLIC AND PRIVATE

The incarnation appears either as male or female, and each of us is the incarnation of God. You're born in only one aspect of your actual metaphysical duality, you might say. This is represented in the mystery religions, where an individual goes through a series of initiations opening him out inside into a deeper and deeper depth of himself, and there comes a moment when he realizes that he is both mortal and immortal, both male and female.

JOSEPH CAMPBELL
THE POWER OF MYTH

From Chaos to Architecture

The universe remained in a chaotic state called Nun, "a dark, sunless watery abyss with a power, or creative force within it that commanded order to begin."[1] From within that matrix of disorder came organization, what physicists and engineers today would refer to as a decrease in entropy. The forces of control and organization have been battling against an ever-increasing disorder—increasing entropy—ever since. This was how the ancient Egyptians envisioned the beginning of the world, an outlook from which they developed their kingdoms. The spark that ignited the fire of civilization on Earth remains unknown, hidden from view by the scant inferences we have from the evidence collected about human prehistory. At the same time we can see parallels between the Sumerian and Egyptian conceptualizations of the universe before creation, when that watery mass filled space before time.

From their investigation to discover the origins of freemasonry, Knight and Lomas conclude that the first Egyptian builders—ancient precursors of both operative masons and Freemasons—developed their craft from the masonry originating in Sumer.[2] Ziggurats, the stepped pyramidal structures built in Sumeria, were constructed as a means for the ruling and priestly members of society to communicate with and conduct rituals in closer proximity to their gods.

Numerous theories abound for the purpose of the Egyptian pyramids, some of which include a site for the pharaoh's interaction with the gods; a resting place for the pharaoh's body while the ruler's soul and spirit journey to a cosmic union with the celestial father; an initiation site for potential pharaohs or priests in ancient secrets of prehistory and the cosmos; manufacturing energy or receiving Earth's electromagnetic energy to serve Egyptian society; and capturing or creating energy in association with the world's grid of sacred sites. Maps have developed in recent decades depicting grids laid across Earth's surface based on locations of sacred sites and Earth's electromagnetic energies. Figures 8.1a, b, and c are three examples of such grids. Some of the grid designs are more fanciful that others, a few include specific polyhedral geometries, and some are based on hard evidence. But to a large degree the true purpose of the pyramids, as singular entities or within the aggregate of Egyptian or world-wide sacred sites, remains open to question. The debate between Egyptologists, other scholars, and the general public rages on.

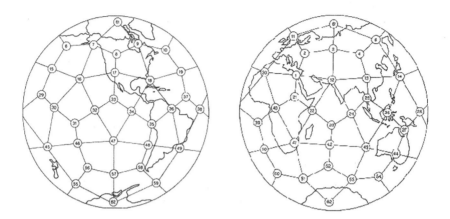

Figure 8.1a: The world grid of megalithic sites, sacred sites and other significant geographic locations on Earth's surface (from David Zink's book The Ancient Stones Speak, Dutton, 1979). The pyramids at Giza are labeled Site 1 on the map (right).

Figure 8.1b: The earth grid map based on studies published in 1973 by Makarov, Morozov, and Goncharov in their article "Is the Earth a Large Crystal?" in *Khimiya i Zhizn'* (*Chemistry and Life*), a journal of the USSR Academy of Sciences. The article proposes that Earth's crystalline power field might yield the geometry of an icosahedron and dodecahedron. The authors theorize that the combination of those two Platonic solids could explain numerous geophysical phenomena. Note location of Giza (center of map).

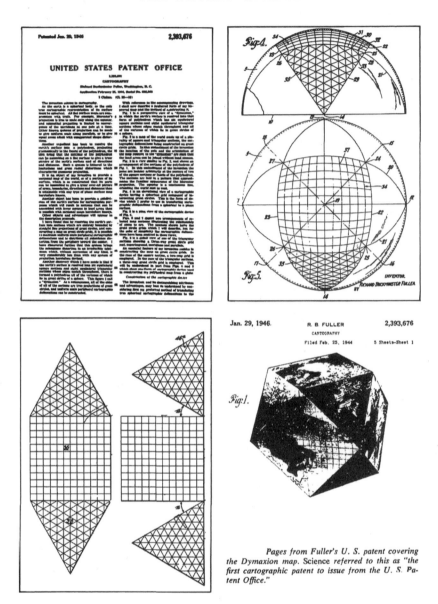

Pages from Fuller's U. S. patent covering the Dymaxion map. Science referred to this as "the first cartographic patent to issue from the U. S. Patent Office."

Figure 8.1c:

R. Buckminster Fuller obtained a copyright for The Dymaxion (dynamic maximum tension) map, a projection of the world's surface onto the surface of a polyhedron, which can then be unfolded to a net in many different ways and flattened to form a two-dimensional map which retains most of the relative proportional integrity of the globe map. The 3-D map takes the form of a cuboctahedron, composed of eight tetrahedra and six half-octahedra. Fuller called the polygon vector equilibrium (VE). Fuller did not specifically identify the location of the pyramids at Giza with respect to form of his map.

Pillars of Egyptian Community

Another architectural feature of ancient Egypt, one that is older than the pyramids, is the pillar or column (Figure 8.2). Obelisks, so readily identifiable with ancient Egyptian architecture, are similar structures that typically stood along structures. These phallic emblems were associated with the Afro-Asiatic *mazzebah* ("pillar"). Serving as a means to connect the ruler and priests on this earthly plane with the gods of the cosmos, the pillar gave spiritual significance to each of the two lands before unification. One sacred pillar was located at the ancient Egyptian city of Inna or Iuna (the biblical city of On, and the Greek Heliopolis), a religious center in what became Lower Egypt. *Innu Mehret* means "the pillar" or "the northern pillar." Innu was sacred ground, inseparable from the group of Heliopolitan gods known as the Ennead, as old as any part of civilized Egypt. It was believed to be part of that first mound of earth that rose from the watery abyss, the sacred land where the gods began ruling Earth during *Zep Tepi*—the First Time.[3] Another column was situated at the city of Nekheb (later Thebes, or Waset). This was the location of the southern pillar in Upper Egypt-*Iwnu Shema*, "the Southern Pillar."[4]

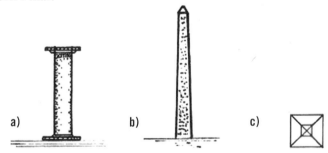

Figure 8.2: a) A pillar or column; b) An obelisk; c) View of obelisk from above.

Based on their analyses of Egyptian beliefs and rituals, Knight and Lomas believe that unification of Upper and Lower Egypt became manifest in those two sacred pillars, symbols of the common interests of the two lands within the one kingdom. They also find important symbolism with the two pillars united by a heavenly lintel represented by the sky goddess Nut, the combination of architectural features producing a doorway opening toward the east and a view of the rising sun. The threshold is seen as the earth itself or the distant horizon.

This structure represented stability of the state and prosperity for the common good as long as the two pillars remained in support of the lintel of the cosmos (Figure 8.3). Knight and Lomas find parallels between ancient Egyptian architectural symbolism and modern-day Masonic symbolism associating concepts of stability, strength, and the Great Architect of the Universe. They note that "each pillar was a spiritual umbilical cord between Heaven and Earth, and the Egyptians needed a new theological framework to express the relationship of their new trinity of two lands and one heaven."[5]

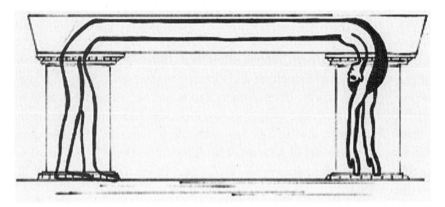

Figure 8.3: The sky goddess Nut forming the combination of architectural features of two pillars and lintel, a doorway opening toward the east and offering a view of the rising sun. The threshold is Earth itself, and the distant horizon.

The engineered pillar and lintel system forms the basis for this framework. Looking westward, toward the gateway:

- The right-hand pillar (signifying Lower Egypt) corresponds to the right-hand pillar Jachin ("to establish"); this is the establishment of the world, as it was at Iuna or Heliopolis that the world was created out of the primordial chaos of Nun.

- The left-hand pillar (signifying Upper Egypt) corresponds to Boaz ("strength," or "in it is strength"), which the two lands would have achieved through unification.

- Unification symbolized by these two pillars represents stability, something that both Upper and Lower Egypt would have achieved together. The stability would support the interrelationship of Earth and the heavens and the tradition of "as above, so below."

Solomon's Temple

Additional details concerning the sacredness of these two pillars may be found in the design and construction of King Solomon's Temple on Mount Moriah at Jerusalem during the tenth century BCE. Sir Issac Newton and others developed plans of the temple based on design information provided in the Bible (Figures 8.4 and 8.5). Solomon's father David brought Israel to the apex of its military power. Through a solemn contract with God, David was granted victories on the battlefield, extending the kingdom's boundary from Gaza to the town of Tiphsah located on the banks of the Euphrates River. In return, David planned for construction of a glorious temple—the House of Yahweh—including the Holy of Holies where the Ark of the Covenant was to be placed.[6] Israel was at peace when Solomon ruled, and the land's unparalleled prosperity led to the erection of the temple that David had envisioned. Solomon was responsible for construction of numerous important buildings across Israel, but this new sanctuary would "outshine all others in the land of Canaan."[6] The structure was to be situated on a hilltop, its porch facing east toward the rising sun.

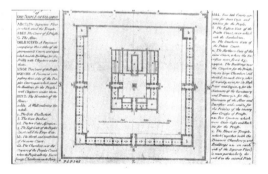

Figure 8.4: Sir Issac Newton's vision of The Temple of Solomon and adjoining grounds based on architectural information provided in the Bible (from Newton's *The Chronology of Ancient Kingdoms*, 1728). Note the near four-fold and mirror symmetries of the structure and gateways.

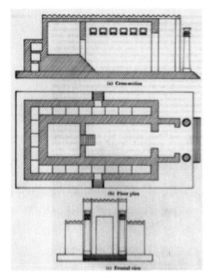

Figure 8.5: Elevation and plan views of Solomon's Temple (from W.F. Albright's *Biblical Archeology*, Philadelphia, PA, Westminster, 1957).

Temples in Egypt and Phoenicia were far more spacious and extravagant, but construction of the Temple at Jerusalem required engineering expertise and craftsmen with skills beyond those of the Israelites. Solomon contracted those services to Phoenicians working under the supervision of a man named Hiram. Artisans over the course of the past 3,500 years have created many renditions of the temple. No one knows for certain the details of its appearance and structural features; the most valuable information concerning its design comes from descriptions in several books in the Bible. However, based on the fact that foreign craftsmen with experience building Phoenician temples worked on the temple, and knowing that architecture in Phoenicia was but a modification of Egyptian design, we can piece together elements described in the biblical accounts to envision the magnificence of what David foresaw, and what Solomon built.

In fact, we might soon have a better understanding of the actual size, location, and features of the temple and associated structures. In February 2010, archaeologist Eilat Mazar announced that an excavation at Jerusalem encountered a tenth-century BC city wall and a royal structure containing Hebraic writing. She suggested that the find would corroborate belief that a palace and fortification around the capital city had been controlled by a Hebrew king at that time. Some archaeologists questioned the stability of a local state at that time, suggesting that Mazar's dating of the structures should be suspect.[7,8] However, features of a number of temples constructed in Egypt, Mesopotamia, and Phoenicia are known through archeological investigation and available literature, and I expect that Mazar's archeological investigation might yield sufficient information to allow a relatively fast determination of the age and purpose of the palace and fortifications.

While Solomon's Temple was not a stone-for-stone replica of any of the structures previously documented in the region, some architectural features are common to all of them, and the general form of the building was "reminiscent of Egyptian sanctuaries and closely matches that of other ancient temples in the region."[9, 10, 11] The Bible describes it as follows:

- The Holy of Holies or Inner House was where the Ark of the Covenant was kept, at the west end of the temple. This room was sacred, corresponding to the adytum of Egyptian temples, and the public was forbidden to enter it.[12] The Bible states that the room was 20 cubits in length, breadth, and height. The rest of the temple (the Holy Place, located between the Holy of Holies and the porch at the east end of the structure) had a height of 30 cubits. It is generally believed that the ceiling elevation was uniform over the temple, such that the floor of the Holy of Holies would have been elevated above the adjoining floor. The temple floor and walls were covered with cedarwood, and then overlaid with gold. Two cherubim constructed of olive wood and standing 10 cubits high were placed in the Holy of Holies; each statue had wings extending 10 cubits from tip to tip, such that together they spanned the breadth of the room and touched at the center. There were no windows.

- The Holy Place, or Greater House, had the same width and height as the Holy of Holies, but it was 40 cubits in length. The floor was constructed of fir wood overlaid with gold. The walls were lined with cedar-carved figures of cherubim, palm trees, and open flowers and then overlaid with gold. Small "loopholes" or windows might have been located near the ceiling.[13] Chains of gold further demarcated the line between the Greater House and the Holy of Holies.

- Peripheral chambers, each 10 cubits wide, were constructed along the north, south, and west sides of the Temple. They included a three-tiered roofing system and might have served as hallways or storage areas. Two additional stories might have been added to the chambers over time.[14] A winding staircase led to the upper floors. These lengthy side chambers would have corresponded to small chambers found in every Egyptian temple.[15]

• The porch was located at the east side of the temple. It was 20 cubits long (north–south) and 10 cubits long (east–west). The extreme height of the porch, described in 2 Chr. 3:4 as being 120 cubits, would be an unusual feature in comparison with contemporaneous temples in the region. The porch is where the two pillars of Jachin and Boaz stood at the north and south sides, respectively, of the entryway. The column height was given as 18 cubits, their width as 12 cubits. Each included a hollow space that was four fingers wide. The pillars were most likely constructed of either copper or bronze (Hebrew: *nehosheth*). Their capitals were 5 cubits in height and decorated with rows of carved pomegranates, chains, and lilies. Columns having similar design features have been documented at the ancient cities of Tyre, Byblus, Paphos, and Telloh. The biblical description of the temple does not indicate a structural connection between the porch and the Greater House. Schick argues that Jachin and Boaz were isolated from the Holy Place but not strictly ornamentation of the building—they did serve some structural purpose for the building.

The architectural style of the temple remains unknown. In *The Biblical World*, first published in 1899, Emanuel Schmidt discusses a variety of possibilities including a design based on the Hebraic tabernacle of the wilderness period; an Assyrian design using a mixture of wood, stone, and metal that was also a common scheme in Egypt; Phoenician architecture, given the source of the supervision and labor for its construction and the use of design elements resembling those found in temples at Byblos and Paphos in Phoenicia; and, of course, the most notable architectural style of the region—Egyptian.[16] Many of the elements included in the temple's design have been lost to time, but the features we know of bear strong resemblance to those found in Egyptian structures. According to Schmidt, such similarities include:

- gateways leading to the temple's exterior lower courtyard, corresponding to large pylons found in Egyptian temples

- a colonnade equivalent to Egyptian peristyle hall

- groves within the lower courtyard reminiscent of Egyptian temple gardens;

- installation of a "molten sea" outside the east entrance to the temple, the counterpart to artificial lakes built at almost all Egyptian temples

- the two large pillars serving a function similar to that of Egyptian obelisks (See Figure 8.5)

- a floor plan for the Greater House and the Holy of Holies that mirrors the Egyptian house of god and adytum; the Ark of the Covenant situated in the middle of the room in the fashion of an Egyptian oracle

- the cherub design and the placement of the Ark of the Covenant, reminiscent of the Egyptian winged sun-disk representing the god Horus and the sacred bark carried in processions and used by pharaohs

- wall decorations, including illustrations of men, animals, and trees, common motifs in the region

- inclusion of numerous small chambers, which "betrays the Egyptian architecture, for they are found not only in Karnak and Luxor, but in all the temples of Egypt"[16]

Based on this comparison of architectural styles, there is no doubt that the design of the Phoenician-built Temple of Solomon was influenced significantly by Egyptian planning, architecture, and engineering. Like the two pillars located at Inna and Nekhab, two columns in the temple, one on the north side of the entryway and another at the south side of the porch, together symbolize the two lands of Israel. With the temple centered at Jerusalem, the left-hand (south) pillar of Boaz signified *strength* and represented the land of Judah, and the right-hand (north) column of Jachin signified *establishment* and represented the land of Israel. The portico thus incorporated the idea of Judah and Israel unified as one state. In tandem with the lintel representing God—Yahweh—the entrance to the temple was a vision of stability, harmony, and balance within the kingdom. The symbolism corresponds completely with the social, spiritual, and political idealism that unified Upper and Lower Egypt: stability through unity. Knight and Lomas state that this conceptualization of strength through the unity of two pillars was adopted by later cultures, including the Hebrews. This must have been the case, as the motif was also used in the teachings of Christ.

Figure 8.6 illustrates this important twin-pillar paradigm. In complete correspondence with Egyptian and Jewish symbolism, Freemasons know the right-hand pillar as Jachin, the first high priest of the temple. For the Qumranians of Jesus's time, the Jachin column symbolized holiness as a fundamental characteristic of *tsedeq*, commonly translated as "righteousness" or "rightness." *Tsedeq* is the principal foundation of divinely appointed order,[21] equivalent to the Egyptian *ma'at*.

Figure 8.6: A keystone arch above two pillars or columns. Egyptian and Jewish symbolism (as well as speculative Freemasonry) consider the right-hand pillar as "Jachin" and the left-hand pillar as "Boaz."

The left-hand pillar symbolized *mishpat* ("judgment") and the figure of Jesus, the prophesied Messiah or king. Such a meaning was certainly acceptable to the Qumranians, but it would not have been acceptable for Jesus to be seen as the right-hand pillar—the high priest—and the left-hand pillar at the same time.[22] As Knight and Lomas point out, it is said that Jesus will sit at the right hand of God the Father; this would be represented by the left-hand pillar, as viewed looking west toward God from the temple door, with God facing east and the *mishpat* pillar situated on his right. But if Jesus, Son of God, may indeed be represented by the two pillars of Mishpat and Tsedeq, then the structure becomes one of Jesus serving as both spiritual and secular connection to God in heaven, who sits on His throne at the ridgeline, the crown of this conceptual temple.

Upon finding that the cubic dimensions of Utanapishtim's ark, as described in the Epic of Gilgamesh, suited the construction of that seaworthy vessel in the shape of a sphere, my attention was drawn to the similar design found in Solomon's Temple. Specifically, the length, width, and height of the Holy of Holies are each 20 cubits. Therefore, it is certainly reasonable that the Holy of Holies, if not constructed in spherical form, may have included a domed roof, similar in form to the Dome on the Rock, the Islamic shrine currently located on the Temple Mount. And, in fact, we do find historic illustrations portraying Solomon's Temple with a dome overlying some portion of the structure.

Of even greater interest are the dimensions given for the Greater House that was situated between the east-side portico and the west-side Holy of Holies. Recall that the dimensions of that room were 40 cubits long by 20 cubits wide, and the corridors extending along the north, south, and west sides of the temple were 10 cubits wide. The total width of the central portion of the temple, including the north and south corridors, was 40 cubits, the same as the length of the Holy Place! Is it possible that the roof of the Greater House was constructed as a dome? Certainly there is no reason why this could not have been so. I would not be surprised to find that the peripheral corridors located along the sides of the temple served as structural supports—buttresses—for the temple's roof, particularly in the case of an open dome design.

A dome can be constructed from a great variety of materials—wood, stone, metal, or other building materials familiar to the designer and contractor. It can be made of a single mass of material or comprised of a series of concentric layers. Unless the space to be covered is too expansive for the strength of the material and structural support of the building, a simple dome (opposed to a compound dome) can be constructed such that the roof and pendentives (the supports underlying and connected to the dome) are unified. The result is a dome of a height that approximates, or may be slightly higher than, another curvilinear roofing system, the barrel vault. Both the simple dome and the barrel vault are ancient architectural features of sacred buildings. Domed and vaulted churches and cathedrals have been common for two thousand years. And domed temples were common in ancient Mesopotamia and Egypt.

With these facts in mind, Figure 8.7 provides two schematic elevations of Solomon's Temple as it could have been built almost three thousand years ago using the materials and technical knowledge available at that time. In these illustrations I assume that a simple dome covered each of the two main rooms, the Greater House and Holy of Holies. The square dimensions of each of those rooms would support their respective domes using four pendentives, one at each corner of the room. Additional structural support is provided by the peripheral corridors and the east-side portico with its two massive pillars. As with many other renditions artisans have generated over the years, this vision of the temple is an illustration based on the nominal information provided to us through biblical texts and the historical architectural and archeological data.

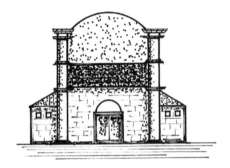

Figure 8.7a: Schematic front-elevation view of Solomon's Temple based on architectural information provided in the Bible. I believe that a dome might have been constructed over the Greater House.

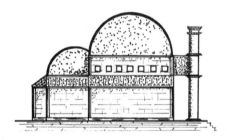

Figure 8.7b: Schematic side-elevation view of Solomon's Temple. Domes might have been constructed over the cubic Greater House and Holy of Holies.

Nine Plus

A lineage of Egyptian gods and goddesses impacted not only Egyptian traditions, but also theology, cosmography, and geometrical relationships found in the symbolism of Judaism, Christianity, and Islam. In the Heliopolitan theology of Lower Egypt, the creation myth taught that the universe begins in the utter disorder of Nun, a uniformly dark, watery body of nothing. Contrary to the laws of physics, entropy decreased as out of this unstructured environment came organization in the form of a mound of dry land upon which Re (Ra) the sun god was self-created in the form of Atum, the Great He-She, the Complete One. But while complete, Atum is conscious of being alone in the universe, and so creates two offspring: Shu, god of the dryness, and Tefnut, goddess of moisture.

The Egyptian sibling gods and goddesses are consorts to each other, and so Shu and Tefnut eventually produce two gods of their own: Geb, god of Earth, and Nut, goddess of the sky and the Milky Way. In turn, Geb and Nut produce the god Osiris and goddess Isis, as well as the god Set and goddess Nepthys (Figure 8.8).

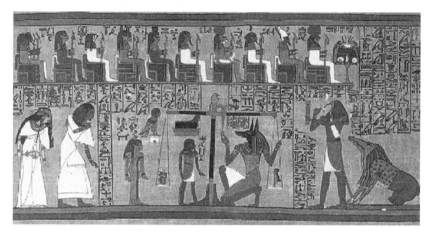

Figure 8.8: The Ennead (above) and the weighing of the soul (below) (from the *Egyptian Book of the Dead, the Papyrus of Ani).*

This pantheon of Egyptian gods and goddesses, known as the Ennead, was the center of worship in Iuna. Ancient legends say that four of the deities—Re (Ra), Shu, Geb, and Osiris—were the first kingly rulers of Egypt. Horus, Osiris's son, is next in succession, followed by Thoth, a god of much practical significance. Thoth represents wisdom, the mind and tongue of Re.[23] He is the arbiter of disputes between the gods. As such, he is the heart of understanding and intention brought to bear through communication. From these qualities he is later associated with magic, writing, science, and judgment of the dead. Thoth's counterpart is Seshat, goddess of wisdom, knowledge, and writing, a scribe or recordkeeper. Consequently she is also the goddess of technical skills associated with geometry and mathematics: astronomy, astrology, architecture, surveying, and building.

There is another goddess with an important role connecting Egyptian mythology with spiritual symbolism. She is Ma'at, Thoth's wife. While Thoth may have been a later addition to the pantheon of Egyptian deities, together he and Ma'at are responsible for ensuring order in the universe, maintaining and regulating the movements of the stars and the order of time (seasons) on Earth. They stand on either side of Re's boat, or barque, which Re rides as he moves across the sky and crosses the netherworld before arriving in the east horizon at dawn each day. Ma'at represents divine wisdom and a universal ethical principle (*ma'at*) understood not only by the ancient Egyptians, but also by proponents of all theologies around the world as it concerns qualities of order, balance, law, justice, morality, and truth. These were the qualities toward which all ancient Egyptians were to strive. Another important role played by Ma'at is that of weighing of souls that took place in the underworld, or Duat. Her symbol is an ostrich feather that she wears on her head (Figure 8.9). It is this feather that is used to measure the soul residing in the departed's heart—if measured to be lighter than the feather, then the soul would attain paradise in the afterlife.

Figure 8.9a: Sketch of the symbol of Ma'at, an ostrich feather.

Figure 8.9b: Detail from Figure 8.8 showing the feather of Ma'at on the balance for "weighing of the soul."

Even the gods were expected to obey this principle. The pharaoh was Lord of the Ma'at, protector of this principle. The rulers and priests held that if the people lived by *ma'at*, then Egypt would be aligned with the natural order of the universe. This may be seen as a method of social control enforced by the rulers over the greater society, and yet it served as a truth that ensured that the world corresponded with the structure of the universe. The pharaoh's activities and proclamations were accepted as products of his personal or priestly revelations and, in turn, the people were to accept a certain loss of freedom and liberty if they were to achieve some measure of peace and protection under the pharaoh's guidance. After all, the pharaoh was responsible for upholding the natural laws of the Creator and preventing disorder, or chaos (*Isfet*), from developing within the state.

In time the concept of *ma'at* was applied to all aspects of earthly existence and understanding of the cosmos, including the broader Egyptian theology. There was a deep and pervasive understanding of the sacred and orderly nature inherent in the universe. Cosmic harmony was an imperative. Disturbing this cosmic imperative could disassociate an individual, group, or the state as a whole from the harmony that was naturally intended. Other features of Ma'at (and *ma'at*) lead me to conclude that a vital but typically unmentioned quality of this goddess and the social concepts applied in ancient Egypt was that of *mercy*.

Whether the context is associated with social justice, an impartial application of the law, pharaonic decision-making, or the workings of the celestial sphere, I believe that universal order and control necessitates some measure of mercy manifested in all actions to protect people, the world, and the cosmos from chaos and wanton destruction. The concept of mercy as an important function of *ma'at* is readily apparent in one Middle Kingdom text in which the Creator declares, "I made every man like his fellow."[24] *Ma'at*'s calling for the wealthy to assist the less fortunate is evident in declarations found within tombs, such as "I have given bread to the hungry and clothed the naked," and "I was a husband to the widow and father to the orphan."[25] Through *ma'at*, the unity of the universe is ensured by the court of public opinion, the court of law, the heart of the pharaoh, and the actions of the Creator.

Osiris

Osiris became associated with the ancient Egyptian underworld. Osiris is the protagonist in the story of the Egyptian Tree of Life (Figure 8.10). Recall that Geb (the earth god) and Nut (god of the sky and the heavens) are the son and daughter, respectively, of Shu and Tefnut and the grandchildren of Atum (Re) the Creator. Re is also the personification of the sun.

Figure 8.10: Osiris seated with Ma'at in his support behind him (from *Papyrus of Nehkt*, a Royal Scribe, in the British Museum).

Osiris, Nut's oldest son, was both a man and a god. He was the first great ruler of the country. Osiris ruled a prosperous nation with the help of Thoth's great wisdom. As mentioned earlier, Isis was Osiris's consort. Their brother Set was jealous of Osiris and schemed to be rid of him. Set held a banquet for Osiris at which he displayed a box and proposed that whoever could lay in the box with a perfect fit could keep it. Set, of course, had previously measured Osiris to be sure that his size would fit the box perfectly. Set had also invited seventy-two other guests who became coconspirators in the plot. After attempts by other guests to fit into the box, Osiris stepped in and Set immediately closed the lid over him. The box became Osiris's coffin and Set and his helpers threw the box into the Nile where it drifted downstream and into the eastern Mediterranean Sea. Eventually the box came to rest along the shoreline at Byblos in Phoenicia. A tamarisk (salt cedar) bush next to the coffin grew to become a magnificent tree that eventually encapsulated the box. The king of Byblos saw the tree and admired it so much that he had it felled and taken to his palace to become a great pillar supporting the roof.

Meanwhile, Isis discovered that the box containing Osiris had floated out to the sea, and she went to find her beloved and bring him back to Egypt for proper burial. At Byblos she encountered the pillar in the king's palace. She pleaded to the king to give her the box within the tree, and he conceded. Isis returned with the coffin, but Set discovered it and cut Osiris's body into fourteen pieces, then discarded them across Egypt so that Isis would never find them. Since Osiris had no heir, Set was soon to receive the right to rule the state. Isis and Nephthys searched for the pieces and located all but one—Osiris's phallus. Isis used magic to make him another one. She rebuilt Osiris's body, transformed into a falcon, and set herself to stand guard over him. In doing so she conceived a son, Horus. Now with an heir apparent, Osiris left the earth to rule the underworld, kingdom of the dead. Isis later gave birth to Horus, who became the prince of Egypt and dueled with Osiris's murderers. In the process, Horus lost an eye but cut off Set's testicles. In the end, Set could not father a child, and Horus became the first king of Egypt.

There is significance to this myth on several levels. As we've seen before, myths typically can be understood to have meanings with physical, intellectual, emotional, and spiritual implications. On a purely emotional level we have here a good story filled with action, adventure, intrigue, and a cast of characters that we relate to; we encourage *ma'at* to prevail. We feel concern, doubt, fear, adoration, and surprise, cheering as Osiris, Isis, and Horus rise above the chaos. They do prevail. And so the people of Egypt win, the country wins, and the Nile continues to bring prosperity throughout the land.

On a purely physical level these events play out across Egypt and the farther reaches of the Near East. Since the beginning of Egyptian civilization, the kings and pharaonic lineages were considered to be both gods and humans. They ruled by divine right. Each was the son of god, a Horus. Upon death each ruler left his body but remained a god to become Osiris, one with the father in heaven—the cosmos. The son of the deceased ruler became the state's new king or pharaoh, the new Horus. The rationale for this nepotistic cycle of political and religious oversight is found in the myth of Osiris. Indeed, Osiris eventually proceeds to the netherworld of the cosmos, the land to the south (in this case down, meaning southward beyond the earth's south pole), and to the kingdom of the dead near the nearest polestar or what may currently be considered as the polar constellation of the Southern Cross.

This netherworld included a Ship of the Dead. Osiris was the pilot of the ship, and the star Canopus was the pilot star at the stern of the ship. The mortal ruler of Egypt becomes the immortal overseer of the Kingdom of the Dead (Figure 8.11). Thus there is a rational basis for the succession of ancient Egyptian rulers as well as a mythological purpose for the southernmost stars in the celestial sphere.

Figure 8.11: Osiris pilots the Ship of the Dead.

De Santillana and von Dechend note a particularly important connection between Osiris as the pilot of the Ship of the Dead and Sumerian mythology, in which Urshanabi is characterized as the pilot of Utanapishtim's ark and the boatman who takes Gilgamesh across the sea to meet Utanapishtim, who may tell Gilgamesh the secret to everlasting life.[26] Recall that Gilgamesh destroyed the sacred stone figures that Urshanabi used to "enable [him] to cross the deep sea without touching the Waters of Death."[27] But Gilgamesh and Urshanabi are able to cross the sea and find Utanapishtim at the boat landing, whereupon Gilgamesh pleads, "I wish to talk with you about life and death. I know that you have found everlasting life and have joined the assembly of gods. I too wish to live on earth forever."[28] It is here that the Sumerian and Egyptian myths intersect, as Osiris is the overseer of the Ship of the Dead and Utanapishtim is the builder of the ark and receiver of everlasting life like the gods in the netherworld.

De Santillana and von Dechend discuss the potential meaning of the Egyptian *mnj.t.* or *menat*, of which "the transitive verb (*mnj*) means to land, from persons, and from ships, and to die, sometimes supplemented 'at Osiris. . .'"[29] They find it, sufficiently striking to see the mooring-post "married" to Canopus in a similar manner as Urshanabi is "married" to Nanshe, Enki's daughter, to whom is consecrated the holy stern of the ship. . . Admittedly, we know as little as before where precisely the *mnj.t.* of the star clocks has to be looked for, but we have made it more plausible that . . . *mnj.t.* must be the decisive plumb line connecting the inhabited world with the celestial South Pole or, let us say, with the orbis antarctus: Osiris being depicted as a circle . . . the verb *mnj.y*, "to land (at Osiris)," points in this direction.[30]

Clearly the purpose of highlighting Earth's polar axis as an important geometric in world mythology is that the movement of Earth around this axis results in the apparent effect of the cosmos circling around the earth. These myths describe the earth's precessional movement as the result of a rotation caused by a celestial mill, fire stick, pillar, or other device turned by the gods or their subordinates. De Santillana and von Dechend drive this point home:

"[F]ire" is actually a great circle reaching from the North Pole of the celestial sphere to its South Pole. . . . The identity of the Mill, in its many versions, with the heaven is thus universally understood and accepted. . . . The Pole star does get out of place, and every few thousand years another star has to be chosen which best approximates that position. It is well known that the great Pyramid, so carefully sighted, is not oriented at our Pole Star but at alpha Draconis, which occupied the position at the pole 5,000 years ago.[31]

Intellectually, the Osiris myth connects earthly existence with the order seen in the cosmos. Instituting this order in this world ensured that the rulers and priests would have control in the state and that the political, civil, and religious activities therein would accord with the natural order of the universe. The alignment between this world and the cosmos was a fundamental aspect of Egyptian civilization, the basis upon which all other purposes of the state were organized. This imperative ensured the success of Egyptian civilization for thousands of years. I'll show later that the myth provides greater understanding of the orderly movements of the sun, moon, planets, and stars as they appear to cross the sky. It was those movements that, as de Santillana and von Dechend demonstrate, manifest the esoteric reality of the physical universe and the source of spiritual understanding buried in myth.

On a spiritual level the Osiris myth provides an order of succession in which the mortal body serves as a mechanism operating within the limits of earthly existence, and the mind as a mechanism through which the relationships between men, gods, and the cosmos can be understood. The human spirit and soul are seen as pervading the body and surrounding it, helping to support those relationships and ensuring an orderly and kindly individual existence with the help of Ma'at. The spirit within is *ka*, representing the source of the powers of life received from the gods, a spiritual double found in all humans. The manifestation of *ka* is *ba*, the immortal soul or spirit that survived after death of the body. *Ba* includes a person's individuality and personality free to roam beyond its abode, the body, and the confines of mortality. Thus the *ba* of the dead pharaoh could travel beyond Earth to partake in the necessary order of the universe and return to the body at will.

Ptah

In Heliopolis and Hermopolis, located in Middle Egypt, Re (Ptah) the sun god was perceived to have been self-created in the form of Atum, bringing order to the universe. In the city of Memphis 18 miles (30 kilometers) south of Heliopolis, Re was identified as Ptar, the earth god. Both Re and Ptar attained self-consciousness when organization in the form of a mound of dry land was created from the waters of Nun. Each was the beneficent source of physical and mental skills necessary for an orderly and progressive civilization. They inspired creativity in the arts and the knowledge and maturity needed for planning and constructing civil works. As we have seen, these were also the affairs of Thoth, the wisdom, mind, and tongue of Re, while his counterpart Seshat supported the geometrical and mathematical skills necessary for building the temples and other great structures that continue to capture our imagination today.

Sir Ernest Alfred Wallis Budge, in his book *The Gods of Egyptians*, describes the common portrait of Ptah as a balding, bearded man wearing close-fitting clothing around his neck (Figure 8.12).[32] His hands project frontward, holding a scepter and emblems of life and stability (the *ahnke* and the *Djed* pillar), while a *menat* representing happiness and pleasure hangs from his back. When he is shown standing, his feet rest upon a pedestal representing *ma'at*; if seated, his throne is set upon a similar pedestal. At his back is the obelisk or *Djed* pillar, again symbolizing stability, with the *Djed* making reference to the tree trunk in which the body of Osiris once hid. From this symbolism we gain a picture of a god who appears quite comfortable in his role of planning, designing, and building. In others words he is an architect, an engineer, a surveyor, and a builder—a true craftsman.

Figure 8.12: Ptah holding a scepter, and an *ahnke* and *djed* pillar, emblems of life and stability.

Once again, parallels can be found in the gods of other mythologies. De Santillana and von Dechend find that:

> it is possible to trace back the significance of the *blacksmith* in Asiatic shamanism, particularly the celestial blacksmith who is the legitimate heir to the "divine archi-tekton" of the cosmos. Several representatives of this type, whom we call Deus Faber [God the Maker], still have both functions, being architects and smiths at the same time . . . the Egyptian Pharaoh also celebrated his jubilee after thirty years, true to the "inventor" of this festival, Ptah, who *is* the Egyptian Saturn, and also Deus Faber.[33]

How does Saturn fit into this scheme of God the Maker? *Saturnian* means "happy" or "peaceful," and thus a Saturnian ruler would reign over a successful and prosperous kingdom. As noted by de Santillana and von Dechend, "The cube was Saturn's figure . . . this is the reason for the insistence on cubic stones and cubic arks. Everywhere, the power who warns 'Noah' and urges him to build his ark is Saturn, as Jehovah, as

Enki, as Tane, etc."[34] With deference to de Santillana and von Dechend, and as established in chapter 5, while the specifications for Utanapishtim's ark expresses cubic geometry, the vessel was not to be built as a cube, but as a sphere. The case for this conclusion is well founded throughout this book.

There appears to be no doubt that Saturn (Ptah) has provided some benefit to this world, including the world of the ancient Egyptians. From a mythological point of view it is Saturn who establishes time. De Santillana and von Dechend state that Ptah, as Saturn, "has been 'appointed' to be the one who established it because he is the outermost planet, nearest to the sphere of fixed stars . . . Saturn does give the measures: this is the essential point . . . it is essential to recognize that . . . Ptah of Egypt . . . is the 'Lord of Measures'—maat in *Egyptian*. . . . Saturn, giver of measures of the cosmos, remains the . . . 'Star of Nemesis' in Egypt, the Ruler of necessity and retribution, in brief, the Emperor."[35] The ancient Egyptian word *rnp* is equivalent to "year," and in this regard "the Lord of the Year" is an additional reference to Ptah.

Mechanics of Space

One of the most important ancient measures of time was the cyclical change observed in the obliquity of the ecliptic. Graham Hancock and Robert Bauval go into great detail on this topic, as it concerns the rise and fall of the constellation Orion (Osiris) during the past thirteen thousand years,[36] while John Major Jenkins describes similar effects as they pertain to ancient Mayan astronomical observations and their prophecy of the coming end of a World Age.[37] For de Santillana and von Dechend, "The fact is that the 'separation of the parents of the world,' accomplished by means of the Ouranos [Uranus] stands for the establishing of the obliquity of the ecliptic: the beginning of measurable time . . ."[38] which is, as they note, "[t]he very same 'event' . . . understood by Milton as the expulsion from paradise."[39]

With regard to the aforementioned precessional movement of Earth around its axis, mythically described as the turning of a celestial mill or fire stick, it was Ptah who oversaw that very rotary motion through use of the potter's wheel, the cosmological instrument for precession. De Santillana and von Dechend declare, "Decisive is the Ancient Egyptian instrument for drilling out stone vessels, which was perhaps even cranked, but there is no unanimity among the historians of technology as to the real nature of this device. . . . The cosmic machine (mill, drill, or churn) produces periods of time, it brings about the 'separation of heaven and earth,' etc. . . . churn, mill, and fire drill [are] . . . machines which were meant to describe the motions of nested spheres."[40]

It is evident that the ancient Egyptians rationalized their concepts of universal order, the mechanics of the cosmos, and the relationship between humans and the gods by placing gods at the helm of creation and organization necessary to fuel civilization. Human dependency on gods and movement of the heavenly spheres was understood and accepted. This relationship continued to fuel conceptualizations of the mechanics of the universe until very recent times.

Claudius Ptolemaeus was a Roman citizen of Egypt. We know him as Ptolemy. His treatises on astronomy, geography, and astrology influenced Islamic and Christian European studies of those sciences for 1,400 years. For astronomers, Ptolemy's *Almagest* (The Greatest) is most important. In it he presents and defends his geocentric view of the cosmos. For astrologers the book is regarded as the prologue to *Tetrabiblos* (Four Books). In his review of important historical markers concerning astrology, David Berlinski notes the detailed analysis of this second of Ptolemy's three important works (the other treatise being *Geography*, presenting the geographic knowledge of the Greco-Roman world).[41] *Tetabiblos* was Ptolemy's melding of horoscopic astrology with the natural philosophy of Aristotle. For its part, *Almagest* is not a light, leisurely read. As Berlinski finds, it "is organized as a geometrical tract, and while the mathematics is never difficult, it is always dry. The book's drama lies with its subject, which is the structure of the universe; its execution appeals to those with a taste for desert landscapes."[42] However, it was the apex of Western astronomical understanding until the sixteenth century, when Copernicus and Johannes Kepler dared to expound ideas considered heresies.

Kepler considered himself to be first and foremost a mathematician and a mathematical "rhapsodist." Berlinski writes, "He was indifferently disposed toward numbers, although necessity forced him to become a superb calculator; geometry had seized his soul and he very early came to the conclusion that 'geometry is one and eternal, a reflection out of the mind of God.'"[43] Of particular interest is Kepler's idea that models of each of the known planets could be nested between regular polygons to represent the order of the solar system. He laid out this theory in his *Mysterium Cosmographicum*:

> The Earth's orbit is the measure of all things; circumscribe around it a dodecahedron and the circle containing it will be Mars; circumscribe around Mars a tetrahedron, and the circle containing this will be Jupiter; circumscribe around Jupiter a cube, and the circle containing this will be Saturn. Now inscribe within the earth [Earth's orbit] an icosahedron, and the circle contained within it will be Venus; inscribe within Venus an octahedron, and the circle contained within it will be Mercury.[44]

The shapes Kepler identifies between the orbs are the five regular Platonic solids of Euclidean geometry. We will have more to say about these solids in later chapters, but for now it is important to realize that there are only five of them, no more, no less. However, the number of known planets numbered six. In keeping with the relationship he found between the Platonic solids and planetary orbits about the sun, Kepler proposed that if Earth's orbit was accepted as the unit measure—the fundamental unit of distance—then his construction of nested spheres and polygons defined the apparent ratio of the other planetary orbits. He assigned the following platonic solids to their respective heavenly spheres (Figure 8.13):

Saturn—Cube

Jupiter—Tetrahedron

Mars—Dodecahedron

Venus—Icosahedron

Mercury—Octahedron

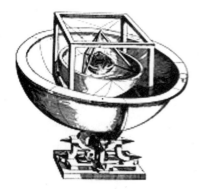

Figure 8.13a: Johannes Kepler's illustrated model of the Solar system based on the geometry of the Platonic solids (from *Mysterium Cosmographicum*, 1600).

Figure 8.13b: Detail of nested spheres and platonic solids contained within the model.

Kepler's follow-up to *Mysterium Cosmographicum* was *Harmonicum Mundi* (*Harmony of the World*). Berlinski calls it a "masterpiece. It is far-reaching in scope and rich in mathematical invention. . . . Kepler's aims in this book are inseparable from his means, and his means are geometrical. In the *Mysterium Cosmographicum*, he had heard a melody rustling throughout the cosmos; he had explained one part of the melody by an appeal to the five Platonic solids. In the *Harmonicum Mundi* that melody becomes contrapuntal."[45]

Kepler found that there are two universal forms of harmony—one musical and one spatial. He concluded that "[m]usical harmonies proceed by means of the division of a straight line into ratios, spatial harmonies by the division of a circle into sections."[46] The straight line and circle were fundamental constructs, and he reasoned that as "shapes given in nature" they must be "shapes resident in the human soul."[47] It is these very same geometries that are used in esoteric fashion within world mythologies. As in nature, so within the human soul. As above, so below. We must understand the mechanics of Earth's movement across space if we are to find the key that unlocks the mystery of the Egyptian pyramids constructed so long ago.

Ever since human beings first peered up toward the sky and saw the spectacular display of the sun, moon, planets, and stars circling overhead, we have wondered what our role is in this heavenly dance. But of course we know that, except for the moon, the cosmos do not revolve around Earth, nor around us. The geocentric view is an illusion. Imagine standing somewhere on Earth's equator. You face toward the east just as the sun begins to appear at the horizon. Perhaps you imagine that the pillars of Boaz and Jachin are at your sides. An orange glow begins to cross the landscape as the sun moves slowly upward. The air is sweet and still. But now consider a greater reality. You are in fact traveling at a speed of about 1,000 miles per hour (almost 1,700 km per hour) around Earth's axis. Earth orbits the sun at a mean speed of about 18.6 miles per second (29.8 km per second), travelling 589 million miles (942 million km) in one year. Meanwhile, the solar system rotates around the center of our galaxy at about 136 miles per second (220 km per second). Yet you feel no hyperhurricane-force winds, no angular momentum that would toss you out into space. We hardly even feel the pull of Earth's gravity.

Obviously there are tremendous forces at work here. It is the spin of the earth that produces the effect of the cosmos crossing the sky, night to become day and day to become night. The moon orbits Earth to produce the lunar waxing and waning we observe each day and night. Earth orbits the sun, giving us just over 365 days per year and four seasons defined by the two solstices and two equinoxes. And the planets orbit the sun, appearing to give chase to one another while running their own circuit along the ecliptic path. Ancient civilizations of Babylonia, Egypt, India, Europe, Central America, and other locations across the earth perceived something special occurring in the sky, an order of things hidden behind what first appears to be a chaos of stars across the night sky, other stars (planets, comets, and meteors) moving against this backdrop, some appearing in the evening sky and later in the morning, returning again in the evening, while the moon waxes and wanes, waxes and wanes. But there are additional complications to this music of the spheres, part of what Kepler called *musica universalis*. And the ancients strove to understand it all.

Reality and Illusion

Like our ancient forefathers, we seem to stand in the stillness of night as the cosmos dances before us. Astronomers of old accepted this perception of Earth's passiveness within a churning ocean as fact, but in truth it is fiction, illusion. As de Santillana and von Dechend demonstrate, myth presents cosmological facts and eternal truths buried in a fictional storyline. Today we live in a culture and a time when myth appears less important. Why did the ancient Egyptian civilization, among others, all but disappear, leaving behind only their monumental megalithic stone works, a few papyri scrolls, and earthen cities in ruin? There a number of fundamental facts regarding Earth's movements that are vital for understanding many of the concepts—physical, intellectual, emotional, and spiritual—buried in myth. All of these concepts are important, so let's review the geometries and motions involved.

1. Earth spins on its polar axis in a counterclockwise direction, from west to east.

2. Earth completes one revolution about its polar axis every twenty-four hours; this defines one day.

3. Earth completes one orbit of the sun every 365.242199 days.

4. Earth orbits the sun in a counterclockwise direction as viewed from "above" the plane of the solar system (looking toward Earth from a point in space along the earth's north polar axis).

5. In space, the concepts of "up" and "down" have no meaning.

6. Earth's orbit changes over time, ranging from near circular to mildly elliptical; the major component of this orbital variation (eccentricity) has a cyclical period of 413,000 years, but the period of other components varies between 95,000 and 125,000 years, combining into a 100,000-year cycle.

7. The period of duration of Earth's eccentricity in orbit imposes long-term yet cyclical variations in climate.

8. The ecliptic is the geometric plane defined by the mean orbit of the earth around the sun; as a plane defined by a mean value that does not change, the ecliptic does not change. Currently the ecliptic plane is inclined by about 1.5 degrees with respect to this orbitally defined plane. The name *ecliptic* is derived from eclipses that occur when the full moon or new moon is located in proximity to the apparent path of the sun.

9. Earth's polar axis is tilted with respect to Earth's orbit around the sun. The angle of the axial tilt (obliquity) varies by about 2.4 degrees over time, with one full cycle of variation (shifting between a tilt of 22.1 and 24.5 degrees and back again) once every forty-one thousand years.

10. The rotational axis of the earth is not perpendicular to its orbital plane, so the equatorial plane is not parallel to the

ecliptic plane; the angle between them currently is about 23° 26.4' (23.44 degrees). This is the axial tilt, or obliquity of the ecliptic; this value is about halfway between the extreme values. The tilt is in the decreasing phase and will attain 22.1 degrees in about 10,000 CE.

11. The orientation of Earth's polar axis relative the positions of fixed stars is in constant change; this change in direction of the polar axis is called axial precession (precession of the equinoxes) and has a current period of about 25,800 years; this precession results from tidal forces exerted by the sun and the moon on the earth, which is not a perfect sphere.

12. Earth's orbital ellipse also precesses as a result of gravitational effects with Jupiter and Saturn; this orbital precession results in a variability of the precession of the equinoxes resulting in a range of about 25,800 to 21,600 years.

13. Earth's equator is defined by the plane perpendicular to the polar axis, crossing through the planet's center point.

14. The four cardinal directions with respect to Earth's axial spin are:

   • north (toward the north pole, opposite of south)

   • south (toward the south pole, opposite of north)

   • east (the direction of earth's spin and perpendicular to the north-south line, opposite of west)

   • west (opposite direction of Earth's spin and perpendicular to Earth's spin, opposite of east)

15. The other planets in our solar system orbit the sun at different rates and along different elliptical paths.

16. The moon completes one elliptical orbit around the earth in approximately 27.3 days, rotating around the earth in a counterclockwise direction as viewed from above the earth's north pole.

17. Our home galaxy, the Milky Way, has a diameter of about 100,000 light years and is about 1000 light years thick; it travels through space at about 345 miles per second (552 km per second) relative to the universal cosmic microwave background.

18. Earth is about 26,400 light years from the galactic center of the Milky Way; Earth currently is about 88 light years above (or below, you decide) from its "equator," more properly referred to as its equatorial symmetry plane.

19. The solar system and the Milky Way in general rotate around the galactic center in a clockwise direction (viewed from a point in space beyond the limits of the galaxy and along the earth's north polar axis).

20. The solar system completes one rotation around the galactic center once every 225 to 250 million years, passing through the galactic equatorial symmetry plane every 33 million years, or about seven times per revolution.

21. The solar system's ecliptic plane is oriented at an angle of about 60 degrees to the galactic equatorial symmetry plane.

From these facts our ancestors observed a number of apparent, if illusory, effects here on Earth. At least a handful of ancient cultures, including the Egyptians and Maya, tracked and recorded their astronomical observations and interpreted some of these effects as signs of great impending change on Earth. The cause for their concern remains in the realm of speculation, but ours is not to judge. These people observed and recorded the information and interpreted the data. While some of the observations were illusions based on effects related to space, time, matter, and energy that were at the time indeterminable, we can conclude that their interpretations were attempts at rationalizing local experience with the observed order of the cosmos. In other words, events occurring on Earth were apparently tied in some way with cosmic *chronos*, time personified in gods such as Kronos, Saturn, Quetzalcoatl, and Osiris. Listed here are ancient and modern perceptions that are pertinent, if also illusory:

1. Celestial bodies generally appear to move across the sky from east to west (*Illusion*).

2. The sun and stars appear to orbit Earth exactly once every twenty-four hours; over the course of about thirty days the sun appears to slowly float past one of twelve zodiacal constellations generally located along the ecliptic path (*Illusion*).

3. The sun appears to complete about 365.24 rotations around the earth in one year (*Illusion*).

4. Heaven and the cosmos are above us, "up there," while hell and the underworld are below us, "down there" (*Illusion*).

5. Confined here on Earth by gravitation, *up* is away from the center of the planet, and *down* is toward the center of the planet.

6. Earth's orbit around the sun is not readily apparent without detailed observation and recording; the geocentric model of the universe was in good standing in Western culture until the sixteenth century.

7. Cyclical variations in climate, imposed by the hundred-thousand-year period of Earth's eccentricity, are generally unnoticeable without long-term observation and recordkeeping, or modern technology.

8. The ecliptic appears as the intersection of the geometric plane defined by the apparent path of the sun across the map of fixed stars of the celestial sphere; it is a great circle; the sun's course among the stars appears to move forever westward (*Illusion*).

9. Long-term yet cyclical variation (with a maximum of about 2.4 degrees) in the tilt of the earth's polar axis might impact climate but is generally unnoticeable without long-term observation and recordkeeping, or modern technology.

10. Movement of the rotational (polar) axis of the earth can be, and has been, observed and recorded over long periods of time, documenting that the north polar axis points to one of several stars over a cycle of about twenty-six thousand years; no star is ever in line with the south polar axis; world mythologies document these events.

11. The concept of world ages is, in part, based on long-term observation of the change in direction of the polar axis.

12. Ancient observation of orbital precession appears to be accounted for within the more general observation and recording of the precession of the equinoxes.

13. Earth's equator can be approximated by locations on the earth's surface where the sun is directly overhead at noon during the vernal and autumnal equinoxes.

14. The celestial equator appears as a straight line viewed from any point on Earth.

15. The points of equinoxes on the celestial sphere appear to rotate 1 degree along the ecliptic every 71.57 years, completing one rotation around the ecliptic in about 25,765 years.

16. The four cardinal directions with respect to the sun are the four directions extending from the viewer at any location on the earth and the point of sunrise and sunset at the time of the winter and summer solstices; these directions generally are oriented toward the northwest, northeast, southwest, and southeast.

17. The orbits of the planets were plotted with only minor error by ancient astronomers; however, those plots typically considered a geocentric view of the solar system.

18. Moon phases are not readily understood without realization of a heliocentric solar system.

19. The solar system's slow rotation around the center of the Milky Way is imperceptible within the limited time span of humans on Earth; the limits of the Milky Way and its apparent speed through space remained unknown for millennia and could not be detected without modern technology.

20. While solar and galactic distances were indeterminable until the advent of modern technology, the angles and relative movements between Earth and heavenly bodies have been tracked for millennia.

21. Earth's true movement relative to the galactic center is unnoticeable without very long-term (over hundreds of thousands, if not millions of years) observation and recordkeeping, or modern technology.

22. Viewed from Earth, the conjunction of the sun with the galactic center was believed to signify the end of one world age and the beginning of another, an occurrence generally associated with

the precession of the equinoxes having a periodicity of about 25,765 years. Most ancient cultures recognizing precession calculated four world ages per precessional cycle, or about one per 6,480 years and one precessional cycle taking 25,920 years to complete. Many ancient cultures identified the year 2012 CE as the end of a world age and beginning of another and related them to zodiacal constellations

23. Currently the sun appears to move along the ecliptic plane and is in the process of crossing the galactic equatorial symmetry plane at the galactic center, the two planes oriented about 60 degrees from each other. The most commonly estimated date for this occurrence as may concern Earth is the winter solstice of December 21, 2012.

This last perception, that the sun moves along a line at a 60-degree angle to the galactic equator, is an important observation because the point at which the ecliptic crosses the galactic equator moves slowly back and forth, first along that equatorial line until it reaches and then passes by the galactic center. In time, the intersection point appears to reverse course, only to cut across the galactic center again and eventually start the cycle anew. While this was happening, ancient cultures observed the morning sun rising before a continually changing background of stars perceived in groups as constellations that represented earthly forms—animals, humans, mythic creatures, and elements of nature. In time, the sun was observed to rise in one constellation, eventually passing through it and beginning the process again and again until the cycle was understood to continue at a constant rate of about 1 degree of rotation along the ecliptic per seventy-two-year interval, a full cycle calculated to be about 25,920 years (72 years per degree × 360 degrees). With twelve constellations (signs) of the zodiac identified along the ecliptic, the sun on average passes through each constellation in about 2,160 years (25,920 years ÷ 12 signs).

Obviously the movement of the earth is complex, the result of an amalgamation of gravitational and inertial effects from the moon, the planets, the sun, and forces beyond. Precession is one of the effects of this cosmic tug-of-war. Earth's polar axis changes its orientation in space, pointing in the general direction of a star we call the north polestar, currently Polaris but predicted to be Gamma Cephei by 3000 CE. The closest stars on the celestial sphere to the south polar axis are in the constellation of the Southern Cross, but there the south portion of the axis never points directly to a single star. Imagine holding the middle of a straight and vertical wood dowel about 2 feet (60 cm) long (Figure 8.14). Rotate the dowel so that it is out of plumb by about 22.5 degrees (half of a 45-degree angle). Then, holding the pole in place, gently and slightly rotate your hand so that the ends of the pole each move in a small circle, perhaps 1 inch in diameter. You are envisioning a model of axial precession, but the earth needs almost twenty-six thousand years to complete one cycle.

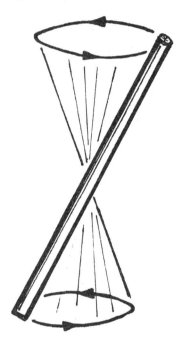

Figure 8.14: Modeling earth's axial precession with a wood dowel.

All of this applied geometry is important because the ancient Egyptians and many other cultures noted the effects of precession, developed an explanation for those effects (some of them illusory, and some very real), attributed those effects to the acts of gods, experienced life and interacted with the natural events thereof, and built a mythology that would allow their knowledge and experience to be communicated to future generations, including ourselves. This is the very essence of tradition, the transfer of knowledge that has no beginning that we can yet determine. The fact that this information is wrapped in mythologies from around the world suggests that it is vital to this world, so necessary to our well-being that it was encoded in Egyptian pyramids, Mayan pyramids, and numerous other architectural structures located at sacred sites around the world.

From ancient Egyptian mythology we find that the pillar of Osiris, his coffin contained within the trunk of a tree at the king's palace in Byblos, is metaphor for the earth's polar axis. The four legs of Nut (yes, four, because she was transformed into a cow that Re rode into the sky after destroying much of human civilization, the cow trembling from the great height, Re's son Shu providing underlying support for Nut, who then spread her legs to become "the four props of heaven at the four cardinal points"[48]) represent locations of the vernal and autumnal equinoxes and the winter and summer solstices at the horizon. Osiris freed from his coffin and with an heir is helmsman for the Ship of the Dead in the netherworld, where the south polar axis intersects the celestial sphere.

Certain numbers found in many world mythologies, such as 4, 5, 7, 8, 9, 12, 30, 72, 104, 360, and others, are associated with astronomical facts and illusions listed above. Graham Hancock concludes:

> But can chance account for the fact that these and other prime integers of precession keep cropping up in supposedly unrelated mythologies from all over the world, and in such stolid but enduring vehicles as calendar systems and works of architecture? Santillana and von Dechend, Jane Sellers and a growing body of other scholars rule out chance, arguing that the *persistence of detail* is indicative of a guiding hand.[49]

Chapter 9
The Great Pyramid

This King is Osiris, this pyramid of the King is Osiris, this construction of his is Osiris . . .

Pyramid Texts, Utterance 600

"What's your guess?" I asked Bauval. "What do you think the purpose of the pyramid builders really might have been?"

"They didn't do it because they wanted an eternal tomb," he replied firmly. "In my view . . . They succeeded in creating a force that is functional in itself, provided you understand it, and that force is the questions *it challenges you to ask."*

Graham Hancock and Robert Bauval
The Message of the Sphinx

We turn now to the pyramids of Egypt and in particular the Great Pyramid at Giza. Graham Hancock was spot on when he questioned if it is coincidence that precessional data is found in world mythologies, calendrical systems, and architecture developed throughout the ages. This persistence of detail might indeed indicate that those cultural traditions—prime examples being the design and construction of megalithic monuments such as the Great Pyramid—are effects of a guiding hand. In this chapter we explore the interior and exterior geometries of the Great Pyramid, and discover the universal symbolism that is both hidden and in plain view for all to see. And just like the mythologies from Sumer, ancient India, and Celtic Europe, the key to understanding this symbolism is recognizing the spherical geometry placed before us.

Pyramidology

The Great Pyramid is held in awe. It continues to be scrutinized from its bedrock foundation to its naked stone apex. Loss of its limestone veneer many centuries ago has not lessened the pyramid's impact on human imagination. Architects are captivated by the structure's simple, timeless form. Engineers remain bewildered by the complex planning and construction methods it required. Mathematicians marvel at the surprising numerical relationships integrated into its design. Geographers find meaning in the pyramid's location and orientation on the landscape. If there is but one manmade geodetic marker of any consequence, this is it: the Great Pyramid of Khufu (See Figure 9.1).

The geographical knowledge and engineering capabilities of the ancient Egyptians are demonstrated by the following facts, each important to hidden geometrical symbolism at Giza.

- Earth's longitudinal meridian passing through the pyramid equally divides the Nile delta region into halves.

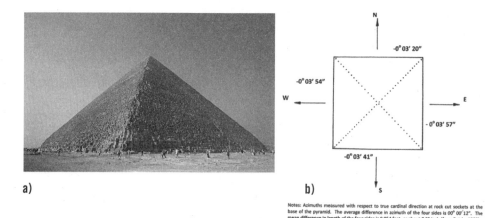

Notes: Azimuths measured with respect to true cardinal direction at rock cut sockets at the base of the pyramid. The average difference in azimuth of the four sides is 00° 00′ 12″. The mean difference in length of the four sides is 0.054 feet, or about 0.65 inch (from Petrie, 1883).

Figure 9.1: a) The Great Pyramid at Giza (photo by Nina Aldin Thune). b) Orientation of four sides of the pyramid (from Petri, 1983).

- The entire delta of the Nile is delineated within a triangular area by extending the northeast and northwest diagonals from the pyramid's apex to the Mediterranean Sea.

- The apex of the river's delta system is located at 30° 06'N 31° 14'E within the Nile's bedrock-lined floodplain; the pyramid is located at latitude 30°N (corrected for atmospheric refraction) and longitude 31° 09'E. Hancock refers to the 00° 05' discrepancy between the apex of the delta system and the center of the pyramid as "an error of just a few minutes of terrestrial arc to the south and west,"[1] but it is possible that: a) the pyramid's built location was the best location for its intended purpose and was acceptable then to the builders, no error envisioned; or b) the location of the apex of the delta system has changed over the course of the last 4,500 years, particularly in light of changes to the flow of the river resulting not only from local human-engineered development in and around Cairo, but also all along the river's natural channel. If this is indeed the case, then we cannot attribute nature's hydrogeological adjustments to human error. In either case the location is an intentional artifact of the pyramid's construction.

- Designers of the Giza necropolis knew that Earth was spherical and calculated its dimensions with surprising accuracy. Hancock finds evidence for this by a symbolic seven terrestrial degrees for the length of the country, the geodetic location and orientation of the Great Pyramid to the cardinal directions, the pyramid's base to height ratio of 2π, and the 1:43,200 scale of the monument's size compared to Earth's northern hemisphere.[2] Recalling that the tetrahedron is constructed using four triangles to form the most fundamental polyhedral space, it isn't suprising that the ancient Egyptians constructed the Great Pyramid having four inclined faces (with a fifth face simply being the square based upon which the monument was constructed), each face oriented toward the polar axis (north or south) or equinox sunrise or sunset. As metaphor, the apex of the pyramid represents the north pole while the base of the structure is symbolic of Earth's equator. Peter Thompkins clarifies the symbolism expressed by the pyramid's form:

- Each flat face of the pyramid was designed to represent one curved quarter of the northern hemisphere, or spherical quadrant of 90 degrees. To project a spherical quadrant on to a flat triangle correctly, the arc, or base, of the quadrant must be the same length as the base of the triangle, and both must have the same height. This happens to be the case only with a cross-section or meridian bisection of the Great Pyramid, whose slope angle gives the pi relation between height and base . . .[3]

It is well known that ancient Egyptians believed in the resurrection of the soul. Egyptologists have evidence for this in the contents of pyramids and other funerary and burial structures: sarcophagi, mummies, and the various accoutrements necessary for pleasant and prolonged afterlives for the pharaohs, priests, and their families. Additional evidence was discovered in the various hieroglyphic texts and engravings on walls inside temples and pyramids. Perhaps the most important texts were scripts covering the walls of chambers inside pyramids built after Khufu's reign. Known as the Pyramid Texts, they are the oldest body of religious texts found in Egypt. The texts describe the intertwining of the cosmic order, the roles of the gods, the relationship of the pharaoh with his subjects and the gods, and the machinations that would ensure his resurrection as an Osiris. One of the works that evolved out of the Pyramid Texts of the Old Kingdom and the Coffin Texts of the Middle Kingdom was *The Book of the Dead*, a funerary text that the ancient Egyptians referred to as "Spells of Coming (or Going) Forth By Day." The text was written on a scroll of papyrus to be included with other items within the burial chamber or inside the sarcophagus itself. The various spells, hymns, and instructions help the deceased work his way through a series of obstacles on the path to the afterlife.[4]

Ancient Egyptian society was more than willing to expend cultural resources on artisans, planners, architects, engineers, and labor—thousands of people working many years—to construct the funerary and burial facilities and conduct ceremonies necessary for a pharaoh to achieve and enjoy life in the otherworld. But we must ask questions the pyramids force upon us. Perhaps the most pressing question is: *are the pyramids nothing more than large, attractive, and expensive burial chambers for an ancient ruling elite?*

Almost two decades have passed since Robert Bauval recognized the importance of "observational astronomy and its material expression in the symbolic architecture of the pyramid structures and their orientations."[5] Results of Bauval and Gilbert's investigation, presented in their book *The Orion Mystery*, demonstrates that the pyramids at Giza are part of a grand construction related to the configuration of stars in the constellation of Orion—"that the celestial form of Osiris . . . corresponded with our modern constellation Orion, the pyramids were indeed Orion too."[6] Yet the mystery of the pyramids continues to attract the attention of Egyptologists and other scholars, as well as amateur historians, archeologists, anthropologists, engineers, and authors whose concern is the unlocking of the design and purpose of these greatest of all monuments. They seek hidden chambers, hoping to discover treasures or answers to the mystery, or confirmation of which pharaoh actually oversaw the pyramid's construction and an explanation for its purpose. Surely form follows function, and the intended purpose of the pyramid must be related to its architecture.

And yet the structure is devoid of any ancient ornamentation and inscription, inside and out. The nearby pyramids of Khafra and Menkaura are similarly unadorned. Pharaohs did not claim their ownership of pyramids at Giza via kingly ornamentation and hieroglyphic statements of sacred intent. Why didn't they include a statement of ownership with each structure, and ensure that a testament of pharaonic ownership and intent for each pyramid would last as long as each monument stood upon the Giza plateau?

Is it possible that this lack of opulence and personal breast-beating by Khufu, Khafra, and Menkaura were intentional? Perhaps the imposition of the three pyramids at Giza is as purposeful as the mysteries they contain. Certainly there is reason for the lack of adornment, and there is a purpose to the megalithic geometry. There is geometry to be found in the overall pyramidal form as well as in the size, shape, and configuration of shafts and chambers located in each structure. Why are those features of the Great Pyramid oriented as they are, and on such grand scale? The Queen's Chamber is lined with limestone. The floor, walls, and ceiling of the King's Chamber are faced with smooth black granite. Only one item is located in the King's Chamber, an unadorned granite sarcophagus abused by looters, explorers, and throngs of tourists.

The series of granite slabs forming relieving chambers, distributing structural stresses around the King's Chamber, have held the most mystery for me. Why are they configured to form the hidden image of a four-tiered top hat above the ceiling slabs, similar to the upper portion of a *Djed* pillar? Were the relieving chambers actually intended to prevent potential collapse of the ceiling overlying the King's Chamber under the weight of overlying stone? These curiosities have remained unresolved while investigators accumulate further evidence for an astral connection to Khufu's pyramid.

Bauval and Gilbert suggest that air shafts connecting the King's and Queen's chambers were associated with fertility ritual, the shafts allowing the pharaoh's soul to extend up toward the phallic region of Osiris-Orion (the Belt stars) and a symbolic seeding of a Horus-king.[7] They note sacred relationships between the ancient Egyptian people, the stars, and the sun, preserving the body, and providing some means for the soul to release from the confines of mortality and journey to an afterlife in the Duat, located within the vicinity of the Orion constellation. Osiris as Orion provided the bridge between our earthly plane and the cosmos.

Hidden but Not Forgotten

We can see in these findings vital concern for preparing and preserving the body of the deceased to ensure translocation of the soul to the Duat and transformation through rebirth to an afterlife in the stars. These funerary concerns, including necessary ritual and ceremony, generally have been perceived as benefitting the pharaohs, their families, and other important officials such as priests. However, there is evidence suggesting the possibility of an afterlife in a broader context. In fact, Bauval and Gilbert allude to two important texts that may suggest the potential for rebirth by all people, although this idea appears to have been missed during their investigation. It is important to view portions of these texts as they pertain to the purpose and symbolism hidden in the geometrical form of the Great Pyramid. The function becomes apparent once we understand the intent of the form, and as you might surmise by now, all is not what first appears to be. As we proceed, keep in mind the symbolism of the pyramidal form representing a hemisphere, and the various levels of meaning (PIES) to be found in myth and architecture.

The first of the two texts is Plato's *Timaeus*, in which dialogues between representatives of Greece and Egypt address ancient knowledge including the lost civilization of Atlantis. In the *Timaeus*, Plato says that the demiurge made "souls in equal number with the stars and distributed them, each soul to its several star . . . and he who should live well for his due span of time should journey back to the habitation of his consort star."[8] The important point here is belief that the number of souls is equal to the number of stars, such that a person who lives well (not in a material sense but in relationship with other people and the world) may return to his or her star. This implies a place for *all* souls in the stars, not only the kingly or priestly among us.

The second text (Leiden Museum Papyrus No. 344) includes passages found in a number of papyrus texts dated to the Middle Kingdom reign of Amenemhet I (c. 1990 BCE). They describe upheavals in the political and social conditions of the state that at that time lead to pessimism on the part of the Pharaoh and other officials. During a discussion concerning how best to handle this dire situation, a scribe by the name of Ipuwer mentions to the pharaoh and the officials a hope for a great messiah (possibly the son of the pharaoh) who might bring order back to the state. Ipuwer recalls when qualities of ma'at (order, balance, law, justice, morality, truth, and mercy) existed under the leadership of an "ideal king."

> Remember . . . it is said he is the shepherd of all men. There is no evil in his heart. . . . Where is he today? Does he sleep perchance? Behold his might is not seen.[9]

Then Ipuwer refers to something hidden inside the pyramid, but might no longer be found there: ". . . that which the pyramid concealed has become empty . . ."[10] The text suggests that there is indeed something vital to the culture of ancient Egypt, something sacred concealed in the Great Pyramid. Ipuwer provides clues. It is associated with the "ideal king," the "shepherd of all men." Are these references to Khufu, pharaoh of the Great Pyramid, or perhaps someone of far greater import? Bauval and Gilbert recognized that the pyramids at Giza were regarded as symbols of Osiris.[11] Is it possible that the Great Pyramid contains Osiris? We will need several keys to open the concealed door that is the Great Pyramid itself.

The First Key

In the early years of this millennium, Vincent Brown developed a theory after reviewing some very enlightening papers written by several Egyptologists and other scholars addressing the content of the Pyramid Texts and the *Book of the Dead*. He was curious about the interior design of numerous pyramids, not the least of was the Great Pyramid.[12] It seems that Egyptologists Mark Lehner and James P. Allen noted particular features of the pyramid reflecting information found in myths and legends, all essentially religious aspects.[13] The Great Pyramid contains no ancient hieroglyphic writing at all. In fact, exterior form of the pyramid at first appears to provide no indication of its intended purpose. Many theories concerning the purpose of the pyramid are based on records, artifacts, and data associated with megalithic structures that have absolutely no direct connection with the pyramid. It is therefore fortunate that completely disconnected texts happen to describe specific features of the Great Pyramid, or the guessing game would continue unabated. As Sir Gaston Maspero said, "The pyramid and the Book of the Dead reproduce the same original, the one in words, and the other in stone."[14]

The first key to unlocking the purpose of the Great Pyramid and its symbolic geometry is to understand ancient Egyptian artisans' use of scale and proportion when designing and executing hieroglyphic drawings and engravings. Architects and engineers used these particular canons of proportion at the megalithic scale of tombs and temples. One such temple is located in Luxor, a city on the banks of the Nile about 320 miles (500 km) south of Cairo (Figure 9.2).

Figure 9.2: Temple of Luxor from the east (by Vivant Denon, 1800).

Luxor's ancient name was Thebes, founded in 1400 BCE. The Temple of Luxor was built during the New Kingdom and actually consists of a complex of buildings referred to as the southern sanctuary. It was dedicated to the gods Amun (Sun god, the Complete One), Mut (Nun, cosmic disorder), and Chons (Khensu, Moon god, keeper of time). The temple was the site of the Opet Festival, a celebration of fertility.

Completion of the temple required several hundred years. The earliest remaining parts of the temple are chapels built by the female pharaoh Hatshepsut, fifth pharaoh of the eighteenth dynasty in the fifteenth century BCE. The chapels housed barques possibly associated with the festival. The main portion of the temple—the megalithic colonnade and the sun court—was built by Amenhotep III. Additions including an entrance pylon and obelisks were constructed by Rameses II to connect Hatshepsut's buildings with the primary temple. Additional chapels were built by Tuthmosis III and Alexander the Great, who modified several of the barque chapels, but the main structure was completed by about the mid-thirteenth century BC. Figure 9.3 shows the temple as it stands today.

Figure 9.3: Temple of Luxor.

Interestingly, the great court of Rameses II located at the north end of the temple has a longitudinal axis that does not align with the orientation of the remainder of the building, giving a bent, almost haphazard look to the structure's footprint. But there was a method behind such apparent design madness, even more surprising considering that the temple took about two hundred years to complete under the direction of at least three pharaohs. René Adolphe Schwaller de Lubicz (1887–1961) is well known for his lengthy study of the art and architecture of the Temple of Luxor. With his keen interest in architecture and hieroglyphics of

Egyptian temples, fueled by his interest in philosophy and metaphysics, he relocated with his family from France to Egypt in 1938. For the next fifteen years he studied the pyramids, the temples, and the Sphinx. He concluded that the Sphinx was thousands of years older than other massive Egyptian monuments and buildings, noting that the extremely weathered Sphinx and surrounding bedrock walls were results of water erosion.

Schwaller de Lubicz was most interested in the Temple of Luxor. His books *The Temple of Man* and *The Temple in Man* detail his findings about the Temple of Luxor, exploring the relationship between ancient traditions of Egyptian mathematics, science, philosophy, and spirituality—much of the discourse derived from the study of mythology, arts, and architecture of that structure. His thesis is that proper understanding of Egyptian traditions requires an appreciation of numerology and sacred geometry. The book's ideas remain contested in some Egyptological circles, while other scholars find merit in Schwaller de Lubicz's conclusions.

Schwaller de Lubicz found that ancient Egyptians developed a detailed understanding of "the laws of creation" and that they presented this understanding in hieroglyphic texts and sacred architecture we can observe today. He referred to the Egyptian conceptual modus operandi as "intelligence from the heart," and "functional consciousness."[15] This was the tradition from the "first time" of creation, continuing through the process of resurrection, and applied with the help of Ma'at until the end of time. At times metaphysical and at times starkly realistic, the ancient Egyptian understanding of natural universal order and cosmic harmony are self-evident, while their traditions symbolize the human quest to relate with, and find comfort in, humanity's perceived reality.

Lucie Lamy, Schwaller de Lubicz's stepdaughter, assisted him with the study of the Temple of Luxor. Another avid proponent of sacred geometry, Lamy's investigations reinforced her stepfather's surprising arrangement of various compartments of the temple and the building's overall form. De Lubicz and Lamy divided the temple into nineteen sections along its length and found that the building's various rooms and colonnades corresponded to specific locations of the human body, from head to toe. Overlaying the temple footprint with the image of a human skeleton emphasized this discovery (Figure 9.4).

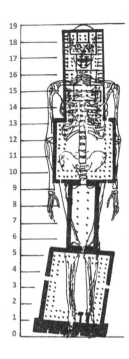

Figure 9.4: Plan view geometry of the Temple of Luxor with overlay of a human skeleton; hair line of body at one end of the temple, bottom of feet at the other end (based on Schwaller de Lubicz, 1998).

Gay Robins's research interests include ancient Egyptian composition, style, and proportion, and she has authored several books about Egyptian art. In her book *Proportion and Style in Ancient Egyptian Art* she notes, "It has long been known that much Egyptian art executed in two dimensions as painting or relief was conceived and carried out on a square grid, which helped determine the proportions of the human figure. . . . In this book I . . . show that the squared grid had an important influence on the position of scenes as a whole. . . ."[16] Robins shows that square grids made by ancient Egyptian artisans were fundamental for preliminary drawings of scenes and persons.

Contrary to the typical Western method of employing an illusion of depth in two-dimensional art, Robins finds that Egyptian artisans "accepted the drawing surface as flat. . . . Objects were shown in what was regarded as their most characteristic form, independent of time and space."[17] A picture of a box would show a rectangle. If something was inside the box, it would be shown above the box. An illustration of a pool surrounded by trees would show a plan view of the pool with surrounding trees lying flat, extending away from the pool. Similarly, the plan view of a building would show doors lying flat on the ground surface next to the threshold, and a shady courtyard might show trees again laid flat on the ground (Figure 9.5).

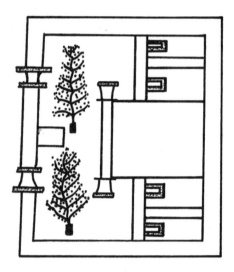

Figure 9.5: Typical ancient Egyptian plan view of a building and courtyard. Note doorways and trees drawn as though flat on the ground surface (based on Robins, 1994).

The fascinating thing Robins demonstrates about Egyptian drawings of human figures is use of a standard grid comprised of nineteen squares, one atop the other, from the soles of the feet to the top of the head. This was the finding of the German scholar Karl Richard Lepsius, author of the 1884 book *Die Längenmasse der Alten.* In an appendix to that work Lepsius describes his observations regarding proportions of human figures shown in artwork of the Old Kingdom. Robins summarizes the findings.

> If, as has become the custom, we take the baseline as 0 and count upward, horizontal 5 runs beneath the kneecap, 6 above the kneecap, 7 beneath the tips of the fingers of the hand hanging by the body, 8 under the thumb, 9 beneath the buttocks, 11 through the navel, 12 through the elbow, 14 through the nipple, 16 through the junction of the neck and shoulders, 17 beneath the nose, and 18 through the hairline. Thus, the horizontal lines of the early system found in the Old Kingdom that ran through the knee, beneath the buttocks, through the elbow, through the junction of the neck and shoulders, and through the hairline in fact correspond with the grid horizontals 6, 9, 12, 16 and 18. . . . For Lepsius, his observations left no doubt that human

figures were drawn according to an established set of proportions obtained originally by a set of horizontal guidelines and then secondarily by the use of the square grid.[18]

Although several other scholars challenged Lepsius's conclusion regarding "an established set" of human proportions applied by Egyptian artisans during the Old Kingdom, Robins finds that up to eight lines passing through observable figures from that period, with the hairline present and the top of the head often not marked, vindicate Lepsius's work. This grid was uniformly applied with only minor variation over the millennia, and two particular ratios were important in the layout of the human body when a person was shown in a fully erect standing or walking position. The baseline of measurement began at the bottom of the feet, equivalent to the bottom of the first square. The navel was located at the top of the eleventh square, and the hairline at the forehead was located at the top of the eighteenth square. Why the import of the hairline? Because there were times when the person drawn was wearing a hat or crown and the top of the head could not be seen. In such cases the height of the person was implied in the drawing, but the top of the eighteenth square was always the height above ground surface for the hairline.

Lamy outlines the same relative proportions in the image of the human skeleton superimposed on the plan view of the Temple of Luxor as described by Robins in the numerous drawings and engravings of people that abound on the walls of ancient buildings. Refer to Figure 9.5. With the north temple wall as the baseline, the bottom of the feet are at the bottom of square one, the navel is at the top of square 11, and the hairline is at the top of square 18. The sun court located in the midsection of the temple, shoulders aligning with the north end of Hatshepsut's chapels, navel at the center of the megalithic colonnade, genitals align with the north side of the sun court, and knees match up with the south end of the entrance pylon constructed by Rameses II.

Schwaller de Lubicz and Lamy conclude this *cannot* be coincidence, given the canon of proportions applied by ancient artisans. Simply put, over the course of about two centuries the architects and planners of the Temple of Luxor used the nineteen-square grid to create a "Temple of Man"—a "Temple *in* Man"—using the same proportions used by artists for depicting human forms on walls. Robins's and Lamy's work strongly support Schwaller de Lubicz's ideas of "intelligence from the heart" and "functional consciousness," at least in terms of the obvious application of ancient intelligence to express the connection between human form and architectural form, affirmation of order of the cosmos found within the body of mankind.[19]

But there are additional secrets to be discovered here. Figure 9.6 shows an overlay of a human skeleton on the Temple of Luxor, and a depiction of Hesire, a Third Dynasty high-level official during the reign of Netjerikhet (Dosjer), based on a wood engraving and also overlaying an illustration of the Temple of Luxor. The image of Hesire based on the engraving shows Hesire in stride to our left, carrying a rod and several other items in his hands. By plotting eighteen squares of equal size from the bottom of his feet to his hairline, we find that the proportions of the figure coincide with those described in Robins's book, and, indeed, with Lamy's illustration for the skeleton at the Temple of Luxor. The engraving of Hesire matches those same proportions even though he is stepping off to his right with his official equipment. The top of his head is in line with the top of the nineteenth square, his hairline the eighteenth, navel the eleventh, knees the fifth, and so forth. Most surprisingly, the angle of his right shin as he makes his stride corresponds exactly with the angle of the Great Court of Rameses II.

Recall that construction of the Temple of Luxor began under the direction of Hatshepsut, fifth pharaoh of the Eighteenth Dynasty and completed by Rameses the Great during the Nineteenth Dynasty. This means that the engraving Hesire prepared during the Third Dynasty is about 1,200 years older than the earliest construction of the temple. The importance of the Egyptian canon of human proportionality is self-evident from this fact. We now have the first key to unlocking the purpose behind the Great Pyramid:

Ancient Egyptian artists, architects and engineers applied a canon of scale and proportion modeled on the human body in the design and construction not only of hieroglyphic drawings and engravings, but also megalithic scale of tombs and temples.

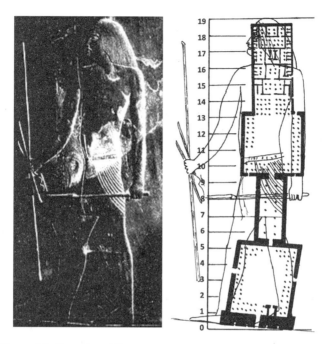

Figure 9.6: Engraving of Hesire (courtesy Vincent Brown, 2002) next to plan view of the Temple of Luxor overlain by an illustration of Hesire. Note similar proportions and angles, including the angle of stride and the Great Court of Ramses II.

The Second Key

Let's return to the pyramid of Khufu and see if this theory holds true for this greatest of monuments constructed almost 4,600 years ago. The figure of Hesire was constructed during the Third Dynasty. Isn't it possible, even reasonable, to expect that scale, if not a canon of proportions, might have been used during the design and construction phases for Old Kingdom building projects as well? Robins states, "There is no doubt that grids had already been employed for other purposes in the Old Kingdom; for instance, they were sometimes used for drawings of expanses of water, the zigzags representing water being obtained by joining diagonals of the squares. . . . Further, tombs are sometimes decorated with panels of geometric patterns."[20] She also notes that, "In the Fourth Dynasty, proportions in the tomb of Khafkhufu I (G7140) come very close to the classic."[21] This canon of proportion seen in the Fourth

Dynasty should be expected, given that there was but one artistic center—Memphis—during the Old Kingdom period. So, let's look in some detail at Khufu's pyramid and see how the classic human proportions may have been used in the construction of that monument.

Vincent Brown suggests that a typical cross-sectional view of the Great Pyramid is one in which the viewer is located either east or west of the structure such that the internal chamber system—a series of passages, chambers, a voluminous gallery, rock excavations, and vents—appears to zigzag its way through the north and central portions of the pyramid (Figures 9.7 and 9.9).[22] It is quite apparent that very little of the pyramid's volume is taken up by those spaces. Most of the pyramid, to the best of anyone's knowledge, is comprised of megalithic stone: an estimated 2.3 million limestone blocks that at one time were covered by a casing of decorative Tura limestone blocks. Few of the white casing blocks remain in place, but when the monument was originally completed the casing created an incline from the pyramid's base to the apex of the Tura pyramidion (capstone) of 51° 50' 34", certainly too steep and slick to climb. There is speculation that the pyramidion might have had a metal veneer to accentuate the apex, but no record provides such detail. It seems as though Khufu's pyramid was meant to be viewed not from its base, but from a distance, if its meaning were to be best understood and appreciated.

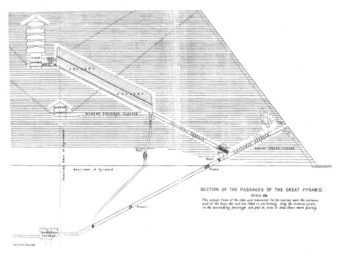

Figure 9.7: Interior of the Great Pyramid showing passages, Grand Gallery, King's Chamber, Queen's Chamber, air shafts (stellar alignments), and the Pit constructed in bedrock below the pyramid (from Petrie, 1883; courtesy Brown, 2002).

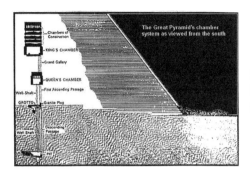

Figure 9.8: Detail of the interior of the Great Pyramid (from Petrie, 1883; courtesy Brown, 2002).

Up close or not, the Great Pyramid's size and precise construction are more than impressive:

- Upon completion it stood 280 Egyptian royal cubits, or 146.5 meters (480.6 ft), tall (1 cubit = 0.524 m).

- Weathering and erosion, and the absence of the pyramidion, have reduced its height a mere 25 feet, to 138.8 meters (455.4 ft).

- Each base side was 440 royal cubits, or 230.4 meters (755.9 ft) long.

- Its total mass is estimated to be 5.9 million metric tons, with a gross volume of about 2.5 million cubic meters (3.3 million cubic yards).[23, 24]

- The estimated average joint width between casing stones was 0.5 millimeters (.02 in).[25]

- The pyramid remained the tallest manmade structure for over 3,800 years (ultimately succumbing to medieval technology and the Roman Catholic Church in about 1300 CE with the completion of the spire of the 160-meter-tall Lincoln Cathedral; the spire blew down in a storm in 1548).

- The average error of the bases is 58 millimeters (less than .25 in).[26]

- The sides of the square base are almost exactly aligned with the four cardinal directions to within four minutes of arc.[27]

- The finished base is virtually square, with a mean corner error of only twelve seconds of arc.[28]

- Based on Sir Flinders Petrie's *The Pyramids and Temples of Gizeh*, the original design dimensions were 280 cubits high with a base length of 440 cubits along each of the four sides.[29]

That this massive structure was built to such fine tolerances speaks of the importance of the pyramid to everyone involved in its planning, design, and construction. The blocks of stone were cut and placed such that a knife could not be inserted between them, even after almost 4,600 years. And all the while the monument remains founded on a square base oriented perfectly to the four cardinal directions. Here are the respective lengths, orientations, and apparent as-built errors of the four base sides:

Side	Design Orientation	Actual Orientation	Apparent Error of Orientation	Design Length	Actual Length	Apparent Error of Length__
North	N 90° E	N 890 57' 32" E	00° 02' 28"	755.9 ft	755.42 ft	- 0.48 ft
East	N 180° E	N 1790 54' 30" E	00° 05' 30"	755.9 ft	755.97 ft	0.07 ft
South	N 90° E	N 890 58' 03" E	00° 01' 57"	755.9 ft	756.08 ft	0.18 ft
West	N 180° E	N 1790 57' 30" E	00° 02' 30"	755.9 ft	755.76 ft	- 0.14 ft
Average Error of Orientation			00° 03' 06"			
			Total Base Length	3023.6 ft	3023.23 ft	-0.37 ft (4.4 in)

It is obvious that the ancient Egyptians intended to align the four sides of the pyramid, all virtually equal in length, with the cardinal directions (Figure 9.2). The east and west sides are oriented along the meridian extending from pole to pole, while the north and south sides extend along the pyramid's latitude, toward the sunrise and sunset during the vernal and autumnal equinoxes. Entry and exit are provided by a passageway located along the centerline of the pyramid's north face.

The proportion of the design length of each base (440 cubits) to the design height of the pyramid (280 cubits) is 440/280, equivalent to 11/7, or about 1.57143; the ratio of two design lengths (halfway around the pyramid base) to the height is 22/7, or about 3.14286; the total perimeter to height ratio is 22/7, approximately 6.28571. These values leave little doubt that the designers intended to imply the value of pi (π) within the structure, even though there is nothing circular or spherical about its shape. Is this truly the case, or just wishful thinking? The facts are that the referenced values (11/7, 22/7, and 44/ 7) are equivalent to π/ 2, π, and 2π to within at least three significant digits, or better than 0.9995 percent of the actual value of pi, which is defined as the ratio of the circumference of a circle to its diameter (π = C/d). This simply cannot be coincidence with regard to the pyramid's design and execution, and here's why.

You may recall from high-school geometry that there are 360 degrees in a circle, a vestige of the ancient Babylonian sexagesimal (base-60) counting system. The Babylonians did not have a digit representing zero, so the number 60 essentially represented both the ending and beginning of their basic unit of counting, just as today we count up to nine and then use the numeral 1 with a zero tacked onto it to start a new round of counts, using the digits all over again. The lack of a zero has been regarded as a problem for the Babylonian counting system, but in fact that concern is moot when the system is applied to dividing a circle into equivalent parts. This is true because the circle has no starting or ending points. If we begin a clockwise count of degrees at the top of a circle, then 1 degree is the angle from the top to the point 1/360 of the circle's circumference from the starting point. We continue counting until we arrive at 359 degrees and one more degree and find ourselves back at the starting point, the top of the circle. But this is not the zero point. It marks the end of the 360th degree and the start of the 361st degree as we proceed with another round of counting. The concept of zero is unnecessary.

Let's switch our focus from degrees to radians of a circle. Recall that the 360 degrees of a circle are equivalent to 2π radians. This seems confusing at first, but remember that for a circle of unit radius (radius = 1) the circumference is equal to pi times the circle's diameter ($C = \pi d$). Since the diameter is equal to two times the radius, we find that $C = \pi 2r = 2r\pi$. We defined the radius as equal to one, so the length of the circumference is $C = 2 \times 1 \times \pi = 2\pi$, defined as the number of radians around a circle.

Now for the fascinating part. Note the following relationships between the geometry of the Great Pyramid of Khufu and a circle or sphere:

Number of Base Lengths	Ration of Base Lengths to Height	Degrees Around Circle or Sphere	Equivalent Number of Radians
1	$11/7 = 1.57143$	90	$\pi/2 = 1.57080$
2	$22/7 = 3.14286$	180	$\pi = 3.14159$
3	$33/7 = 4.71429$	270	$3\pi/2 = 4.71239$
4	$44/7 = 6.28571$	360	$2\pi = 6.28319$

The relationship that we discover from this exercise is the second key to unlocking the purpose behind the Great Pyramid:

Each side of the pyramid represents one fourth of the circumference around a circle.

The Third Key

There is growing acceptance by scholars of the deliberate use of pi in ancient Egyptian design.[44] Miroslav Verner suggests, "We can conclude that although the ancient Egyptians could not precisely define the value of π, in practice they used it."[30] However, this geometric relationship in two dimensions, within the plane of the circle and the square, gives us pause to consider if there might be a similar relationship in the third dimension. Let's look at the four faces of the pyramid, from their base length to the apex. We know that the design length of each base was 440 cubits, and we can calculate the altitude of each side (the length perpendicular to the base, measured from the ground surface to the apex) because we know the design height of the pyramid, 280 cubits. The altitude (A) calculates out to approximately 356.0899 cubits. The area of each triangular face is equal to the altitude times half of the base length: $A = 356.0899 \times .5(440) = 78340$ square cubits. So, the area of the pyramid's four faces is $4 \times 78,340$, or 313,359 square cubits.

Using our second key we know that the total length of the base of the pyramid (1760 cubits) is to be interpreted as equivalent to a circumference of a circle. But what circle? What is the radius or circumference? We have two obvious choices, one in which the radius is equal to half the length of a base (meaning the circle fits exactly within the square footprint of the pyramid, and the other in which the circumference equals the total length of the four base lengths). Looking at the first case, the radius will be $r = .5(440) = 220$ cubits. The area of the hemisphere of such radius is $A = 4\pi r^2/2 = 4\pi(220)^2/2 = 304,106$ square cubits. In the second case, the total base length of the square of the pyramid is 1,760 cubits. The radius of a circle with a circumference of 1760 is equal to $r = .5C/\pi = \frac{1}{2}(1,760)/\pi = 280.11$ cubits. This is approximately 0.04 percent (2.3 inches or 5.9 cm) more than the estimated design height of the pyramid, essentially confirming what we discovered with the second key, that the proportions of the Great Pyramid were meant to indicate an equivalency to a sphere having a radius of 280 cubits. The area of this hemisphere is $A = 4\pi r^2/2 = 4\pi(280)^2/2 = 492,602$ square cubits.

Let's compare these two results with the total area of the pyramid's four faces (313,359 cubits). In the first case the surface area of a hemisphere having a radius of 220 cubits is 304,106 cubits; this is about 97.0 percent of the total area of the four faces. In fact, the near equivalency of these two areas should not be surprising, since our second key suggests that the parametric dimensions of the pyramid are meant to signify the dimensions of a sphere contained within it. But in the second case, it is obvious that a hemisphere of radius 280 cubits would be sizably larger than the 440-cubit base length and width of the pyramid, and in fact the surface area of the hemisphere is just that. The surface area of a hemisphere having a radius of 280 cubits is 492,602 cubits; this is about 157.2 percent of the total area of the four faces. Results of this second case appear to well exceed the value we had hoped for.

However, as stated at the outset of these last calculations, there are two obvious choices in the size of the circles (or spheres) to evaluate, one in which the sphere fits exactly within the confines of the square footprint of the pyramid, and the other in which the sphere's circumference equals the total length of the four base lengths. An interesting relationship arises from the proportions relative to these two circles: The ratio of the area of the larger sphere (the second case) to the smaller sphere (the first case) is 492,602/304,106, which equals approximately 1.619837. This value is surprisingly close to the value of phi ($\Phi = 1.618034...$), an irrational number that has relevance to natural processes in innumerable ways and is found in the geometry of the five-sided polygon and pentagram.

I won't delve here into the many surprising ways in which phi is found in nature and geometry, as there are more than sufficient resources on that topic already available to the reader. However, it is notable that the proportional relationship of the surface areas of these two hemispheres is within about 0.01 percent of the value of phi. Additional calculation will show that a change in radius of even one foot in either of the two spheres will increase the deviation between the calculated proportion and the value of phi. To find an equivalency to phi by keeping our value from the first case in good standing, the radius of the sphere in the second case would have to be 279.84 feet, less than 2 inches (4.7 cm) short of the design height of the pyramid. Certainly this "error" might be attributed to the value of the pyramid's design height calculated in modern times, or actual construction tolerances for which we have scant evidence. Therefore the apparent error should be re-evaluated as potentially within acceptable design or construction tolerance by the builders, rather than dismissing the argument that the ancient Egyptians not only had an understanding of the value of pi, but of phi as well. We have found the third key:

The ancient Egyptians knew and used pi and phi in their constructions, and they ensured that this tradition was communicated to us via the Great Pyramid.

The Fourth Key

Returning to cross-sectional views of the Great Pyramid, Brown suggests, "While designing the internal works of the pyramid, a number of different perspectives from all directions would have undoubtedly been drawn up. This concentration on the alignment of the pyramid with the northern polar region of the sky, however, would suggest that the view from the south looking north was considered the principal perspective."[31] This perspective is shown in Figure 9.9. Brown argues that the northerly view of the system of passageways and chamber inside the pyramid "forms an image resembling that of a standing man wearing a crown. This man stands facing west, the entrance to the 'Underworld.' His crown formed by the set of small chambers situated above the King's chamber could be viewed as the top section of the *Djed* or *Tet* pillar, a symbol that later became synonymous with the resurrected god, Osiris (Figure 9.10). The custom of Ancient Egyptian Kings to identify themselves with Osiris in death dates back to a very early epoch, and the ritual of 'Raising the Djed Pillar' dates back to even earlier times."[32]

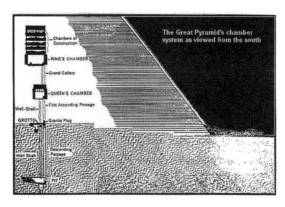

Figure 9.9: View of the interior of the Great Pyramid looking toward the north, showing the image of a crowned figure standing and looking toward the west (illustration courtesy Brown, 2002). The crown is formed by several "relieving" chambers immediately above the King's Chamber, appearing similar to the upper portion of Osiris as the *Djed* pillar.

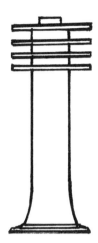

Figure 9.10: The Djed Pillar, symbol of the resurrected Osiris.

We have returned to discussion of the Egyptian pillar, in this instance a column known as a hieroglyph and engraved symbol on temple and tomb walls and ceilings. It is a symbol that is not foreign to our discussion, since it represents stability, the backbone of Osiris. One special form of the *Djed* is the *Banebdjedet*, which I interpret as *ba* (immortal soul or spirit, also *ram*), *neb* (lord), *Djed* (pillar of stability), and *et* (truth or peace): That is, "stability in the true spirit of the lord." There has been some confusion concerning the word *ba*, the result being that *baneb* could have meant one of the gods having ram-like qualities or simply referred to the soul of a particular god.[33] Banebdjedet was personified as a god with the head of at least one ram and usually four.

The center of the cult was at Djedet (Greek: Mendes), located in the east portion of the Nile delta. The city's name might indicate that the residents considered the area one of peaceful stability. Sounds pleasant enough. Banebdjedet's ram heads could represent the four souls of Re, but might also refer to four gods of the Ennead—Osiris, Geb, Shu, and Ra-Atum—all of whom had large shrines constructed in their honor at Mendes.[34] Another cult center was located at Djedu (renamed Pr-Asir, or "house of Osiris"; Greek: Busiris), another city located in the Nile delta. It was there that Isis found Osiris's backbone after Set had cut up and thrown his body across Egypt. In Djedu's Renewal Festival, a *Djed* symbolizing a phallus was raised as acknowledgment of the "potency and duration of the pharaoh's rule."[35]

We now have a vivid idea of what the *Djed* hieroglyph represents. A reasonable approximation would be the modern phrase "the backbone of society," but this doesn't provide the depth of physical, intellectual, emotional and spiritual meaning the word held. *Djed* represented the common bond between all people of ancient Egypt, part strength, part stability, playing a central role in a culture that prized the concept of *ma'at*—divine wisdom, a universal ethical principle emphasizing the qualities of order, balance, law, justice, morality, and truth. Ancient Egyptians also cherished a deep, almost reverent sense of social responsibility with everyone in the community. This was what the backbone of Osiris represented. The Egyptian kingdoms could not thrive without the principle of *Djed*. It was a culturally accepted truth put into action. It was tradition. The people didn't *have* to accept it as a part of life. They *wanted* to live it, experience it, and pass it on to others. If the pillar of Jachin stood on the north side, and Boaz stood at the south, the *Djed* stood in the center—not overt and boastful, but perhaps set back a bit from the other two as if to say in a very sacred manner, "I am the backbone of the human soul." This symbolism *is* tradition in every sense of the word.

Vincent Brown suggests that the cosmic order the Ancients envisioned demanded that funeral ceremonies held on Earth be "meticulous and elaborate and [that] great care was taken to ensure that the King was entombed with the necessary equipment to guarantee him everlasting life. The pyramid itself functioned as a part of this funeral equipment."[36] The canon of proportion was applied in the hieroglyphs and sculptures that lined the ceilings and walls of the pyramids and other tombs. Brown described the interior of the tombs:

> Pyramids built after Khufu's, such as that of Unas, had their chamber and passage walls covered in hieroglyphic scripts [the Pyramid Texts] conveying the King's relationship with the cosmic forces and the heavenly gods. These inscriptions contain spells and directions for the rituals that were carried out to allow the King to be transformed and resurrected. In them the King is likened to the Djed Pillar and is repeatedly identified with Osiris. In fact, Osiris features heavily in the Pyramid Texts with over 170 separate spells or utterances referring to him, suggesting that even at this early point in history Osiris had become a key player in the funeral texts.[37]

But this was not the case for Khufu's Great Pyramid. No hieroglyphs were drawn on exposed walls and ceilings, and no necessary equipment was included for use by the deceased. No sarcophagus, and no mummi-fied pharaoh of Khufu awaiting new life. In fact, it looks like a pyramid and has passageways and chambers like a pyramid, but it provides no direct evidence of it having been used as a pyramidal tomb. However, Vincent Brown found a remarkable coincidence, or is it coincidence at all?

> When the canon is applied to the layout of the chamber system it can be confirmed that the positioning of the chambers and aspects of their dimensions comply with the proportions of figures depicted in papyri and tomb reliefs from the same period. This innovative arrangement of the pyramid's chamber system to form a hidden "statue" of the anthropomorphized Djed pillar may reflect an early experimental stage of pyramid design. . . . The resemblance of the god of resurrection inside Khufu's pyramid would indicate that the architecture was designed to incorporate key aspects of Ancient Egyptian myths and legends, elements that form the basis of their religion.[38]

James P. Allen, Wilbour Professor of Egyptology in Egyptology and Ancient Western Asian Studies at Brown University, analyzed texts found within the Fifth-Dynasty Unas Pyramid and a number of other Old Kingdom pyramids and tombs. He discovered that particular texts were grouped together and placed at similar locations within each pyramid. Allen concluded that "the direction of the texts lead the soul's path through the tomb, flowing from the innermost parts of the sarcoph-agus chamber, through the Antechamber to the outside of the pyramid."[39] The whole canon of the interior architecture supported the king's soul in moving out of the body and sarcophagus located in the west portion of the burial chamber and proceeding to the east side, mirroring the sun's movement at night across the netherworld, only to be reborn in the east. The soul would continue through an antechamber before entering a number of passageways, and would cross the threshold of the exit to be bathed in light. This was the king's *ba* "Going Forth by Day."

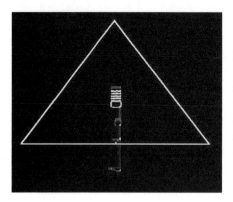

Figure 9.11: The Great Pyramid, symbolic burial palace of Osiris, built for the purpose of preparing (constructing), entombing, and resurrecting the Djed Pillar (illustration courtesy Brown, 2002). This is sacred symbolism in the highest regard, expressing belief in the resurrection of mankind.

From this architectural canon we can see that the zigzagging passageway and chamber system of the Great Pyramid, like those of the other pyramids, is intended to address sacred concepts, designed with a specific purpose in mind, and built according to design (Figure 9.11). Given the apparent magnitude of effort that went into the design, engineering, and construction of the monument, I'm not surprised that the pyramid's interior has a holistic design. However, the intent of the symbolism that Vincent Brown finds in that design is stunning! Egyptologist Mark Lehner concludes:

> Physically entombed in the pyramid, the dead King became identified with Osiris, the divine father of Horus. The pyramid complex was, in one sense, a temple complex to the Horus-Osiris divinity, merged with the sun god in the central icon of the pyramid. . . . [the pyramid was the] embodiment of light and shadow: and the union of heaven and earth, encapsulating the mystery of death and rebirth.[40]

We can now understand why Brown sees the size and configuration of space within the Great Pyramid as the image of a man wearing a crown and facing the west. This man is the symbol of Osiris, the central pillar. From top to bottom, the whole series of passageways and chambers is configured to represent Osiris as the *Djed*, scaled in accordance with the same traditional canon of proportion as that found in texts and engravings and size and configuration of temples. Even the relieving chambers located above the so-called King's Chamber resemble the four tiers of the *Djed*. Vincent Brown notes, "The custom of Ancient Egyptian Kings to identify themselves with Osiris in death dates back to a very early epoch, and the ritual of 'Raising the Djed Pillar' dates back to even earlier times."[41] So with the symbolic *Djed* firmly set within Khufu's pyramid, we can stand back and look north toward the pyramid and envision the entombed Osiris prepared to enter the underworld and undergo resurrection in sequence with the same protocols undertaken by later pharaohs upon their deaths. Brown states, "This image contains the same elements of the Legend of Osiris: the body of Osiris inside the Pillar in the House of the King."[42]

This King is Osiris, this pyramid of the King is Osiris, this construction of his is Osiris.

Utterance 600 Of The Pyramid Texts

We have discovered the fourth key:

The Great Pyramid is the symbolic burial palace of Osiris, built for the symbolic purpose of preparing (constructing), entombing, and resurrecting the Djed pillar.

The Big Picture

Vincent Brown suggests that "if the King is Osiris and his Pyramid is Osiris, then by constructing the chambers of his pyramid in such a way that they form an image of the *resurrected* Osiris, Khufu had in effect ensured his own resurrection."[43] Might the construction of Osiris in the form of the *Djed* have a universal meaning beyond simply the king's own resurrection? Isn't it reasonable to suspect that Khufu's pyramid, the greatest pyramid and the prototype of later pharaonic tombs, was meant to express the possibility, or even enhance the potential, of the resurrection of humankind? Recall that there are no hieroglyphs, engravings, or other means of communication inside the pyramid to associate Khufu with the monument. Of course, there are hieroglyphs on stones lining the relieving chambers that include the name of Khufu, but those markings are hidden and not expected to be viewed, should anyone have entered the structure after its completion. The hieroglyphs were simply the means by which quarry masons and onsite laborers communicated while construction was proceeding. I conclude that the Great Pyramid stands as a monument symbolizing the resurrection of all human beings. This certainly conforms with tradition of *ma'at*.

At first we would expect that Osiris in the form of the *Djed* would stand upon the bedrock surface underlaying the stone blocks of the pyramid. But this is not the case. In applying the canon of proportion, with the height of the *Djed* measured from the ceiling of the uppermost relieving chamber, Brown suggests that "[Osiris] appears as though half buried, rising out of the mound of bedrock upon which the pyramid was built."[44] The following table summarizes the height of each chamber associated with the *Djed* based on measurements taken by Petrie.[45]

Location in the Great Pyramid	Height from the Bedrock Passage at the Bottom of Djed Pillar
Campbell's Chamber Apex	179.72 cubits 3,706 inches ± (Petrie, 1883)
King's Chamber Ceiling	150.51 cubits 3,103.6 inches ± (Petrie, 1883)
Queen's Chamber Apex	109.62 cubits 2,260.5 inches ± (Petrie, 1883)
Pyramid's Entrance Ceiling	91.5 cubits 1,886.9 inches ± (Petrie, 1883)

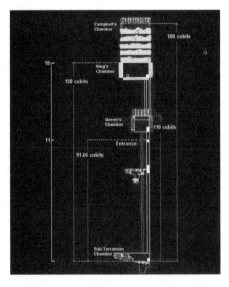

Figure 9.12: The symbolic burial palace of Osiris showing dimensions from the subterranean chamber to significant locations within the pyramid, crowned by the top of the Djed pillar, in conformance with the ancient Canon of Proportions (illustration courtesy Brown, 2002).

The meaning of the symbol of the *Djed* in Khufu's pyramid becomes all the more transparent when Brown illustrates these dimensions with respect to the pyramid's interior, as shown in Figure 9.12.

The four keys can now be used to unlock the mystery that has for so long prevented us from understanding the purpose of the Great Pyramid. Recall:

Key #1: Ancient Egyptian artists, architects, and engineers applied a canon of scale and proportion modeled on the human body in the design and construction of not only hieroglyphic drawings and engravings, but megalithic scale of tombs and temples.

Key #2: Each side of the pyramid represents one fourth of the circumference around a circle.

Key #3: The ancient Egyptians knew and used pi and phi in their constructions, and they ensured that this tradition was communicated to us via the Great Pyramid.

Key #4: The pyramid of Khufu is the symbolic burial palace of Osiris, built with the symbolic purpose of preparing (constructing), entombing, and resurrecting the Djed pillar.

The purpose of the Great Pyramid is now plain for all to see.

The Ancient Egyptians applied their canon of scale and proportion in tandem with the universal principle of ma'at for the symbolic entombment and resurrection of the Djed pillar. The design of the pyramidal form included use of pi and phi to signify that the monument exposed above the ground surface represents spherical geometry, Earth's northern hemisphere. Within the womb of the earth the Djed (all of humanity) is raised from the depths of the underworld to be resurrected, joining the eternal order of the cosmos.

Contained for millennia in the architecture of an ancient culture, hidden from view yet recorded in texts and on walls of temples and tombs, the purpose of the Great Pyramid is the same as that of Osiris, Horus, Mithras, Krishna, Dionysus, Jesus, and so many other mythical and historical heroes sacrificing themselves for the sake of humanity. They are united by a common purpose. Their differences are artifacts of individual cultures, products of the variety of theological traditions. Osiris is not Mithras, and Jesus is not Krishna, but they each represent the same essential concepts symbolized by the Great Pyramid.

The measurements rounded to whole-number values in Figure 9.12 vary only slightly from those made by Petrie, who himself remarked upon the difficulty of accurately measuring the heights of the chambers. The small chambers situated above the King's Chamber would be particularly difficult to measure, and for the apex of Campbell's Chamber he relied on the measurements taken by Vyse. The pyramid's entrance height is notably not a round figure when measured in cubits. Keep in mind, however, that the canon of proportion seen in ancient Egyptian figures varied over the thousands of years of the kingdom. As Robins notes, "When a figure differs slightly from ideal proportions, it is not because it was produced by a bad draftsman, but because there was room for some variation."[46] This is certainly true during the construction phase of any major engineering project, even today. And so, Vincent Brown concludes that the ratio of the pyramid's entrance height to the King's Chamber height above the bedrock-excavated subterranean passage can be described as 11/18.[47]

Lastly, it is worth noting again the paucity of archeological evidence concerning Khufu's reign and what this might mean. We know that he was the first pharaoh to have his pyramid constructed at Giza, and what an imposing edifice it is upon the landscape. Yet, there is hardly a statue of the king to be found. Egyptologists believe that many of the written and artistic records were destroyed. However, there are virtually no artifacts or hieroglyphics found on or within the Great Pyramid and, in fact, it seems possible that Khufu expressed his self-deification in the pyramid itself. What is self-evident is that Khufu refrained from having the Great Pyramid marked with his own name. The symbolic raising of the soul after death, as evidenced by the size and configuration of unsigned, unadorned chambers and passageways of the pyramid's interior, signifies the raising of the *human* soul—the pharaoh's and the slave's, yours, and mine (Figure 9.13).

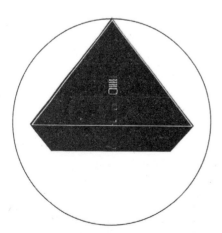

Figure 9.13: The form of the Great Pyramid is symbolic of a hemisphere of Earth. Osiris as the Djed (stability) is within the pyramid. The resulting monument is a vision of rebirthing the human species, joining the eternal order of the cosmos.

It is no wonder that architects, engineers, mathematicians, and geographers are in awe of the Great Pyramid. By putting together the interior and exterior symbolism we can appreciate this monument to all humankind with even greater reverence, for the pyramid stands as a statement of the birth, life, death, and rebirth of every human soul. It symbolizes the relationship between us and the Creator within the context of the universal spherical geometry expressed by the pyramidal form—symbol of the world as the rebirthing chamber for humanity. The effort needed to construct the Great Pyramid as a monument to the universal soul must be seen as great sacrifices by a pharaoh and his people.

The tradition of sacrifice was a characteristic of other ancient cultures as well. Those people also built temples and pyramids, yet were far from Egypt. We turn next to the Maya of Central America.

Chapter 10

Mayan Canamayte

Something incredible occurred in the center of the Americas that has persistently intrigued and baffled European colonizers. The discoveries and achievements of American Indian civilizations reveal an unparalleled genius.

JOHN MAJOR JENKINS[A]
THE 2012 STORY

This deduction, which was the simple truth for the Maya, has been something that modern science has only just begun to recognize; that even the tiniest components of matter conform to certain geometric patterns whose dimensions are maintained by an intelligent energy— an energy that fuels all human intellectual and creative energy.

HUNBATZ MEN[B]
THE 8 CALENDARS OF THE MAYA

235

Like the Lakota, the Maya of pre-Columbian Central America understood that time and space express the circular nature of the universe. Their cosmology recognized a strong correlation between mundane experience and the eternal macrocosm. It is this relationship between humans, nature, the cosmos, and indeed the very center of our galaxy that we review in this chapter. Our guides at the beginning of this journey are Adrian Gilbert and Maurice Cotterell, whose book *The Mayan Prophecies: Unlocking the Secrets of a Lost Civilization* provides important information concerning the source of the geometrical principles supporting Maya perceptions about the human condition.[1] This understanding is more than a millennia old, possibly much older than Maya civilization itself. We then follow up with a conceptual model of these relationships founded on geometrical principles as understood by contemporary Maya.

Background

Pre-Columbian Maya traditions were far different from those of Western culture. Nonetheless, while owning few personal possessions and cultivating few staple crops, corn being the most important, the Maya managed to produce some of the world's most treasured works of art and architecture. They ensured that their kings and priests were well cared for, in terms of everything from attire befitting the great leaders of their culture to a supply of humans chosen for sacrifice in accordance with ritual and ceremony. The rulers themselves, and likely their family members, also participated in painful rituals of sacrifice to help ensure fertility of the land, health and well-being of the people, and their own entry into the realm of the afterlife. Their scientists were expert astronomers, discovering cycles of time and forecasting the end times and the beginnings of new worlds based on astronomical observations and detailed recordkeeping—records that were retained and referred to over many generations.

The Maya golden age occurred between about 600 and 800 CE. Only a few centuries later, Maya civilization collapsed across the Yucatan. The cause of this collapse remains unknown, though researchers have suggested environmental degradation, disease, other socio-cultural catastrophes, and combinations of such events. In any event the cities reverted to the jungle. Left behind were houses, markets, pyramidal temples and observatories, infrastructure, agricultural fields, and pastures dating from the Terminal Classic period, about eighth century

CE. Within about two hundred years the Yucatan was invaded by the Toltec, who replaced the Maya as the most powerful people of the Mexican region. Later, the Aztecs became the leading socioeconomic force in central Mexico, only to be decimated by Hernando Cortez and his conquistadors.

Like the Maya, the Toltec and Aztec peoples practiced ritual sacrifice, but the sheer brutality of the rituals and the numbers of sacrificial victims appear to have risen to almost unbelievable proportions over time. Gilbert and Cotterell note, "As part of their solar religion [the Toltec] carried out human sacrifices on a regular basis. Using obsidian knives, they cut open their victims' chests and tore out their still-beating hearts as an offering to the sun god. They believed that by doing this they were feeding it with its favorite food, human life-force, which would ensure that the sun would keep on rising."[2] In their time, the Aztecs "adopted these superstitious beliefs and customs, and took them to absurd limits. Human sacrifice, and especially removal of hearts, became the central mystery of their religion."[3]

Aztec legends told to Spanish victors in the 1500s state that the Toltec ruler who took over the lands of the Maya was a god-king named Topilzin-Ce-Acatl, a ruler of Tula who had been exiled by his rival Tezcatlipoca or "Smoking Mirror." The Aztec legends say that Topilzin-Ce-Acatl, a man who sought peace after having taken over the Yucatan and establishing a capital at Chichen Itza, gave himself the title of Quetzalcoatl or Kulkulcan. This was likely as much a political maneuver rather than a demonstration of authority. *Quetzalquatl* is the name of an Aztec a god-king, described as a white man with a beard who would return from the east to reclaim his kingdom after some period in exile. Then prosperity would return to the Aztec and all people would live in peace. Kulkulcan is the Maya equivalent to the Aztec Quetzalquatl (Figure 10.1). The Aztec legend indicates that Topilzin-Ce-Acatl may have adopted the name Kulkulcan to placate his Maya subjects by taking the role of their long awaited god-king.

Figure 10.1: The head of Quetzalcoatl projecting from the Aztec Temple of the Feathered Serpent at Teotihuacan. The Mayan deity of *Kulkulkan* is the equivalent of the Aztec feathered serpent, Quetzalcoatl.

Who or what was this Quetzalcoatl of the Aztec, this Kulkulcan of the Maya? Certainly he was held in high regard by the people of Central America for centuries, his importance transcending cultures. He was banished from Mexico and the Yucatan, yet his return to the people on a predetermined date was eagerly anticipated. The name *Quetzalcoatl* is derived from two words, *quetzal* and *coatl*. Those words are from the Nahuatl group of Nahuan Uto-Aztecan languages. Nahuatl was the language of the Aztecs and today is spoken by more than 1.5 million people in Central Mexico. Quetzals are birds in the family *Trogonidae*, their habitat including forests, woodlands, and humid uplands in and around Mexico. The Nahuatl word *quetzalli* means "large brilliant tail-feather" and comes from the Nahuatl root *quetz*, "stand up," referring to a plume of upstanding feathers. *Coatl* is the Nahuatl word for snake. Together, the two words mean (*quetzal*) bird snake or, more commonly, "plumed serpent."

The term references the eternal struggle by terrestrial man to relate to the heavens, the cosmos, and creator, to rise above our confinement within the physical universe. If we had wings, we could lift ourselves up and fly to heaven. The brilliant plumage of the quetzal would be grand and appropriate attire for such a journey. If the plumed serpent could return to our world, then the joys of heaven might be experienced here, Eden on Earth. Kulkulcan, then, is the Maya translation of the intent of Quetzalcoatl.

A Universe in Maize and Sand

The *Bacabob* are four gods, all brothers, worshiped by the Maya. Said to have escaped destruction caused by a world-wide flood (the deluge), they stand at the four corners of the world on the back of a turtle (Earth), upholding the heavens. The four corners of the Bacabob are the four directions lying between the cardinal directions. Each corner has a characteristic color: white for the north, yellow for the south, red for the east, and black for the west. A giant tree, the World Tree, is located at the center of the turtle's back. The World Tree is Earth's central axis, the *axis mundi*. This quadripartite structure of the Maya world was mirrored in the "sand" diagram drawn by Maya elder and spiritual guide Carlos Barrios during a fire ceremony held at the ancient Maya city of Cahal Pech in 2005 (Figure 10.2).[4]

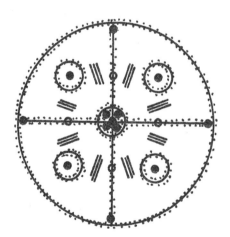

Figure 10.2: The quadripartite structure of the Maya world, as diagramed by Maya elder and spiritual guide Carlos Barrios during a fire ceremony at Cahal Pech in 2005 (based on Gilbert, 2006).

The primary purpose of the fire ceremony was to ask forgiveness of Mother Earth and to ask her for help so that humans can create harmony and balance between the energies.[5] The diagram, created using maize flour, included:

- a circle divided into quadrants with four orthogonal lines extending from the center of the circle to its perimeter and terminating with arrow points

- four small circles, one in the center of each quadrant of the large circle

- a fifth small circle centered in the middle of the large circle

- five copal balls, one placed in each of the small circles

- four copal balls, one placed on each of the arrow points

- four additional copal balls placed in the central small circle

- eight groups of three candles placed symmetrically at eight locations between the quadrant lines and small circles

- four larger, red candles, one placed on each of the quadrant lines

- several drawings, including a pyramidal structure, dots, a crescent shape, and other figures, placed east of the large circle

During the ritual Barrios noted this fire diagram had been used for a thousand years, and the story surrounding it included the following events.

- Four *balonev* (jaguars) came from the Pleiades after the destruction of last World Age; their purpose was to teach humanity.

- Another balonev by the name of Akabal (Venus), the most important healer, arrived as the new light, and then the five balonev called Father Sun.

- Each of the four jaguars brought incense for a ceremony and stood at one of the corners.

- The large circle is Mother Earth.

- The arrows point to the four directions of the universe.

- The center circle is the place (anywhere) where the fire celebration is conducted, and represents the ether elements.

- The other four small circles represent the four corners; the red, black, white, and red races; the four elements of fire, earth, air, and water; and the eyes of the Great Spirit before us (We are directly connected to the Great Spirit).

Throughout the ceremony there is important symbolism in the circle, acknowledgment of space in terms of direction and other parameters that we can then view as vectors, and energy of the ceremony represented by the light and heat of the fire. In fact, each of the five elements (fire, earth, air, water, and ether) participated in the ceremony. The quadripartite symmetry of the diagram is an integral part of the ceremony, and this symbolic imagery is seen time and again in Maya traditions.

A Snake in Stone

Ancient cultural traditions are expressed at Chichen Itza and at other Maya cities by the orientation, form, ornamentation, and function of many buildings that remain. A prime example of this dedication to form and function of Maya architecture is found with the Pyramid of Kulkulcan (Figure 10.3). Fray Diego de Landa described the pyramid in his *Relacion de las de Yucatan*, written in the 1560s:

> This structure has four stairways looking to the four directions of the world, and 33 feet wide with 91 steps to each that are killing to climb . . . When I saw it there was at the foot of each side of the stairway the fierce mouth of a serpent, curiously worked from a single stone.[6]

Figure 10.3: Pyramid of Kulkulkan at Chichen Itza.

Gilbert and Cotterell note the serpents carved from stone and mentioned by Landa "are far from unique, however, and are to be found at the foot of virtually every stairway of every building of any significance in Chichen Itza."[7] With ninety-one steps located at each of the four sides of the pyramid, each side oriented toward one of the four cardinal directions, and one additional step to the top of the temple, the four approaches to the apex are analogous to the 365 days of the year. And now well known to anyone interested in Maya architecture, there is a cosmological significance to this pyramid. As the earth spins during the afternoons of the spring and autumn equinoxes, the sun causes shadows cast on the balustrades of the Pyramid of Kulkulcan to move, making the serpents appear to slither down the pyramid—the living Quetzlcoatl returning to terra firma. Gilbert and Cotterell conclude, "Given the calendrical significance of the steps on the [Kulkulcan] Pyramid and the orientation to the cardinal directions, it seems obvious that this building was meant to symbolize the central pivot or shaft of the world around which the universe turns."[8]

In 1994 Gilbert found a series of booklets written by Jose Diaz-Bolio, an amateur archeologist and anthropologist with a fascination for the arts and architecture of the Maya. Born in Mérida, Yucatán, in 1906, Diaz-Bolio was educated in Mexico and the United States and became a successful pharmacist in Mérida. His interest in the past and present of Yucatán resulted in his publication of several booklets, including *Why the Rattlesnake in Mayan Civilization*, *The Geometry of the Maya and their Rattlesnake Art*, and *The Rattlesnake School*, each of which Gilbert obtained in a bookshop in Mérida during his research on Maya astronomy and mythology. Another of Diaz-Bolio's publications, *The Feathered Serpent-Axis of Cultures*, was written in 1955. Gilbert notes that Diaz-Bolio wrote and self-published more than twenty booklets and books based on his studies. Diaz-Bolio's basic thesis is that all representations of the Mesoamerican feathered serpent are derived from the colored pattern of scales of a rattlesnake-*Crotalus duress*.[9] This snake is indigenous to the Yucatan Peninsula and surrounding region, and had cultural significance from the United States to Argentina. According to Diaz-Bolio, there were several aspects of *Crotalus*, or Ahau Can, that fascinated the ancient Mayans, but of most interest was the pattern on its back. The pattern is said to have provided the Maya "a biomathematics," and "a Euclidian projection like a bio–abacus in [the snake's] skin pattern."[10, 11]

Figure 10.4: The rattlesnake Crotalus duress, the Mayan Ahau Can. Note the pattern of diamonds and crosses along the length of its back.

The particular design on Ahua Can's back, consisting of interlocking squares and crosses as shown in Figure 10.4, is reflected in the art and architecture of not only the ancient Mayans, but also in cultures across Central and South America. Gilbert considers the Ahua Can pattern the most important inspiration for Native American art. He recalls,

> [Diaz-Bolio] showed me a picture from a temple at El Tajin, Veracruz, that exactly matched the pattern of Crotalus and it was quite clear that this was no accident . . . mosaic-friezes from Mitla . . . composed of zigzag lines made out of little stones . . . were based on the same snakeskin patterning. The tiny carved stones that made up the mosaics were just like the scales of a snake and matched the way in which the pattern was made out of regular diamonds of color. . . . The Zapotecs had taken the basic design and produced variations on the theme in the way that a composer might write a symphony based on a simple melody. . . . Don Jose assured me that this was so, and that not only was this the case in architecture but in all the other arts, too, including even the simple embroidery on the costumes of the peasants. This explanation made sense to me, for the preponderance of zigzags in Indian design had to owe its origin to something important.[12]

Diaz-Bolio described to Gilbert how the *Crotalus* pattern results from a series of squares, each containing a cross shape. Diaz-Bolio called this the *canamayte* ("snake square") pattern (See Figure 10.5). This cubic symmetry found on the back of a serpent was the basis of Mayan science, particularly geometry. "Studying the *canamayte* pattern," Gilbert says,

[we] could see how this might certainly have been an important inspiration behind the development of not just Mayan but all American Indian architecture. Laid flat, the most natural thing would be to orientate the square in such a way that its sides pointed towards the four cardinal points. This would mean that the cross would also be directed in this way. The natural way of expressing this in three dimensions would be as a pyramid, oriented to the points of the compass and with a staircase running down the center of each of its three sides. This is exactly how many of the pyramids of the Maya and others were in fact built, the most obvious example being the Chichen Itza Temple of Quetzalcoatl itself. . . . More complicated sites such as the hilltop of Monte Alban, were designed on *canamayte* principles, again involving orientation to the cardinal points but this time using multiple squares.[13]

Figure 10.5: The Mayan canamayte ("snake square") pattern derived from the scales of Crotalus duress.

Diaz-Bolios publications include illustrations of the *canamayte* defining the floor plans and elevations of Mayan buildings. Oriented on its side, the pattern resembles the profile of steps extending up a pyramid. Other architectural features can be seen in the *canamayte* as well, including cross-sections of doorways and the orientation of temple roofs.

In the Epic of Gilgamesh, Gilgamesh collects a plant that was to give him the secret of eternal youth, but later he loses it along a stream bank to a serpent that "smelled the appealing fragrance of the plant. It glided out of the water, slithered up the bank, took hold of the plant with its mouth, and carried it back into the water. As it returned to the water it shed its skin, emerging younger and fresher looking."[14] And so it was that the snake received the secret of eternal youth. In a similar fashion,

Diaz-Bolio found that *Crotalus* sloughed its skin once a year, in mid-July, near one of the two times when the sun reaches its apex in the Yucatan sky.

This "natural correspondence between sun and serpent, that annually renew themselves together,"[13] did not go unnoticed by the Maya. For all of the cultural significance that Ahau Can had for the Maya, including and well beyond the *canamayte* pattern, Gilbert finds that "[f]or these people, the Ahau Can and everything to do with it was as much of a religious and cultural emblem as the Cross is to a Christian. It filled their lives and they proclaimed their adherence to its values at every opportunity, whether it was the building of a temple or the embroidering of a smock. According to Don Jose there was another initiation that is still practiced in some parts of the Yucatan involving a live rattlesnake. In this the right hand is passed over the snake nine times to the left and then the left nine times to the right. This was done to confer artistic talent, particularly in embroidery."[15]

Mayan Cosmology

It is clear that the Maya were aware of very complicated geometries of the movements of the earth, moon, sun, planets, stars, and the Milky Way. They successfully applied mathematics to recording and predicting those movements. Their mythology reflects an understanding of the cosmography presented to them each day and night, and that understanding was expressed in the geometry of the cities they left behind in the Yucatan and in southern and central Mexico. Detailed discussion of those ideas and meanings of the cycle-ending date can be found in numerous publications containing a broad range of interpretations.[15]

According to Hunbatz Men, a contemporary Maya *hau'k'in*, or daykeeper, the esoteric meaning of *Kulkulcan* "is one who not only knows the seven psychospiritual forces that govern our bodies but who also uses them and understands their intimate relationship with natural and cosmic laws."[16] The word *Teotihuacan* comes from: *teo*, "god"; *ti*, "place"; *hua*, "emerge"; and *can*, "wisdom"; and so the translation of the name is "The Place Where God's Wisdom Emerges."[17] Thus the temple is held in great reverence.

John Major Jenkins' interest in Mayan astronomy was, in part, piqued by Giorgio de Santillana and Hertha von Dechen's discussion of ancient myths from around the world, which relates belief in a time "when cosmic harmony would return and an earthly paradise would resurface."[18] He studied Mayan astronomy, mythology, and architecture, as well as the dynamics of Earth's processional movement and the impact it has on our perception of the night skies. Jenkins recognized that the December solstice sun will align with the Milky Way's galactic center in 2012, at the end of what the Maya considered to be one-quarter of a precessional cycle, or 6,540 years. Such an alignment with the March equinox, June solstice, September equinox, or December solstice results as Earth cycles through a quadrant of precession once every 6,540 years. So the alignment scheduled to occur in 2012 "during the traditional 'beginning' point of Earth's yearly cycle"[19] has not occurred in about the last 25,800 years and will not do so again until completion of another 25,800-year cycle. This is cause for excitement as it relates to the calendric cycles of the ancient Maya and the current World Age shift that the Maya and many other cultures expressed through their mythologies, perhaps since well before the last alignment of 6,540 years ago. The basic cosmo-geometrical facts leading Jenkins toward his theory of Maya Cosmogenesis are:

- The ecliptic path of the sun, moon, and planets across the sky, and the twelve constellations located along that path, with the sun passing through all twelve constellations over the course of one year. Jenkins notes that the ecliptic crossed the Milky Way at an angle of 60 degrees in the vicinity of Sagittarius, and the ancient Maya referred to this cosmic cross (the "crossroads") as the Sacred Tree (Figure 10.6).

- Our galaxy, the Milky Way, appears as "a bright, wide band of stars arching through the sky . . . many dark, blotchy areas . . . observed along the Milky Way's length . . . are "dark-cloud" formations caused by interstellar dust. The most prominent of these . . . looks like a dark road running along the Milky Way, and it points right at the cosmic crossing point, the center of the Maya Sacred Tree. . . . The Maya called this dark-rift the Black Road, or the Road to the Underworld. They seem to have imagined it as a portal to another world, and the December solstice sun can enter it only in AD 2012."[20]

- Viewed from Earth, these celestial objects as symbols of Mayan tradition will converge at the center of the galaxy. According to

Jenkins, the Maya believed that the galactic center "is the cosmic womb from which new stars are born, and from which everything in our Galaxy, including humans, came."[21]

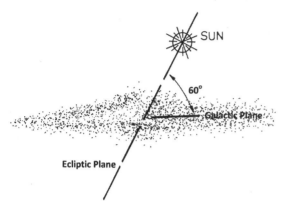

Figure 10.6: The ecliptic crosses the Milky Way at an angle of 60 degrees in the vicinity of the constellation Sagittarius. The Maya referred to this cosmic crossroad as the Sacred Tree.

The shape of the Milky Way is, in fact, generally that of a macro-cosmic pinwheel (Figure 10.7). We see the galaxy on its edge, such that it appears as a broad band of stars crossing the night sky, its galactic center located about 60 degrees from the ecliptic plane. The central portion of the galaxy actually bulges above and below the disk, and a black hole is located at the very center, pulling all matter within its event (space-time) horizon, beyond which events such as light cannot affect an outside observer. The angular momentum of matter moving through the gravitational well of the black hole forms a disc-like structure around its center as, in many cases, relativistic jets extending from the poles of the spinning black hole emitted energy. These jets form the galactic axle around which all other galactic matter rotates (Figure 10.8). Jenkins sees the central portion of the galaxy analogous to a womb, the cosmic source pregnant with solar children, a vision he believes the Maya perceived as well.

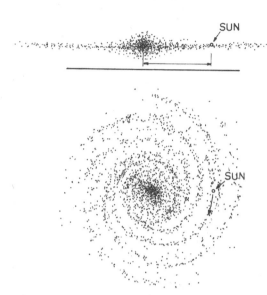

Figure 10.7: Our galaxy the Milky Way has the form of a macrocosmic pinwheel.

Figure 10.8: Jet-like emissions from the polar regions of a black hole (photograph from NASA).

The essential idea of Jenkins' theory is "that the ancient Maya chose the 2012 end-date because this is the date on which occurs a rare alignment of the solstice sun with the Galactic Center."[22] His no-nonsense, detailed methodology and conservative interpretation of the facts render his ideas a credible interpretation of Maya mythology. Their incredible understanding of geometry and mathematics and the cosmic relationships with form and function in their cities (let alone individual buildings) all serve a higher purpose. And just as importantly, research by Jenkins and others shows the importance many ancient cultures placed on precession. Even the Olmec, who built the earliest large pyramid in Central America in about 100 BC, six centuries before the classic period of the Maya, are known to have understood and applied the geometry of precession in their architectural and social orientation with the cosmos.

Cosmic Calendars

As many as three thousand years ago, the Olmec correlated cycles of time and organization in the physical world with observable celestial cycles. The Maya, too, noted this apparent harmonious relationship, something that Jenkins describes as an indigenous holographic paradigm he details in his books *Maya Cosmogenesis 2012* and *The 2012 Story*.[23] The ancient Maya used calendars to map the structure of human experience, whether it be mundane or of a spiritual nature. Modern Maya continue to use calendars for this purpose, such as the Tzolkin 260-day calendar uniting the processes of the cosmos with the earth. The Tzolkin includes twenty day-signs combined with a number from one to thirteen, creating a total of 260 unique day units. The thirteen numbers are reference to the waxing moon (the actually observed phase change from new moon to full moon). The number of the twenty day-signs is based on the sum of human fingers and toes, but it is also associated with the twenty-day *uinal* period of the Long Count calendar. The day-signs and numbers are expressed and counted alongside each other (e.g., 4 Ahau), similar to the way that month days and weekday names are tracked (e.g., Friday, December 21). Maya midwives continue to calculate birth dates with the Tzolkin, and Maya daykeepers apply the 260-day cycle to identify personality characteristics based on the date a child is born.

The Maya Haab calendar essentially reflects the 365 days of the solar year but does not include the approximate one-quarter of a day needed to account for the actual time necessary for Earth to complete one full revolution around the sun. However, the lack of precision in this 365-day vague year of the Maya was intentional. The Haab is based on the cycles of Earth and was used in daily life and agricultural work. It includes eighteen months with twenty days each. When the Tzolkin and Haab calendrical systems are combined, they define a cycle of 18,980 days, just thirteen days less than a full fifty-two years, and this period is referred to as a calendar round. Two calendar rounds are called a Venus Round, the name being associated with the eight-year cycle during which Venus rises in the east five times (every 584 days) as the morning star and returns to the same place on the celestial sphere, only to begin the cycle again. One important result of the Tzolkin-Haab system is that New Year's Day of the Haab calendar falls on one of only four possible day-signs of the Tzolkin; these four important day-signs, "year-bearers," are *Kej* (deer), *Eb* (tooth), *N'oj* (thought), and *Iq* (wind). When placed in a twenty day-sign circle, the four year-bearers divide the circle into quarters (Figure 10.9). Each one of the year-bearers was associated with one of the four cardinal directions as well as a mountain generally located in that direction and one of the four seasonal quarters (an equinox or solstice).

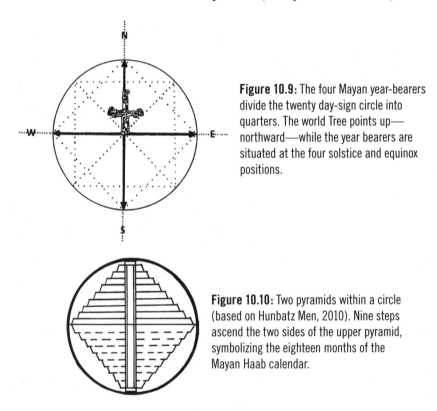

Figure 10.9: The four Mayan year-bearers divide the twenty day-sign circle into quarters. The world Tree points up—northward—while the year bearers are situated at the four solstice and equinox positions.

Figure 10.10: Two pyramids within a circle (based on Hunbatz Men, 2010). Nine steps ascend the two sides of the upper pyramid, symbolizing the eighteen months of the Mayan Haab calendar.

Two pyramids are illustrated in Figure 10.10. One is in an upright, normal position, while the other is an inverted reflection of the first. Note the nine steps ascending each side of the upper pyramid. These steps are symbols for the eighteen months of the Mayan year, in accordance with the Haab calendar.

The Tree of Life was important to the Maya and many if not all indigenous people of the Americas. For the Maya, the Tree of Life connects the galactic center (Wacah Chan) to Earth and the underworld. The Haab represents the root system of the Tree of Life, and the Tzolkin provides the foliage extending out and communicating the earth's energy, a harmonic frequency, to the universe. In this respect the Haab and Tzolkin calendars can't be used on their own, but together they can connect "the spirit and the body, bridging the microcosm and the micro-cosm, creating our reality."[24]

The Mayan Long Count is unique and used for much longer periods of time. It uses five numerical place values based on the following relationships:

1 day	=	1 *kin* (day)
20 days	=	1 *uinal* (vague month)
360 days	=	1 *tun* (vague year, not to be confused with the Haab calendar)
7,200 days	=	1 *katun* (19.7 years)
144,000 days	=	1 *baktun* (about 394.26 years)

Note that each successive unit is twenty times the previous unit of time, except for the *tun*, which is eighteen times one *uinal*. Finally, a World Age cycle amounts to thirteen *baktun*, or 5125.36 years. Rather than using the glyphs that the Maya used to record Long Count dates, scholars use a simple numerical system with (in order) each of the *baktun, katun, tun, uinal,* and *kin* values separated by periods. The start date of the calendar, then, is 0.0.0.0.0, equivalent to 13.0.0.0.0 due to the cyclical nature of the system (the initial or zero date is also the last day of the previous cycle). Results of Jenkins's analysis indicate that the ancient Maya calculated the last zero date as occurring on August 11, 3114, and so the next thirteen-*baktun* cycle end date (1,872,000 days later) is 13.0.0.0.0, or December 21, 2012, the following thirteen-*baktun* end date will occur in 7138 CE, and so forth. But regardless of the date according to the Gregorian calendar, 13.0.0.0.0 will always fall on Tzolkin day 4 Ahau.

The Maya developed and applied additional calendar systems that I will not describe here. Suffice it to say that their calendars amalgamated their understanding of astronomical events with solar seasons, moon cycles, agriculture, human gestation, hours of the day and night, and mythological tradition. While complex, it also had untold value with regard to daily, monthly, and yearly activities and traditions, as well as a very deeply held concern about World Ages. While some may argue that Mayan astronomy was really about a poorly veiled cultural fetish for astrology, we should keep in mind that their ability to observe the skies and understand and record time was far superior to the ability of any other contemporary culture. And, lest we forget, the Gregorian calendar of days and weeks is filled with references to a celestial and mythological cast of characters.

A Modern Perspective

How are the relationships between humans and the universe, space and time, applied by the Maya today? Given that the ancient Mayan culture is so far removed in time from our own, is there value yet for the detailed study of space and time that they so carefully studied and recorded? What relevance does this information have for us today? We can find answers to these questions from the works of the *hau'k'in* Hunbatz Men, a shaman, teacher, and authority on Mayan history, chronology, calendrics, and knowledge, from the mundane to the cosmic.

Our focus in this chapter is on the application of geometry to expressing Maya principles in religion, philosophy, the sciences, architecture, medicine, and other aspects of their culture. Hunbatz Men states that these principles arose from one spiritual idea:

> that the entirety of the cosmos is permeated by sacred energy, and as the cosmos unfolds in countless permutations it constantly reveals the sacred—and thus determines everyday life. In the Mayan way of thinking, human beings are harmonized with divine energy, such that the divine manifests in the myriad forms and beings of the physical world, the world of nature, while the physical world and all its manifestations in nature are reflections of the divine. Being master astronomers, the Mayan sages naturally extended this conception throughout space, such that the whole universe—just like the individual human being—was perceived as a manifestation of divine energy, constantly moving and changing.[25]

Hunbatz Men explains that nature was the Maya mother and guide, Mother Earth, the goddess Ixmucane. Nine is a sacred number with reference to Ixmucane, who was a creative force behind the cyclical processes of life. As we have seen, the Mayans applied their knowledge of astronomy and mathematics to develop calendrical systems that reflected the divine cycles of time at scales from macrocosmic to the microcosmic. Humans were within the microcosm but also part of the macrocosmic Father: All is related and in each of us. Hunbatz Men says Maya believe that astronomical calendars can explain the path humanity must follow to transcend ignorance, that we should study Mayan philosophy, commit to following the path, and live in accordance with the principle of *Panche Be*—seeking the root of truth—then seal our commitment with the words *In Lak'ech*— "You are I and I am you."[26]

The universe operates from an intelligent energy. Hunbatz Men gives an example of the relationship between humans and this universal energy by way of the Mayan discovery of the twenty-three-year sunspot cycle and the cosmo-biological relationship between the sun and nature here on Earth.[27] That this relationship exists and has both qualitative and quantitative effects on human life has been demonstrated by Cotterell and others.[28] Recognizing the effects of this relationship with the cosmos, the Maya found it vital to record their observations of nature and the cosmos so that they could comprehend and appreciate their past and present and prepare for the future.

Recognizing that they were a part of nature, the Maya concluded that they, like everything in nature, had souls, and that the soul had a physical form, since all of nature has form. Spirit, on the other hand, is pure energy: *k'inan*, from the word for sun, *K'in*, and suffix -*an*, which expresses a conditional form of the verb "to be." *K'inan*, then, is spirit or spirit energy. Soul is manifested in Soul. This simple truth for the Maya is a recognition that the microcosm conforms to certain geometric patterns with dimensions maintained by an intelligent energy, the same energy that promotes human intelligence and creativity. The Maya concluded that there is a supreme being in unity with the physical universe. Hunbatz Men says this supreme being is found in "the measure of the soul and the movement of the energy that is spirit—the universal dynamism that stimulates and motivates life in its manifestation of spirit and matter; the principle of intelligent energy that pervades the entire universe, animate or inanimate."[29]

This is the same concept of plurality in unity and unity in plurality detailed in the Upanishads. Hunbatz Men explains that forces of nature (perceived as gods, humans, even numbers) were part of the whole, physical expression of Hunab K'u—architect of the universe, supreme energy, Giver of Movement and Giver of Measure, "because *there can be no movement that does not have measure.* Knowing that God is energy and energy is God, the Mayan philosopher-scientists established the oneness of human beings with Hunab K'u, who is communicated numerically as the union of the numbers 13 and 20, which represent movement and measure, energy and form, soul and spirit. This numerical representation is found in the geometric form of a square within a circle—a synthesis of universal geometry based on the human body."[30] This is illustrated in Figure 10.11.

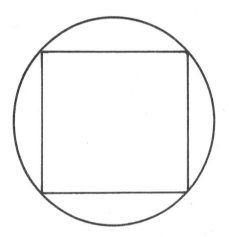

Figure 10.11: The oneness of human beings with Hunab K'u symbolized as the union of the numbers 13 and 20 representing movement and measure, energy and form, soul and spirit (based on Hunbatz Men, 2010). The form of a square within a circle represents a synthesis of universal geometry based on the human body. Note that the numerical relationship is approximate; if the area of the circle is 20 square units, then the area of the square is about 12.73 square units, a little less than 13. A value of pi ~ 40/13 would result in an areal ratio between circle and square of 20/13.

In his essay "Hunab K'u, Synthesis of Mayan Philosophic Thought," anthropologist Domingo Martinez Paredes compares the teachings of Pythagoras with the Maya understanding of sacred numbers:

> For both Pythagoras and the Maya, numbers were held sacred. Pythagoras taught that the number 12 was the infinite, while the Maya, for their part, believed the number 13 to be the infinite, since a meaning had already been given to 12: this number represented the 12 deities who guarded the corners of the heavens, the ideal boundaries within which all things of life take place.

Pythagoras conceived of the number 5 as the vital order, and the Maya considered it so as well; it represented no less than the solar path, the greatest expression of the omnipresence of Hunab K'u—the bearer of Movement and Measure—revered at 2 solstices and 2 equinoxes, plus the central point of the sun, totaling 5. Likewise, he was represented as the human torso, with the two cavities from which the arms rotate, the two joints of the legs, and the navel in the center.

If the Pythagoreans claimed that all things are numbers and that the world is formed by numbers, the Maya, for their part, determined that all things are realized through the course of thirteen numbers, and are subject to two fundamental geometric figures: the circle and the square, or the spherical and cubic forms.[31]

The Maya conceived of a Tree of Life (*yaxche* "*ceiba*") located at the center of the world as a means to communicate between the various spheres located above and below Earth. The ancient Mayan word for tree, *te*, is symbolized by the configuration of the letter *T*. The Maya Tree of Life is a metaphor for spirit, wind, or the divine breath, once again similar to the teachings in the Upanishads and the Lakota conception of *Wakan Tanka*. As the *axis mundi*, the Maya Tree of Life connects earthly existence to the spirit world below and heaven above. It is a representation of Hunab K'u. Hunbatz Men notes that the structure of Maya is similar to all forms of indigenous architecture across the Americas, including the *T* shape in doorways, windows, and other mundane uses of the form of the cross. Use of the *T*-symbol united the Maya "with their true origins."[32] He also suggests that the Tree of Life was a symbol the ancient Maya took to Asia, Africa, and Europe.

Maya concepts of the Tree of Life, the shape of the earth, the four cardinal directions, and the sacred nature of the universe include the basis for understanding cosmology, mathematics, and geometric relationships that make up the wisdom of the Mayan Hunab K'u. Hunbatz Men concludes that this is the same wisdom understood by the Arahuaco of Colombia, Aztec/Mexicas, Nahua, Inca, and many other indigenous cultures of Mesoamerica. This geometric wisdom is illustrated in Figures 10.12 and 10.13.[33] Even simply constructed wood-frame Maya architecture reflects this geometry, as illustrated in Figure 10.14.

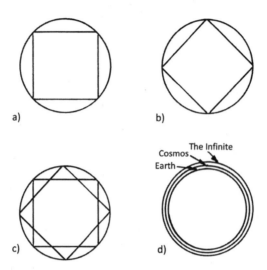

Figure 10.12: Mayan wisdom expressed in geometrical symbolism (based on Hunbatz Men, 2010): a) Father of Thought; b) Mother of Fertility (Earth); c) Father and Mother, the Sacred Ones; d) Mundane existence is found on Earth; to experience the infinite we must work in the spiritual domain.

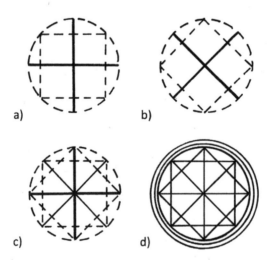

Figure 10.13: Teachings of the Arahuaco of Columbia shown in geometrical form: a) Father and Mother (Earth Goddess); b) Father and Mother in equilibrium; c) Father, Mother, and emergence of the human being; d) Symbol of the need for human beings to live harmoniously with each other and the environment to ensure perfect equilibrium.

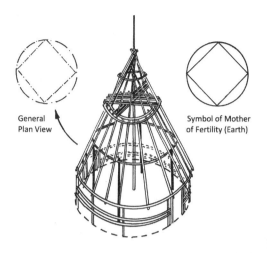

General
Plan View

Symbol of Mother
of Fertility (Earth)

Figure 10.14: The geometry found in traditional Mayan wood-framed architecture reflects the sacred symbolism illustrated in Figures 10.12 and 10.13 (based on Hunbatz Men, 2010).

The key to understanding the diagrams is the combination of circle and square as the symbol of Hunab K'u, the Sacred One, ruling all mundane and spiritual existence. Hunab K'u already existed within the infinite void while the essence of everything else was suspended in the calm, silent heavens. Thus every part of the universe pervaded every other part before becoming manifest. The Maya believe that Hunab K'u is the one source of all energy—divine consciousness—manifested in the infinite number of forms throughout every dimension. Every being is connected to all others through our hearts, and Hunab K'u is the heart of all life. Hunbatz Men suggests that each of the infinite forms of Hunab K'u are frequency vibrations, like the tonalities of sound, "interconnected waves of energy, or vibration. So unlike the modern world, which is obsessed with material reality and linear time, the Maya base their understanding of reality on frequencies and vibrations—the unseen as well as the seen."[34] Numbers—from zero to the infinite—are, of course, created by humans as symbols relating to physical quantities, and represent unique frequencies and tones. (It seems reasonable, therefore, that through divine consciousness we might perceive the physical, tonal nature of unique yet irrational numbers such as pi, phi, $\sqrt{2}$, and so forth, as well as unreal numbers such as $i = \sqrt{-1}$!)

Having mentioned the ability of the Maya to relate numerous calendars into one whole that unifies both microcosm and macrocosm, we can now understand the cosmic knowledge they believed can return mankind to the correct alignment and natural rhythms of creation. Hunbatz Men suggests that the Gregorian timekeeping of Western society and most of the world today doesn't synchronize with natural cycles, but disempowers us as it obscures our connection to nature, the cosmos, and Hunab K'u. We are forced "into the abyss of physical and spiritual destruction . . . [amounting] to a conspiracy against the human race and Mother Nature."[35] But, he says, Maya timekeeping can pull us out of this destructive behavior, and this is so because the Maya "relationship with nature is connected to [their] perception of time."[36] This understanding of and return to "cosmic time," a re-synchronization with Hunab K'u, will direct us out of "thousands of years of darkness" and human beings can then "attain the status of a truly civilized culture."[37]

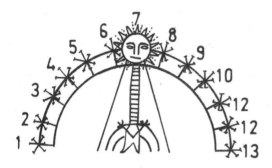

Figure 10.15: A human form with thirteen segments along the neck and a head representing the sun (from Hunbatz Men, 2010). The sun and stars symbolize the thirteen hours of daytime.

The Maya counted thirteen intervals (hours) in each day and each night, a total of twenty-six intervals per revolution of the earth. Figure 10.15 depicts a human being standing on Earth, his body extending outward; his neck includes thirteen segments and his head is represented as the sun, for humans are part of the earth *and* the sun. Thirteen stars, including the sun, form a semicircle arching over Earth and defining the thirteen hours of daytime. The sun is the seventh star. There are also thirteen measures of the night. Hunbatz Men provides the following meaning of the numbers 0 through 13:[38]

Number	Meaning
0	The principle; basis of Mayan counting system; cosmic egg; the spiral form symbolizes the Milky way, source of our existence
1	Unity; beginning; creative spark
2	Knowledge of polarity and duality; number of mystic vision resulting from recognition of conflict that comes from duality
3	Activation and movement; growth stemming from intention
4	Definition and measurement; four elements, direction, and planes of existence; harmony and stability

Number	Meaning
5	Empowerment and integration; potential for choosing conflict or creativity from being part of the whole; number of inquiry, comparison, questioning
6	Universal harmony and balance as we move through the cycles of change; intuitive flow; equilibrium, development, pursuit
7	Number of chakras, colors of the rainbow, and stars in the Pleiades; number at the center of the number 13; wisdom, focus, will
8	Harmony, regeneration, restructuring, awareness of boundaries; discovery, invention, analysis
9	Number of completion; doorway to the next realm; sum of the previous numbers and number of supreme foundation
10	Intention and manifestation; completion of the will; number of perfection of the physical; raw power into reality
12	Crossroads; cycles preparing to end and choices coming from such endings; regeneration, rejuvenation, power of communication
13	Living the truth; perfection of the whole; prophecy and destiny; beginning and end; clarity of the void

Looking back at the geometric wisdom of the Arahuacos in Figures 10.12 and 10.13, we find geometric structures that reflect the meanings of the numbers 0 through 9 as outlined above. The principle of the cosmic egg is, of course, represented by the circle (number 0). An empty circle would also represent the unity principle of Hunab K'u before manifestation of the physical universe, and the circle remains in the all of the other figures—*In Lak'ech*. The square in the circle, Father of Thought, is masculine unity (number 1), while the same figure rotated 45 degrees is the Mother of Fertility, feminine unity. Combined, there is duality (number 2) reflected in the overlapping of lines to from a cross. The pyramidal structure is constructed from three lines, the activation of human intention (number 3) to interact with his environment and the growth of civilization. The number 4 is clearly represented in the squares, diamonds, and crosses, which represent the Father, the Mother, the emergence of humankind, and the equilibrium we must strive to achieve. Interestingly, the number 5 is not presented in the diagrams; this might not be without reason as the underlying cubic symmetry of the figures does not readily support the pentagram, and therefore, questioning and choice of conflict versus creativity are minimized.

Two triangles are created from six lines. The number 6 as harmony, balance, and development is represented by the two triangles (pyramids) shown in Figures 10.12 and 10.13. At first glance the number 7 is not represented in the diagrams. However, note that the two pyramids and the diamond of the Mother of Fertility point to the very top and bottom of the circle, the location of the seventh hour of day and night, respectively. Such is the number at the center of the number 13 and the number of feminine wisdom, focus, and will. By overlaying the diamond with the square we have the eight-sided star (number 8) representing harmony, regeneration, discovery, and analysis. The eight points on the circle touching the corners of the square and the diamond, and the ninth point located at the center of the figure, represent completion (number 9) and entry into the next realm.

Piedra del Sol

There is another calendric icon in the shape of a circle, but it isn't Maya. It was constructed by the Mexica, otherwise known to us as the Aztec. It is the Aztec Calendar Stone, or *Piedra del Sol* ("stone of the sun"), a Mesoamerican megalith carved out of basalt (Figure 10.16). It is almost eleven and a half feet in diameter, three feet thick, and weighs over twenty-four tons. In the introduction to their book, *The Aztec Calendar Stone*, Khristaan Villela and Mary Miller describe the importance of the stone from the time it was sculpted, about five hundred years ago, to today.

Figure 10.16: The Aztec Calendar Stone or *Piedra del Sol* —Stone of the Sun.

Aztec sculptors carved both the top face and the rim of the disk with mythological and historical imagery consistent with the needs of its patron, Motecuhzoma II (1466–1520), and its place in the constellation of sculptures made in Tenochtitlan in the last years before the conquest of 1519-21.

The Aztec Calendar Stone is one of the most famous objects in world archeology, art history, and visual culture . . . The afterlives of the Aztec Calendar Stone suggest levels of "knowing" that span a spectrum in which the object moves from being almost entirely drained on meaning to being elevated as a recondite, mysterious, and impenetrable repository of the secrets of the universe. Despite its ubiquitous, almost numinous, presences, the full meaning of the monument remains just beyond our grasp . . . Those sculptors left behind only the object—no contemporary accounts of its design and creation are known to us. Everything written about the Calendar Stone has always been refracted through the lens of modern inquiry, forever departing and forever arriving.[39]

Why has there been so much interest in the *Piedra del Sol*? What does its matrix of iconography mean? And is it truly a calendar? The answers to these questions are found in the numerous glyphs carved into the stone as well as the shape of the stone itself. There are numerous scholarly and detailed descriptions of the Calendar Stone, many of which were published in English for the first time in 2010.[40] Essentially the form of the stone and layout of the various carvings on its face and sides are as follows.

- The central image is of a human-like head, commonly interpreted as an Aztec sun god, earth god, or similar deity. The god faces front, peering out of the stone, contained within a circular ring.

- The god's face is the central figure within a glyph for the date of 4 Ollin (Nahui Ollin: 4 Motion). The glyph consists of four rhombohedral shapes oriented toward the upper left, upper right, lower right, and lower left relative to the central face, and two curvilinear shapes located immediately left and right of the god. The two curved shapes each appear to enclose a clawed hand wrapped around a heart. Each of these six shapes represents an important date related to the Aztec calendar.

- The arms or wings of the Ollin glyph contain the names of the previous world creations, or suns, according to the order used by the Aztecs. Each is named by the date when it ended, beginning in the upper right and proceeding counterclockwise: Nahui Ocelotl, or 4 Jaguar; Nahui Ehecatl, or 4 Wind; Nahui Quiahuitl, or 4 Rain (this sun ended in a rain of fire); and finally Nahui Atl, or 4 Water. Four Motion is the date in the Aztec 260-day calendar when the Fifth Sun—the sun of our world—is to end. Four smaller glyphs carved within the central circle surrounding Nahui Ollin are the dates 7 Monkey, 1 Rain, and 1 Flint, and a headdress glyph now recognized as the name of Motecuzoma II.

- The 4 Ollin glyph is surrounded by another circular ring which is in turn enclosed by a series of twenty day-sign glyphs presented in order beginning near the top of the 4 Ollin glyph and proceeding around the ring in a counterclockwise direction.

- The twenty day-signs are enclosed by four arrowhead-like carvings and forty quincunx glyphs. Four additional "arrow points" are located immediately beyond the ring of quincunxes. Villela, Robb, and Miller describe the arrow points as "eight large symmetrically disposed points . . . usually interpreted as sunbeams, although the four lesser points may also represent perforators, used to draw blood, since the six square forms with which they alternate have been described as the decorated end of a blood-letter, like a knife handle."[41]

- The next three rings of glyphs include: a) a series of rounded square-like forms commonly interpreted as feathers, b) three thin concentric circles underlying fourteen equally spaced circular glyphs interpreted as greenstone beads and having the meaning of "preciousness," and c) twelve groups of four symbols, each symbol looking like the bilateral end of a tuning fork and interpreted by Hermann Beyer as drops of blood.[42] These groups of "drops of blood" are separated by the aforementioned arrow-points and eight rectangular glyphs that include a quincunx under more "feather" forms and a glyph comprised of two concentric circles.

- The perimeter ring around the stone consists of two *xiuhcoatls*—fire serpents—with their tails located at the top and their heads meeting at the bottom. Their segmented bodies enclose fire-butterfly symbols, and similar fire signs erupt from their backs. Their noses curl back and are decorated with star symbols. At the "top" of the monument their tails frame the date 13 Acatl, or 13 Reed, while at the bottom their mouths open to reveal the faces of Tonatiuh [a sun god] on the left and the fire god Xiuhtecuhtli on the right.

- The edge of the circular stone includes a row of small disks and symbols representing sacrificial knives and the planet Venus. Beyer interprets this grouping of symbols as representing the night sky, Venus, and a beam of light.[43]

While it is true that the Calendar Stone includes specific dates and day names, it does not illustrate complete listings of Aztec-defined days, months, and years that would render it a calendar. Nor does it provide a methodology for calculating dates. And so, we can first conclude that the Aztec Calendar Stone is *not* a calendar. Instead, it is a monument to important dates in Aztec history, the first four worlds, as well as the modern world—the fifth sun.

The identity of the central figure—the Aztec sun god Tonatiuh, the earth deity Tlaltecuhtli, or some hybrid entity—has intrigued and confounded historians and archeologists for centuries. For much of the last two centuries most scholars have concluded that it is Tonatiuh. However, research published in the 1970s suggested that the face included particular aspects that were not characteristic of other renditions of the masculine Tonatiuh, who was considered a daytime sun god, but of a night sun deity, the feminine earth god Tlaltecuhtli, or some combination of gods.[44] Curiously, this figure is situated within the innermost circle of the stone—comprising the central portion of the 4 Ollin glyph. In fact, it is not difficult to imagine this godhead within a circle as representing the body of an animal, with the glyphs of the four previous worlds, or suns, representing the animal's legs. Even the two curvilinear shapes to the left and right of the central figure appear as the head and tail of this quadruped—your choice as to which end is which. Finally, we see an arrow point located immediately above the godhead, and what appears to be a set of two quincunxes and five greenstone "preciousness" beads below the head. Those two carvings can easily be envisioned to represent a perforator that sliced through the neckline of the central figure and, indeed, six or possibly seven disks located along the neckline of the godhead have been interpreted as drops of blood.[45]

My interpretation of these symbols is that they represent an Aztec god of the fifth world surrounded by the four previous suns and forming the body of a turtle. This turtle is Earth, the fifth world. The fifth and current sun is shown in sacrifice. Earthly sacrifices are represented by the organs being grasped at both ends of the turtle. Looking outward from Earth we see the day-signs representing time, the numerous quincunxes symbolizing the four cardinal directions, the *axis mundi*, and the sky supported above Earth. The eight arrow points extend outward, perforating the cosmos filled with feathers (plumes), serpents, stars, and fire. The fire serpents wrap around the universe not once, but many times, as evidenced by the multilayered serpent bodies that can be seen at the top of the stone. These two Xiuhtecuhtli are the gods of the sun and of fire, encircling a universe filled with a multitude of stars and beneficent planets of which Venus is paid the most homage.

We have seen a variation of this imagery before, from another time and another place. Recall that it is the Hindu deity Shiva who as Nataraja dances at the end of each cosmic age. He transforms the world by removing maya. The transformation necessitates destruction as an act of cleansing and creation. He is surrounded by rings of water and cosmic flame, the universal consciousness. The flame of Shiva is the great blackness, Time. Shiva is eternity and wisdom, the dark and the light, the destroyer and creator. He is emancipated, liberated, and self-preserving. He is the world, the Milky Way, and the universe. He may dance upon the demon of ignorance and destroy in a terrible and avenging manner when it is necessary to preserve the soul of the universe. He represents unity within eternity—the cycle of life, death, and rebirth. The Aztecs knew this to be true as well. The central figure of the Calendar Stone is the god of both Sun (sky) and Earth, with both male and female qualities, like Shiva, the creator and the created, the destroyer and the destroyed, sacrificed so that the essence of the universe can be self-preserved.

There is another important similarity between the symbolism of the Aztec Calendar Stone and statuary associated with Shiva, not least of which is the fourfold symmetry found in both. This symmetry appears to be a fundamental effect of our conception of the universe. But more importantly, we should not constrain ourselves to the two-dimensional graphic symbolism found in the stone and statue. If the central figure of the Calendar Stone and of Nataraja is perceived as surrounded by fire and other elements of the universe, then it is not within the

Plate A: Stone tools with circular and spherical forms related to hominid activities have been dated to the early Paleolithic, over 1.5 million years ago. How, why, and when the circle became symbolic of sacred relationships remains unknown. *(Photo by Paul D. Burley)*

Plate B: Cloud Peak Medicine Wheel, Big Horn County, Wyoming. This structure and the surrounding environment are located in the Cloud Peak Wilderness area. The site is sacred not only to the Lakota, but to all Native Americans. It is the source of the author's journey to understand the meaning of this universal symbol, and the site became sacred for the author as well. All people encountering the site should respect the medicine wheel and associated features of the area. The site remains in use for ritual purposes and should be left undisturbed. Help protect and promote the sanctity of all sacred sites. This is everyone's responsibility. Remember: Take only pictures. Leave only footprints. *(Photo by Paul D. Burley)*

Plate C :Dawn at Bear Butte, South Dakota. Bear Butte lies at the center of the region occupied by the Lakota until encroachment by Euro-Americans in the mid-nineteenth century. Traditional Lakota territory is symbolized by the Sacred Hoop envisioned at Earth scale, defined in mythology, and lived by the people. All the land you see is sacred—it is Mother and Grandmother. *(Photo by Paul D. Burley)*

Plate D: The buffalo symbolized the bounty that Mother Earth provided to the Lakota when the people lived on the Sacred Hoop in their traditional ways. Bison represent strength and fortitude, yet they would give up their lives for the Lakota, who would make beneficial use of every part of a buffalo's body. This was a part of the cycle of life. *(Photo by Paul D. Burley)*

Plate E: Wells Cathedral at Wells, Somerset, England. Construction of the cathedral began in about 1175 and was completed in 1495. The structure itself required eighty years to construct, with the grand west front (shown) being the last portion to be completed. It is the first English cathedral built in the Gothic architectural style. Note the bilateral symmetry of the west front including the Vesica Piscis-shaped central entrance. The twin towers are reminiscent of the twin pillars of Jachin and Boaz, allowing a view of the main tower toward the east, the direction of sunrise during the solstices. *(Photo by Paul D. Burley)*

Plate F: View looking east from the nave toward the sanctuary and transept in Wells Cathedral. The scissor arches at the center were constructed in the mid-fourteenth century to prevent collapse of the main tower. Note use of the Vesica Piscis, circle, and two-fold bilateral symmetry in the form of the arches. *(Photo by Paul D. Burley)*

Plate G: View of the tile floor in the apse of Wells Cathedral. Note numerous individual and sets of tiles expressing cubic geometry. Also observe the eight-armed star at the center of the floor, beneath the table. The same geometrical form is sacred in Lakota tradition—Wakan Tanka, the Great Mystery, gave stars the ability to help and bless the people. The eight-armed star representing the Morning Star (Venus) gives name to the star quilt often made for newborns who may then be given the understanding between darkness and light, ignorance and knowledge. Other Native American tribes refer to the design as "God's Eye." The purpose for placing this universal symbol in the apse at the east end of the cathedral is obvious. *(Photo by Paul D. Burley)*

Plate H: This is a view of the vaulted ceiling immediately prior to crossing the threshold between the nave at Wells Cathedral and the south cloister hallway. Note the array of eight arches connected to the uppermost feature, a circle of stone masonry within which a wheel engraved in wood. The wheel includes a rim, eight spokes, and a circular hub. The circular masonry gives the impression of Sun shining down upon Earth. *(Photo by Paul D. Burley)*

Plate I: View of the fluted arches on the ceiling over the organ at Wells Cathedral. Note the four-fold symmetry. Each quadrant includes nine ribs, for a total of thirty-six ribs arching across the sky. Envisioning the scene as a cosmic dome, there would be another thirty-six ribs below the horizon. The total number of ribs, then, would be seventy-two. This is the same number as the curvilinear edges of the disdyakis dodecasphere. Organ pipes are apparent in the lower portion of the view. *(Photo by Paul D. Burley)*

Plate J: The ceiling in the choir of Wells Cathedral. Multiple use of four-fold symmetry is obvious. Note, too, the masonry ribs arching across the ceiling to create six-armed crossing points, and the numerous hexagons created by the basic pattern. *(Photo by Paul D. Burley)*

Plate K: View of the pattern of masonry arches found in the cloister hallways at Wells Cathedral. Note the central rib extending across the center of the photograph. From left to right the central rib encounters an eight-fold intersection, then six, then four, six, and eight again. This pattern continues along the length of the cloisters. Note, too, the octagons each centered by the eight-fold crossing points, and that the ribs form triangles across the ceiling. In fact, this pattern exactly matches the pattern of the DD and DDS. It is as if the pattern of vertices, edges, and triangles across the length of the cloisters was created by rolling the DDS along the ceiling, producing a two-dimension version of the pattern. *(Photo by Paul D. Burley)*

Plate L: This is the pattern engraved on the seat of a wooden chair located in St. Mary's Church at Cerne Abbas, Dorset, England. The church building dates to the early-fourteenth century. However the chair is undated. Note the four-fold symmetry of the central flower pattern at the center of the seat, followed by a circle, an eight-rayed star pattern, another circle, a circular sinusoidal pattern including a total of nine wavelengths, another circle, and lastly a pattern of scrollwork with bilateral symmetry. *(Photo by Paul D. Burley)*

Plate M: This etching is located on the north wall of the apse at St. Mary's Church at Cerne Abbas, Dorset. The etching includes a red circle surrounding the central figure of the Seed of Life. The figure is not dated, and the specific purpose of it at this location is not known. Within Judeo-Christian traditions the Seed of Life is a symbol of the seven days of creation, and protection or preservation. *(Photo by Paul D. Burley)*

Plate N: View of the south entrance to the Lady Chapel at Glastonbury Abbey, Glastonbury, Somerset, England. Lady Chapel was the first of a series of structures constructed beginning in the twelfth century. The buildings replaced the previous monastery that dated from the 900s CE but was destroyed by fire in 1184. The chapel was completed in the fully Romanesque style prevalent before Gothic architecture took hold. Note the semi-circular arches over the clerestory windows and the entrance situated between two structural pillars. Also note the pattern of semi-circular arches forming a series of Vesica Piscis extending along the wall above the entryway. *(Photo by Paul D. Burley)*

Plate O: The museum at Glastonbury Abbey exhibits numerous artifacts from the Great Church and associated structures that were destroyed in 1539 under order by Henry VIII. This photograph shows detail of a carved human figure from the abbey. Note the lozenge or diamond-shaped pattern along the belt or band-like feature. The pattern is similar to the *canamayte* ("snake square") pattern of ancient Mayan tradition. *(Photo by Paul D. Burley)*

Plate P(a): Ornate masonry on display in the museum at Glastonbury Abbey. The five-fold rotational symmetry of the rose is bounded by a curvilinear octagon. *(Photo by Paul D. Burley)*

Plate P(b): Another masonry artifact in Glastonbury Abbey museum. A circular form at left and another circle at right that appears to be a Catherine Wheel (eight spokes with central hub and peripheral rim). They may be symbols of Moon and Sun. *(Photo by Paul D. Burley)*

Plate Q(a): A stone-carved cross pattée from the former Glastonbury Abbey and now displayed at the abbey's museum. The cross pattée has been a sacred symbol in many traditions including the Knights Templar, the Church of England, British royalty, and Freemasonry. *(Photo by Paul D. Burley)*

Plate Q(b): Tiles from Glastonbury Abbey. Note the four-fold (almost eight-fold) rotational symmetry created by the combination of four identical tiles. Artifact displayed ay Glastonbury Abbey museum. *(Photo by Paul D. Burley)*

Plate R: View looking south toward the Abbott's kitchen at Glastonbury Abbey. The stone structure is octagonal. *(Photo by Paul D. Burley)*

Plate S: View inside the Abbott's kitchen at Glastonbury Abbey, looking up toward the center of the vaulted ceiling. Note the eight masonry ribs supporting the roof, the octagonal cupola with sixteen windows providing light to the interior, and the circular masonry at the center that supports the central tower of the structure. *(Photo by Paul D. Burley)*

Plate T: The open cover of Chalice Well at Glastonbury, Somerset. Note the Vesica Piscis on the cover, while the grate below is a pattern of squares. Chalice well is associated with a mineral spring containing ferrous (iron) salts causing the water to exhibit a reddish hue. Joseph of Arimathea is said to have built the first Christian church at Glastonbury, and to have buried or washed the chalice Jesus used at the Last Supper. The spring was an important source of water for Glastonbury Abbey.

Plate U: View of north-side masonry wall supporting rail lines along the River Avon at Bath, Somerset, England. The elevated rail line dates from the nineteenth century. While the wall serves a purely secular purpose as a part of the infrastructure for the city, there are several aspects of the architecture related to sacred symbolism. Note the central doorway and twin towers, the eight-armed window design within a circle, and the mirror symmetry. These attributes are common to entrances into temples, while the current setting is associated with a gateway to the city center. *(Photo by Paul D. Burley)*

Plates V(a) (left) and V(b) (right): Views of the front entrance to the Masonic Temple in Virginia, Minnesota (left) and Resurrection Catholic Church in Eveleth, Minnesota (right). A scan of each structure yields numerous commonalities in these buildings. Note mirror symmetry, the crosses, the circular windows each located along the line of symmetry, and each main entrance flanked by pillars. The temple has a rounded form, while the white limestone and window works at and above the doors, and along the upper portion of the building, are suggestive of the ancient symbol tau (in the form of T) representing a sacred gateway. Note that the main tower of the church is flanked by two smaller towers and the windows on each side direct the eye up toward the circular window which is divided into quadrants, a symbol of a consecrated church. This is the same symbol as the Native American Sacred Hoop, Odin's cross representing the sun, and a common symbol for Earth as well. *(Photo by Paul D. Burley)*

Plates W(a) and W(b): View of B'nai Abraham located in Virginia, Minnesota. The synagogue was built in 1909 and is the only synagogue in Minnesota on the National Register of Historic Places. The building was renovated recently, evident by new concrete steps and entranceway apparent in view (a). Note in view (b) how the architecture of the north side of the building emulates the geometry of the Tree of Life of Kabbalah. The Sefirah located at the lower end of the Tree of Life is Malkhut—the earthly kingdom, the Higher Mother—and, indeed, we see it situated here in the earth beneath the synagogue. Malkhut is seen as the fertile ground in which the nine other Sefirot are rooted. *(Photo by Paul D. Burley)*

Plate X: A main entrance into Trinity Lutheran Church in Virginia, Minnesota. The octagonal window over the triple set of doors is striking; also note the octagonal vent that mimics the form of the window below. Again, mirror symmetry is evident from the roof line, through the brickwork and lighting details, to the landscape plan adjoining the entrance. *(Photo by Paul D. Burley)*

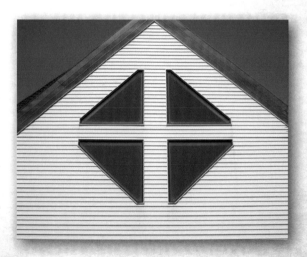

Plate Y: View of the south window above the sanctuary at United in Christ Lutheran Church in Eveleth, Minnesota. While the architecture is contemporary, the symbolism remains united with ancient and indigenous traditions, including the mirror symmetry of the roofline and window placement, and the four-fold symmetry of the window itself, which creates the form of a cross within its midst. *(Photo by Paul D. Burley)*

two-dimensional plane that this is so, but also in the third dimension of space and the fourth dimension of time, as well. And so the concentric circles and glyphs found on the Piedra del Sol represent spheres of space and time that extend across the universe. Similarly, Nataraja dances within the sphere of our universe, each of these universes conceived as our galaxy, the Milky Way. The source of destruction, rebirth, and life is derived from the center. This center sacrifices itself for the sake of the universe and for us. It is a sacrifice that repeats itself through time and has been recognized by an array of cultures over thousands of years.

Life Cycles at Palenque

Adrian Gilbert's thesis is that "the behavior of [the Maya] star- and calendar-based culture was governed by one principle: Events repeat themselves."[46] At first thought this statement seems prosaic, another rendition of "history tends to repeat itself," or Georgio de Santayana's "Those who cannot remember the past are condemned to repeat it."[47] However, Gilbert goes on to explain the belief held by the Maya, Aztecs, and other similar cultures that nature operated in terms of a *cyclical* progression through time, as evidenced in the calendars discussed above. Gilbert describes a surprising geometry that shows up time and again in ancient cultures around the world. It is not a matter of people condemned to repeat a forgotten history, but rather cultures at different times and at different places around the world apparently remembering the same thing, expressing the same concept, and symbolizing this *thing* in the very same way. In other words, humans tend to symbolize shared, vital information in the same way.

An excellent example of this tendency is shown in the symmetry found in Babylonian depictions of the Tree of Life and two engravings of Mayan kings and deities at Palenque. Scholars believe that the Babylonian Tree of Life, a magical tree that grew in the middle of paradise, represented sustenance provided by the gods. The Babylonians typically illustrated the Tree as having a trunk from which a number of branches extended straight up and to the left and right. At the end of each limb was a leaf, flower, or cone. The Tree was tended by gods and goddesses who collected its fruits. In fact, this form of the Tree and illustrations showing two gods tending to it goes back to the thirteenth century BCE in

Urartu—an Iron-Age kingdom centered in the Armenian highlands, where the symbol appeared on fortress walls.

Returning now to the artwork of the Maya, an extraordinary discovery was made in 1952 by the Mexican archeologist Alberto Ruz. It was Ruz who found the burial chamber and carved-stone sarcophagus of Pacal I inside the Pyramid of Inscriptions at Palenque (Figure 10.17). Gilbert notes that the Mayan word *pacal* means "shield," and Pacal I's title, *Makina*, means "Great Sun Lord."[48] Ruz described the sarcophagus' lid.

> Richly carved, most of its upper surface depicts what is clearly intended to be a mythological scene. At its center is what appears to be a strangely dressed man at the foot of a symbolic tree. The man looks as though he is falling backward, while the tree has the appearance of a Roman cross, its crossbeams terminating in what look like a pair of snake's heads. Draped over the tree/cross is what looks like another serpent, this one with what appears to be dragon heads at either end. Perched on top of the tree is a curious bird with a bifurcated tail.[49]

Figure 10.17: The carved stone sarcophagus of Pacal I inside the Pyramid of Inscriptions at Palenque, depicting Pacal I rising toward the World Tree (photograph by David Pentecost).

Indeed this is the renowned engraving that Erich von Däniken in 1968 proposed was an illustration of a man at the controls of a spacecraft.[50] For Gilbert, however, the scene is one of Pacal I falling into the jaws of the death before the Tree of Life. Specifically, he concludes that the lid shows Lord Pacal, dressed as the corn god, falling into the gate of death; the symbolic tree is the Wakah-Chan ("Raised up Sky," the World Tree), a wooden pole connecting Earth to the celestial sphere; the double-headed serpent represents the Milky Way; and the bird at the top is *Vucub Caquix* ("Seven Macaw"), representing the constellation Aquila (Latin: *Aquilae*, or "eagle") of Western mythology.[51] Gilbert's interpretation of the scene at Palenque (also known as Nachan, which he interprets as "sky city") is that it shows Pacal about to die while before him stands the World Tree extending up to the zenith; the Tree is the *axis mundi* terminating at the *Vucub Caquix*, which is Aquila; the double-headed serpent represents the Milky Way and more specifically the area of the constellations of Sagittarius and Scorpio. And as Gilbert shows from Mayan mythology contained in the Popol Vuh narratives, in tandem with Western mythology, Aquila was wounded by a blow dart or an arrow and fell from the zenith point in about 3114 BCE. By the sixth century CE, when Pacal I lived, the zenith point would have been located at the Pleiades. It was the Pleiades where the soul of a good Mayan ruler was believed to be taken to the afterworld, Xibalba.

The components of the symbolism suggested by Gilbert are reasonable, but I believe he missed the mark as it concerns Pacal I. Looking at the carving again, it is very apparent that Lord Pacal is not sprawled out as if falling into the clutches of death, but rather he is hunched forward and looking up with intention while his hands grasp the World Tree. It appears to me that he has indeed already passed from our world and is shown being transported *upward* along the Tree to the zenith point. Thus, the sarcophagus lid illustrates honor bestowed on Pacal I by Mayan belief that his soul would be taken to Xibalba in the Pleiades.

This World Tree takes on additional symbolic importance when it is shown in other reliefs at Palenque, including a tablet at the Temple of the Cross and a plaque at the Temple of the Sun (see Figures 10.18 and 10.19). Gilbert notes that the symbolic meaning of those carvings remains unresolved by scholars, but that most Mayanologists accept Linda Schele's theory that Pacal I's grandson Chan-Bahlum ("snake jaguar") had them constructed "as part of a propaganda campaign intended to bolster his position as *ahau*, or king."[52]

Figure 10.18: The World Tree symbolized in a tablet at the Temple of the Cross at Palenque. The central figure is the World Tree with the cross branch of the ecliptic and the zenith *Vucub Caquix.* Drawing by Linda Schele © David Schele, courtesy Foundation for the Advancement of Mesoamerican Studies, Inc., www.famsi.org.

Figure 10.19: The World Tree symbolized in a plaque at the Temple of the Sun at Palenque. The central figure is GIII, a sun god. by Linda Schele © David Schele, courtesy Foundation for the Advancement of Mesoamerican Studies, Inc., www.famsi.org.

Whether or not this perceived propaganda campaign was real is not our concern here. However, what we see in these two reliefs is a symmetry we have seen before. The central figure of the tablet in the adytum of the Temple of the Cross is the World Tree, the cross branch of the ecliptic, and the zenith *Vucub Caquix*. The Tree is flanked by Pacal I to the left and Chan-Bahlum to the right, each appearing to petition Xibalba. This mythologic and cosmographic symbolism mirrors the Babylonian representations of the Tree of Life with its two attendants. Similarly, the plaque located at the Temple of the Sun shows Pacal I on the left and Chan-Bahlum on the right. Again, these two kings appear to be petitioning the central figure. Gilbert describes the relief as he viewed it in the temple:

> Instead, at the center of the picture was a war shield bearing the face of a god. Schele identifies this god as "GIII," one of the Palenque trio and generally regarded as a sun god. Behind the shield, like arms displayed on the walls of some Scottish baronial castle, were two crossed spears with flint heads. Underneath was a relief of a throne supported on the heads of two aged gods. They were seated cross-legged, their heads bowed forward, and one of them was clearly *L*, or *Itzamna*.[53] [Itzamna is the Smoking God believed to be associated with Xibalba and who ruled over the first day of creation in the present Mayan age, which scholars calculate to have begun August 12, 3114 BCE, ending on December 21, 2012.]

The World Tree is missing in this scene. Or is it? The plaque includes the same symmetry as the tablet at the Temple of the Cross. If fact, this is the same symmetry as that present in Babylonian art depicting the Tree of Life with attendants. It is the same as that of the ancient Egyptians in their representation of the two pillars of Jachin and Boaz that bound the sun on its diurnal journey at dawn at the vernal and autumnal equinoxes. Recall that the central figure of the Piedra del Sol is the Aztec sun god/earth god of the fifth world. An arrowhead-like perforator crosses through this god, while additional perforators extend toward the cosmos in eight directions. Two Xiuhtecuhtli, gods of the sun and fire, encircle the stone, holding together this depiction of universal self-sacrifice. In this scene the World Tree is symbolized by the Aztec sun/earth god looking upward and outward, making a connection with the cosmos.

Look again at the Mayan plaque from the Temple of the Sun—the central figure represents GIII, a sun god; two spears cross through this god, supported by a platform bounded on four sides by glyphs and the bodies of two gods, one of them Itzamna. The two attendants of the sun god are Pacal I and Chan-Bahlum. No, the World Tree is not missing here. It is represented by the sun god who brings life itself. He is petitioned by kings and sacrificed in the name of the world. Indeed, the plaque from the Temple of the Sun and the Piedra del Sol illustrate similar beliefs, similar knowledge common to two non-contemporaneous cultures. Even the symmetries are similar, if not the same, if we envision the plaque from above. If we, the viewers, are the third attendant in this picture, who might be the fourth, located at the other side of the sun god?

As will be shown in the next chapter, this same symmetry is found in biblical visions expressed in ancient Judaic symbolism from the sixth century BC, Solomon's Temple constructed at Jerusalem in the third century BC, and the geometric construct of Kabbalah's Tree of Life, which continues to be studied by mystics to this very day.

Chapter 11
Judaism, Kabbalah, and the Tree

...but we should also recall the quite metaphysical significance that Plato accords to the study of geometry—through geometry the highest earthly being is perceived, and true being, i.e. divine being, is betokened.

THORLEIF BOMAN[A]
HEBREW THOUGHT COMPARED WITH GREEK

Spiritual Energy

The word *Kabbalah* is derived from the Hebraic root word *kibel* (or *qibel*, QBL) meaning "to receive," or "given," used in the context of communicating tradition. By tradition we receive knowledge about our family, culture, customs, and values. People in all cultures and societies receive and accept this knowledge as it has been told to us through the generations, and so it will be passed on for generations to come. Tim Dedopulos describes Kabbalah as "the first and ultimate knowledge, greater than any master. It takes the form of a large body of teachings—written, oral, and practical—about the nature of God himself, the birth of the soul, the process of the creation of the universe, the purpose of life, and what happens afterward."[1]

271

However, Kabbalah is not just knowledge. Understanding can only be achieved through a specific and highly disciplined lifestyle. For the religious Jew, it is living the life of Torah and making a connection between the physical reality and intense spirituality. Still, a person must receive a blessing from the creator of the mystical powers necessary to ascend beyond physical reality. Through a lifetime of study, the aspirant achieves that sense of wide-eyed wonderment and understands the responsibility of living a spiritual life.

Kabbalah provides a structure for building relationships with the creator and all of creation. Its ancient wisdom explains how spiritual energy moves through the cosmos. This energy is the Great Mystery of the North American Indian, the Mysterious Unknown of the Zohar, and as we shall see later, the energy that drives R. Buckminster Fuller's "miraculous web of interacting patterns."[2] This energy, and the patterns it produces, manifest in the universal symbolism we have seen so far, leading us to an understanding of the true meaning of the circle, and the geometry of the Sacred Sphere.

Secret Science

As historical background for understanding the relationship between Kabbalah and its associated geometrical symbolism, we need to briefly review the development of written Jewish laws and traditions over the past two thousand years. This history is varied and complex. We will limit our review to those concepts that help lead us to understanding the Tree of Life as presented in Kabbalah.

Derived from the Hebrew root *l-m-d* ("study" or "teach"), the word *Talmud* generally refers to the authoritative body of Jewish law and traditional beliefs. The Talmudic period lasted from about 200 BCE to 500 CE. There are two Talmuds, the Jerusalem (or Palestinian) Talmud and the Babylonian Talmud. For many centuries the Jerusalem Talmud remained in the shadows of the favored Babylonian Talmud. Jewish law and traditional beliefs, like the knowledge passed down by prehistoric cultures around the world, were communicated orally. The biblical Oral Law, given to Moses on Mount Sinai, was written in the Mishnah by about 200 CE. *Gemara*, the rabbinical discussions of the Mishnah and written down during the latter part of the Talmudic period, became the name used for the Babylonian Talmud. So the Talmudic period is divided into two: the Mishnah and the Gemara.

Jewish law is based on a hierarchy of sources in which older sources such as the Bible have greater authority than laws provided in the Mishnah. Similarly, laws in the Mishnah have greater authority than those in the Gemara. Since the Mishnah addresses conclusions of the aforementioned rabbinical discussions, it generally does not include the arguments, proofs, and debates that upheld biblical texts. However, the traditional method of learning the Oral Law is preserved in rabbinic interpretations of traditional (biblical) legal texts. There are a number of interpretations and critical assumptions within the Kabbalah that were developed by various teachers at the Merkaveh mystical schools. Thus the Kabbalah is complex, and an appreciation of earlier Merkaveh thought is important to understanding its deeper meanings.

The word *merkaba* may derive from the ancient Egyptian *mer ka ba*, "beloved spirit and soul." This phrase is an expression of oneness, the energy of the spirit and soul of being, whether of a god, human, or aspects of the animal kingdom personified in the pantheon of Egyptian gods. In Hebrew we find the following definitions of *Merkaveh* (*Merkabah*, *Merkaba*, *Mercaba*):

1. The Rakefet Dictionary defines Merkabah as "A Chariot, vehicle; used in two senses: first, as a chariot, the Qabbalists saying that the Supreme forms and then uses the ten Sephiroth as a chariot for descending through the various worlds enumerated in the Qabbalah. These worlds are the ten Sephiroth themselves. "Adam Qadmon" (the Heavenly Man) is the same as the ten Sephiroth considered as a hierarchic entity permeated by and inspirited by the divine hierarch or Supreme. Here it is generally equivalent to the Sanskrit vahana."[3]

2. Secret wisdom or knowledge: ". . . without the final initiation into the *Mercaba* the study of the *Kabala* will be ever incomplete, and the *Mercaba* can be taught only in 'darkness, in a deserted place, and after many and terrific trials.' Since the death of Simeon Ben-Iochai this hidden doctrine has remained an inviolate secret for the outside world. Delivered *only as a mystery*, it was communicated to the candidate orally, *'face to face and mouth to ear.'*"[4] Rabbi Laibl Wolf says, "The secret wisdom or knowledge is envisaged as a vehicle or chariot

because what men call esoteric wisdom is the vehicle for the communication to human consciousness of the mysteries of the universe, and consequently of man."[5]

The first of these two definitions includes the word Sephiroth in regard to a divine hierarchy of sacred symbols. This hierarchy and the use of *Sephiroth* in a specific geometrical construct as means to define a perceived relationship between humans and the Creator will be detailed later in this chapter. Sephiroth (or *Sefirot*, Hebrew: plural of *Sephirah* or *Sefirah*) are emanations that proceed from God—the One Source or the Boundless Light.

These definitions of the Hebraic word *Merkaveh* refer to secret knowledge communicated to humans (although not necessarily all people) regarding mysteries of the universe. This knowledge is referred to as a chariot in that the wisdom is transferred either by tradition or divine means. Therefore, from the apparent Egyptian derivation of the word that knowledge is associated with universal energy of the spirit and the soul. Study of this esoteric wisdom was the focus of the Merkaveh schools and the science of Kabbalah.

As a science, Kabbalah is divided into three topics of study: theoretical, meditative, and practical. It is the structure developed from theoretical concepts that allows for proper understanding of the meditative and practical aspects of Kabbalah. The Otz Chiim, or Tree of Life, provides the theoretical framework for kabbalistic mystical teachings. It is thought that the structure it exhibits expresses how God created the universe. All understanding can come from this structure. Through our understanding of the Tree of Life we discover the key to perfecting ourselves and mastering the world. Our souls are born from this understanding.

Eastern meditative practices emphasize proper breathing. Rabbi Laibl Wolf writes of a significant correlation between these techniques and the anthropomorphic association of breath with creation as outlined in the Hebrew Bible: "And He breathed the living soul into his [Adam's] nose." The Hebrew word for "breath" is *neshima*, while the word for "soul" is *neshama*. The connection is obvious. Focusing on breath is focusing on the life force itself. The Hassidic masters of Kabbalah offered the meditational imagery of God's investiture in the world as the process of *memalleh*—the *filling* of the world with Godly presence, just like breath filling the lungs.

Recalling that Ezekiel experienced his vision of the Merkaveh while captive in Babylonia, it is interesting to note that the Sumerian word *melammu* means "glory" or "radiance," something within which the goddess Innana enveloped herself. Is it possible that Ezekiel experienced *memalleh* as he envisioned the *melammu* of God? Might these words have complementary derivations?

Similarly, both kabbalistic teachings and Eastern meditation promote an emptying of the mind. From the opening of Genesis we learn that, "In the beginning God created the heaven and the earth." Kabbalah teaches that the world was derived from a void, an emptiness known as *halal* that existed before creation. By emptying the mind we can begin a process of contemplative filling (*memalleh*). For some people, insight into universal reality is achieved through an understanding of the biology and mechanics of breathing in tandem with learning to empty the mind of the mundane concerns of life. Just as the aspirant practicing Eastern meditation learns breathing techniques to achieve a proper state of stillness and creative thought, so the metaphor of breath is apparent in the Torah's account of creation. The Creator formed Adam's physical body and then *breathed* life into him. Rabbi Laibl states Kabbalah emphasizes God's breath as the life force that sustains existence.

> The "cosmic breath" is referred to as the process of memalleh, thereby animating it. As the Creator meets Its creatures face-to-face, Its breath passes between them by way of a "[cosmic] kiss" . . . The Creator and Its creations share the "cosmic breath." Breathing consciously, with awareness, can become a powerful tool for reclaiming inner balance and personal equilibrium. It can also induce altered states of consciousness, which facilitate deeper self-realization and imprint behavioral transformation.[6]

Ezekiel's Vision of the Merkaveh

The biblical text of Ezekiel's vision of God's chariot is most important to Jewish mysticism—Merkaveh Mysticism—so named for the chariot that he rode to the heavens. As such, it is the primary basis upon which much kabbalistic thought is developed. The prophet Ezekiel lived during the sixth century BCE. Biblical history tells us that he was a priest in the temple at Jerusalem before its destruction, when King Nebuchadnezzar's army exiled thousands of Israelites from their homeland to the kingdom of Babylon. Ezekiel inspired the Merkaveh mystic tradition by his vision during that exile. Aspirants of that tradition sought to recreate similar visionary experiences including ascension into God's realm. Specifically, Ezekiel's importance to the Kabbalah is his vision of the Merkaveh, the chariot-throne of God (Figure 11.1).

Figure 11.1: The Merkaveh, the chariot-throne of God, envisioned by Ezekiel. Engraving of "Ezekiel's Vision" by Veit Dietrich, 1506-1549. Courtesy of the Richard C. Kessler Reformation Collection, Pitts Theology Library, Candler School of Theology, Emory University.

Ezekiel 1:1–28

Now it came to pass in the thirtieth year, in the fourth month, in the fifth day of the month, as I was among the captives by the river Chebar that the heavens opened, and I saw visions of God. In the fifth day of the month, which was the fifth year of king Jehoiachin's captivity, the word of the Lord came expressly unto Ezekiel the priest, the son of Buzi, in the land of the Chaldeans by the river Chebar. And the hand of the Lord was there upon him.

And I looked, and, behold, a stormy wind came out of the north, a great cloud, with a fire flashing up, so that a brightness was round about it; and out of the midst as the color of electrum, out of the midst of the fire. And out of the midst came the likeness of four living creatures. And this was their appearance: they had the likeness of a man. And every one had four faces, and every one of them had four wings. And their feet were straight feet; and the sole of their feet was like the sole of a calf's foot; and they sparkled like the color of burnished brass. And they had the hands of a man under their wings on their four sides; and as for the faces and wings of them four. Their wings were joined one to another; they turned not when they went; they went every one straight forward.

As for the likeness of their faces, they had the face of a man; and they four had the face of a lion on the right side; and they four had the face of an ox on the left side; they four had also the face of an eagle. Thus were their faces; and their wings were stretched upward; two wings of every one were joined one to another, and two covered their bodies. And they went every one straight forward; whither the spirit was to go, they went; they turned not when they went. As for the likeness of the living creatures, their appearance was like coals of fire, burning like the appearance of torches; it flashed up and down among the living creatures; and there was brightness to the fire, and out of the fire went forth lightning. And the living creatures ran and returned as the appearance of a flash of lightning.

Now as I beheld the living creatures, behold one wheel at the bottom hard by the living creatures, at the four faces thereof. The appearance of the wheels and their work was like unto the color of a beryl; and they four had one likeness; and their appearance and their work was as it were a wheel within a wheel. When they went, they went toward their four sides; they turned not when they went.

As for their rings, they were high and they were dreadful; and they four had their rings full of eyes round about. And when the living creatures went, the wheels went hard by them; and when the living creatures were lifted up from the bottom, the wheels were lifted up. Whithersoever the spirit was to go, as the spirit was to go thither, so they went; and the wheels were lifted up beside them; for the spirit of the living creature was in the wheels. When those went, these went, and when those stood, these stood; and when those were lifted up from the earth, the wheels were lifted up beside them; for the spirit of the living creature was in the wheels.

And over the heads of the living creatures there was the likeness of a firmament, like the color of the terrible ice, stretched forth over their heads above. And under the firmament were their wings conformable the one to the other; this one of them had two which covered, and that one of them had two which covered, their bodies. And when they went, I heard the noise of their wings like the noise of great waters, like the voice of the Almighty, a noise of tumult like the noise of a host; when they stood, they let down their wings. For, when there was a voice above the firmament that was over their heads, as they stood, they let down their wings.

And above the firmament that was over their heads was the likeness of a throne, as the appearance of a sapphire stone; and upon the likeness of the throne was a likeness as the appearance of a man upon it above. And I saw as the color of electrum, as the appearance of fire round about enclosing it, from the appearance of his loins and upward; and from the appearance of his loins and downward I saw as it were the appearance of fire, and there was brightness round about him. As the appearance of the bow that is in the cloud in the day of rain, so was the appearance of the brightness round about.

This was the appearance of the likeness of the glory of the lord. And when I saw it, I fell upon my face, and I heard a voice of one that spoke.

The Chariot Revealed

Over the last 2,500 years, interpretations of Ezekiel's vision have been many and varied. Each of the numerous illustrations of the Merkaveh leave little to the imagination with regard to the appearance and functioning of the literal chariot, four creatures, wheels, and occupancy of the airborne craft. Even heat of fire surrounding the Merkeveh and "noise of tumult" are suggested in the artwork. However, for our purpose we will limit our analysis of the Merkaveh to the general character of the chariot and its wheels as Ezekiel's description qualifies them. As before, we take to heart the axiom the simpler, the better. We don't want our imaginations to run wild and imprint the information given with visions of our own, while at the same time, certainly, there is room for interpretation due to the limited description offered. I suggest an alternative interpretation of the vision beheld by Ezekiel.

Note that the Merkaveh arrives on a stormy wind from the north (up) and in the midst of a great cloud. This wind is the "cosmic kiss," *memalleh*, the creator meeting Ezekiel face-to-face, the creator's breath passing between them. There is a fiery brightness "round about" the cloud, suggesting that the brightness appeared to surround the chariot in a spherical form. The chariot was the color of electrum, an alloy of gold and silver with copper and trace amounts of other metals. The ancients knew of this natural alloy, which has a white-gold metallic luster. Ezekiel states that the firmament was located over, and carried by, the wings of the four creatures, and the firmament was like the color of "terrible ice." The likeness of a throne stretched over the heads of the four creatures, and so the throne appeared to be located on the firmament. But what exactly is firmament? Why is the apparent body of the chariot described as a firmament?

The word *firmament* derives from mid-thirteenth-century Latin word *firmamentum*, literally "a support or strengthening," from *firmus*, "firm," which in turn was derived from the Greek *stereoma*, "firm or solid structure," during the fifth-century translation of the Bible. In turn, *stereoma* was translated from the Hebrew *raqia*, a word used for both the vault of the sky and the floor of the earth in the Old Testament; *raqia* probably literally meant "expanse," from *raqa*, "to spread out." But in Syriac it meant "to make firm or solid," and so an erroneous translation resulted.[7] *Raqia* would have been used to denote the space or expanse that is like an arch appearing immediately above us. It would have been regarded as a solid body as well as expansion. *Raqia* was used to describe a division between the waters above and the waters below (Gen. 1:7), it supported the upper reservoir (Ps. 148:4), and it supported the heavenly bodies (Gen. 1:14). *Raqia* is also spoken of as having windows and doors (Gen. 7:11, Isa. 24:18, Mal. 3:10) through which the rain and snow might descend.[8] What this means is that the firmament of Ezekiel's vision appeared as the space below the vault of the sky and over the floor of the earth, between the curving surface of the sky and the plain of the earth.

In the vision we have the firmament as the space bounded by the above and below. What is this space? Ezekiel describes the firmament as like the color of terrible ice (or crystal) stretched forth over the heads of the four creatures. Note that he is not describing color here; rather, he is presenting us with a metaphor: The firmament is *like*, or was as, the color of terrible ice. This is an important difference, and we go back to Hebrew to understand this better. The phrase "was the color of the terrible crystal stretched forth" in Hebrew is "`ayin yare' qerach natah," which would translate as "an eye to fear ice spread out."[8] Now we have a more concrete picture of what Ezekiel saw as the body of the chariot: It was the curved form of heaven over the earth, crystal-like, and appearing as a large eye to be feared.

The eye within the firmament was the likeness of a sapphire blue throne and upon it was the likeness of a man surrounded by a bright luster of electrum behaving as fire. The vision is of a large eye to be feared; looking into the eye is a bright blue canopied seat of a king; and in the center of that location is what Ezekiel sensed was man (Heb.: *Adam*) surrounded by a fiery white-gold luster. Ezekiel also likened the appearance of the area proximal to the chariot to a bright, round rainbow, again suggestive of a spherical symmetry. The rainbow effect indicates that Ezekiel might have envisioned a refraction of light resulting from the bright luster of the Merkaveh and the electrum-colored fire surrounding the likeness of the man upon a throne. This refraction of light indicates the crystal-like nature of the Merkaveh. It is, therefore, reasonable to interpret the refraction of light and the rainbow effect to an icy, crystal-like appearance of the firmament. This is the all-knowing, all-seeing eye of God. This knowledge, that the chariot is the eye of God, is the secret wisdom, Merkaveh, which Ezekiel envisions through the divine. The energy associated with the divine spirit and soul of God is, of course, expressed in Ezekiel's commentary regarding the wind, storm clouds, fire, brightness, and tumultuous noise of the vision. This is illumination. It is the glory of the Lord.

Next, let's look at the information we have for the wheels of the Merkaveh. Ezekiel tells us there are four wheels, and they work (function). The wheels spin. Also, they are like the color of beryl. Beryl is a mineral that can have a creamy white to light green color, but beryl is most associated with green. Other translations of Ezekiel's vision identify the wheels as being composed of chrysolite, the name given to a type of metamorphic rock with a significant amount of olivine, another green mineral. Another possibility, although not in favor in most transcriptions of the vision, is that the wheels had the appearance of chrysotile, the most common of the asbestos-form minerals. Chrysotile is a mineral obtained from serpentinite rocks, so named because the rocks contain mineral formations that resemble snakeskin, with a mottled color, resinous or wax-like luster, and often a curving, polished surface. In fact, chrysotile fibers themselves can have a curly habit (form). Chrysotile was used as a primary material in automobile brake linings before asbestos was banned from such use in much of the world. In the end, although chrysotile might have served as a fire-proof material for the fiery Merkaveh, we can only conclude that the four wheels had the appearance of beryl, chrysolite, or chrysotile.

Ezekiel describes the wheels as looking like a wheel within a wheel, and when the chariot moved they went toward their four sides even though they did not move. This apparent conundrum can be solved by considering the function of the Merkaveh. Essentially, the chariot was moving across the sky, so it would not have been necessary for the wheels to rotate in order for the Merkaveh itself to move. At the same time, we are told that the wheels went toward their four sides when the chariot moved. In other words, the wheels appeared to be reaching out or extending toward the front, back, left, and right of the chariot. Since there was no apparent rotation or spinning of the wheels, we are left with only one wheel configuration that can apply without introducing further complexity to the situation: The wheels were spherical.

With regard to the four creatures, Ezekiel considered them to be likenesses, not the real thing. And so, again, they are symbols. Each has four faces, the hands of a man under their wings on their four sides, straight feet with the sole like the sole of a calf's foot, and the likenesses sparkled like burnished brass. They each had the face of a man, a lion, an ox, and an eagle. They are most often interpreted to be cherubim, celestial beings. And this appears to be so, as the ox, lion, eagle, and man represent the four quadrants of the zodiac—Taurus the bull, Leo the lion, Scorpio the scorpion (also considered to represent the eagle in ancient times), and Aquarius as man. These four constellations at the four corners of the ecliptic plane support the celestial sphere. They are representative of the four world ages that so many cultures have included in their mythologies over the millennia and are indicative of the recognition and importance of the precession of the equinoxes to man's past, present, and future.

Figure 11.2 depicts my interpretation of Ezekiel's vision. It is a representation of the precession of the equinoxes with humans occupying Earth at the center of the diagram. For illustrative purpose the human form is represented here by Leonardo da Vinci's "Vitruvian Man." The firmament is Earth veiled in its sapphire-colored atmosphere and bathed in the light of the sun. Earth is the kingdom of humankind; that we rule the earth is signified by the appearance of spectral fire about the man. A circular path—the ecliptic—surrounds Earth. The ecliptic, of course, is the path of the sun, moon, planets, and constellations of the zodiac as they cross the cosmic dome. The cosmos is filled with stars, like shimmering eyes all about. The course of each heavenly body extending across the ecliptic is like the rim of a great wheel. Together they are like wheels within wheels, and they all move in direction—from east to west across the sky.

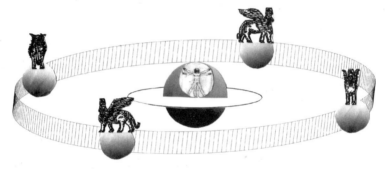

Figure 11.2a: An alternative interpretation of Ezekiel's vision of the chariot-throne of God. Oblique view of the precession of the equinoxes with earth at center and four lamassu rotating through the two equinoctial points where the ecliptic and equatorial plane intersect.

Figure 11.2b: An alternative interpretation of Ezekiel's vision of the chariot-throne of God. Orthogonal view over Earth, the throne occupied by the image of a man whom I represent with Leonardo da Vinci's "Vitruvian Man."

The four creatures are represented by lammasu, the human-headed, winged lion with the tail of a scorpion. Lammasu were guards at the gates of Assyrian temples. Ezekiel could not help but notice them while captive in Babylonia. However, the lammasu also represents four constellations of the zodiac—Taurus the bull, Leo the lion, Scorpio the scorpion, and Aquarius the water bearer. In fact, the Babylonian equivalent of Aquarius is "The Great One" who pours water from his overflowing vases to cause rain to fall from heaven and the rivers to swell, irrigating fields and helping revitalize the Mother Goddess in spring. These four constellations, at the quarter points of the cosmic wheel, are the protectors of Babylonia. When they move, they all move forward, even though they appear to move in a different direction—looking southward we see one moving from our left to right; to the west one falls below the western horizon; behind us another is moving from our right to left beneath the earth; and the last one rises up from the horizon in the east.

For Ezekiel the vision "was the appearance of the likeness of the glory of the lord." It was a recognition of the wonders of the universe from the perspective of humankind on Earth, a vision of power and strength in the eye of God, and the elegance of nature symbolized by precession—*infinitio no appareo*—eternity made manifest.

This symbolism is also shown on an alabaster throne housed at the treasury of St. Mark's Basilica in Venice. The throne was likely carved in Alexandria about two thousand years ago. On the sides of the throne are four carved figures: the bull, lion, eagle, and man. Stephen Huller concludes that, for the person who was seated on that throne, the bull represented the underworld, the lion was the attainment of the Jewish kingdom, the eagle the rising toward heaven, and man as emerging toward his future. [9]

Regarding the importance of Ezekiel's vision Dedopulos states that, "the most critical part of the vision was the implication that the divine presence on Earth was not restricted to the House of God at the temple of Solomon. Instead, God could be seen and felt anywhere and everywhere, for those that had eyes to see . . . This revolutionary implication enlivened scholars and mystics for centuries to come, and kindled the thirsty speculation that gave rise to Merkavah mysticism—taking its name from the chariot of the vision. The mystics Ezekiel inspired literally sought to find out how God could be discovered. This, in turn, gave birth to Kabbalah."[10]

Surprising Correlations

The symbolism in Ezekiel's vision correlates significantly with the Revelation of St. John in the New Testament. While much younger than the Book of Ezekiel, the Revelation text lends credence to the picture we based on our review of Ezekiel's experience. Specifically, note the similarities that follow from Revelation.

Revelation 1:10–16; 4:6–8

> I was in the Spirit on the Lord's day, and heard behind me a great voice, as of a trumpet, saying, I am Alpha and Omega, the first and the last . . . And I turned to see the voice that spake with me . . . and in the midst of the seven candlesticks one like unto the Son of man, clothed with a garment down to the foot, and girt about the paps with a golden girdle. His head and his hairs were white like wool, as white as snow; and his eyes were as a flame of fire; and his feet like unto fine brass, as if they burned in a furnace; and his voice as the sound of many waters. And he had in his right hand seven stars: and out of his mouth went a sharp two-edged sword: and his countenance was as the sun shineth in his strength. And before the throne there was a sea of glass like unto crystal: and in the midst of the throne, and round about the throne, were four beasts full of eyes before and behind. And the first beast was like a lion, and the second beast like a calf, and the third beast had a face as a man, and the fourth beast was like a flying eagle. And the four beasts had each of them six wings about him; and they were full of eyes within. . . .

Also, note the following statements from Revelation corresponding to the mythologies we looked at from ancient Babylonia.[11]

Revelation 22:1–2; 2:7

> And he shewed me a pure river of water of life, clear as crystal, proceeding out of the throne of God and of the Lamb. In the midst of the street of it, and on either side of the river, was there the tree of life, which bare twelve manner of fruits, and yielded her fruit every month: and the leaves of the tree were for the healing of the nations.

> To him who overcomes, I will give the right to eat from the tree of life, which is in the paradise of God.[12]

Most surprisingly, note the similarities between the Merkaveh in Ezekiel's vision and the scene depicted in the Mayan plaque from the Temple of the Sun at Palenque!

- Ezekiel saw in the midst of the cloud the color of electrum; the plaque shows the central figure of the sun god.

- Ezekiel saw the likeness of four living creatures and four wheels; the plaque shows gods squatting, heads bowed.

- Ezekiel saw high, dreadful wheel rims; the plaque shows glyphs extending up and outward from the locations above the gods.

- Ezekiel saw a firmament supporting a throne and the likeness of a man; the plaque shows the sun god and spears supported on a platform.

- Ezekiel saw that wherever the likeness of the four living creatures went, the spirit of the living creature was in the wheels; the plaque shows the platform carried by gods of symbolic importance, wherever the platform went, these gods would be the means of transport.

- Ezekiel said the appearance of the Merkaveh was like a bright rain(bow) brightness round (spherically) about the chariot; the plaque shows the sun god carried by a chariot of gods.

- Ezekiel's vision was the appearance of the likeness of the glory of the Lord; the plaque shows the appearance of the likeness of the sun god attended by kings and other gods.

Physically, intellectually, emotionally, and spiritually—even geometrically—the scenes are surprisingly similar. The two events are separated by 1,200 years and thousands of miles. They are nuanced by culture and tradition, but the symbolic vision is identical. How can this be?

Sepher Yetzirah

The earliest piece of the Kabbalah is the Sepher Yetzirah, Book of Creation, completed by about 300 CE by unknown authors. They are said to have been given to Moses on Mount Sinai. Moses chose a few future *nistarim*—literally "hidden ones"—and taught them the *Sod* ("secret") level of the Torah in which reality is expressly limited to symbolism, numerology (*gemetria*), and the spiritual forces that underlay creation. Rabbi Laibl Wolf notes that Moses educated his people at four levels, or *Pardes*, the "orchard" of Kabbalah mysticism—each level decrypting another layer in the following order:

PARDES–the orchard of kabbalah mysticism

P   *Peshat*   plain meaning
R   *Remez*   hint, allusion
D   *Derush*   derivation via analogy, metaphor
S   *Sod*   secret, the Kabbalah[13]

Moses taught each person at Mount Sinai the overt meaning of the "rules governing personal behavior, social consciousness, code of ethics, and lifestyle and relationship with their Creator. This level of instruction became known as *Peshat*, meaning 'Plain, Unadorned, Stripped of Complication.'"[14] But Moses limited the number of people attending the next set of lectures. This level investigated the deeper allusions that were only hinted at in the text, and the attendees were chosen based on their ability to realize the hidden applications of the *Peshat* level. This second level, *Remez*, concerns a deeper layer of intent not readily apparent in the basic information. Attendance at the third sessions, the *Derush* ("inferred") level, was even more restricted, as this select group was taught to understand the application of appropriate analogies, allegories, and metaphors for communicating complex concepts that otherwise could not be properly expressed. Rabbi Wolf notes,

> At the same time that it illuminates, an analogy obscures, because it is couched in an example this is usually quite unlike the intended information. Yet, ironically, by *hiding* the informa-tion in an analogy, you actually reveal its essence. This is what *Derush* does. It makes the essence of the Torah known via meta-phors and tales. To the sages they become the keys to hidden meanings. A whole body of Torah literature known as Midrash is devoted to the *Derush* form of teaching.[15]

And lastly, the *Sod* level of instruction was given by Moses to the chosen few, the *nistarim*, who received lessons about world reality, of the ethereal energy, the essence that is the foundation of creation itself.

The Tree of Life

As the foundation for Kaballah, the Sepher Yetzirah describes the " . . . thirty-two wondrous paths of wisdom engraved by God . . . thirty-two books, with number, and text, and message . . . Ten Sephiroth of nothing, and the twenty-two letters of foundation. . . . "[16] The Tree of Life of the Kabbalah consisting of those ten Sephiroth, spheres of influence that are linked by twenty-two *Nativoth*, or paths of knowledge (Figure 11.3). The purpose of this structure is to help us understand the process by which we can relate to God, and how He can relate to the world. Although one element of God manifesting into the universe, each Sephira contains infinity within itself. The Sephiroth measure "the depth of the beginning and the depth of the end, the depth of good and the depth of evil, the depth of the heights and the depth of the depths, the depth of the east and the depth of the west, and the depth of the north and the depth of the south—and one master, God, faithful King, rules them all from his holy place."[17]

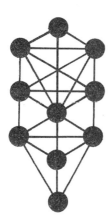

Figure 11.3: The Tree of Life of Kabbalah, including ten spheres of influence (Sephiroth) and twenty-two paths of knowledge (Nativoth).

The basic structure of the Tree of Life is described in the *Sepher ha-Bahir*, the *Book of the Brightness* that first appeared in the rabbinic academies of Provence in France. The Bahir is written in a style resembling the Midrash. Drawing on sources from the Merkaveh mystic schools, it includes references to the Sepher Yetzirah, the Sepher Hadisim, and the Talmudic Raza' Rabbah, or "Great Secret." And it is in the Bahir where the Sephiroth are structured with interconnecting *Nativoth* that diagram God's model for creating the universe, giving form to the eternal laws that allow spiritual energy to move through the universe, and through which humans may re-ascend. This is the Otz Chiim, the Tree of Life. Dedopulos says, "The Sephiroth clearly descended from the crown of perfect divinity (the Sephira Ether) to the earthly kingdom, Malkuth. But the Bahir also strongly introduced the concept of balance—for there to be harmony between the physical and the spiritual, there had to be harmony between other polarities: male and female, severity and mercy, wisdom and understanding, good and evil. Just as the Tree of Life consisted of the spheres of god's limitless bounty, so there was a reverse tree, a shadow side, made up of hollow husks, the Qliphoth, connected by dark paths indeed."[18] Also referred to as Maaseh Merkabah, "Workings of the Chariot," it projected the idea of an "active mystical experience . . . [combining] the ideas of the work of creation, the animating of matter, with the radical concept of a celestial projection as a way to return to the divine source."[19]

My intention here is not to detail the intricacies of Kabbalah. These mysteries remain unsolved although thousands of books have been written on the subject, and the debate continues. However we can't proceed with further study of the Tree of Life without a brief under-standing of the importance of Kabbalah to the study of Judaism and the difficult road that must be travelled if the mysteries of Kabbalah are to be unraveled. Rabbi Wolf explains,

> We discover in the Kabbalah that the world doesn't change of its own volition. Our personal transformation alters the shape of the Cosmos, not the other way around. We are responsible for affecting change in the Cosmos, and it is profoundly enabling when one comes to truly understand that the Cosmos seeks our well-being. . . . When we give to others we create space within for a beneficent Cosmos to replenish us.[20]

Rabbi Wolf finds that Kabbalah provides tools with which to reinter-pret reality, reshaping the world by creating a revolution in our relationships, with the objective being the universal goals of:

FULFILLMENT: The balance of mind and emotion.

SECURITY AND INNER PEACE: The faith with which to vanquish fear. The capacity to embrace inner strength, wholeness, and serenity.

LOVE AND COMPASSION: The emotional connections that bond people in warm, empathic, and committed relationships.

HONEST AND CONFIDENT EXPRESSION: The mastery of words, behavior, and thought, so that they become an expression of our inner truth and uniqueness.[21]

Ancient Kabbalah can be explained and decoded by Chabad Hasidism, a theology and transpersonal psychology that was developed to achieve personal, emotional, and mental well-being. Wolf notes that one of the findings of this theology is, "person and Cosmos are complementary and synergistic aspects of the greater Whole. It teaches that what we think, say, or do leaves a mark on the universe. And the changed universe 'communicates' with us in the new 'language' that includes our contribution."[22]

And so our ability to achieve these universal goals lies in our understanding of the true nature of mind and emotion. Rabbi Wolf notes that mind consists of three spiritual energies (three of the *Sefirot*) mediated by the physical brain, and that these energies create enlightenment and understanding. Emotion consists of the other seven *Sefirot*, energies that manifest in the heart. In sum they are the ten *Sefirot*, spiritual energies that flow throughout the physical body.

With this background let's now look at these ten "spiritual energies," *Sefirot*—spiritual flows. The following description of the meaning of the *Sefirot* is based on the teachings of Rabbi Wolf. Essentially, the *Sefirot* are the time, space, energy, and matter of the universe. They provide the foundation for individual personalities. They have spiritual qualities, but they become mind and emotion. They originate from the infinite source of *Ein Sof*, "without end," flowing from four parallel spiritual realms or worlds (Heb.: *Olamot*). The array of *Sefirot* is like a beam of source energy having ten distinct wavelengths. As the "light" of this array passes through each spiritual realm, the qualities inherent in the *Sefirot* become the spiritual beings in that realm, within that space. These spiritual beings take on the form of angels, holy animals, wheels, and many others influencing the animate, physical universe.

There are an infinite number of these realms, but *Atzilut* is the closest realm to the Divine Source (creator) (Figure 11.4). *Yetzira* (or *Yetzirah*: shape and quality) is the realm of emotion. *Assiya* (actualization) is the physical reality in which we live. It is the reality of time, space, and earthly consciousness. *Neshama* (soul) can be thought of as a spiritual umbilical cord that extends through all realms, one end connected to the creator and the other end found within our spiritual and physical selves. And so the flow of the ten *Sefirot* is through *neshama*, then becoming the ten qualities of personality—the flow of mind and emotion.

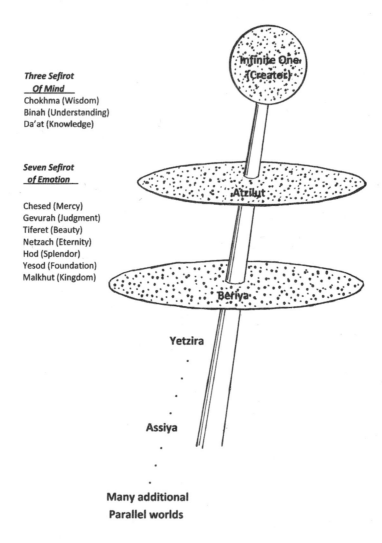

Three Sefirot
  Of Mind
Chokhma (Wisdom)
Binah (Understanding)
Da'at (Knowledge)

Seven Sefirot
  of Emotion

Chesed (Mercy)
Gevurah (Judgment)
Tiferet (Beauty)
Netzach (Eternity)
Hod (Splendor)
Yesod (Foundation)
Malkhut (Kingdom)

Infinite One
(Creator)

Atzilut

Beriya

Yetzira

Assiya

Many additional
Parallel worlds

Figure 11.4: Atzilut, the closest realm to the Creator, which is the Divine Source (based on Wolf, 1999).

The upper realms of the *Sefirot* are literally unembodied mind and emotion, pathways through which we touch the universe. Words cannot express the flow of emotion easily, and that is why we find it difficult to express our true feelings. Metaphors and analogies seem better suited to this purpose. Our thoughts are the framework for envisioning an intelligible universe, but it is feeling that, according to Rabbi Wolf, expresses

> the amplitude, frequency, and depth of our relationship with the spiritual and physical environment . . . Feelings do not "talk" to us—they emote within us. They provide us with a framework within which we experience our reality. Both words and feelings are expressions of the soul. When we touch the Cosmos through Emotion we are connecting in a personal way.

- MIND is the way the SOUL expresses itself through the physiology of the BRAIN.

- EMOTION is the way the SOUL expresses itself through the metaphoric HEART. Compassion is not weakness. It is strength.[23]

Within the hierarchy of the ten *Sefirot*, *Keter* is at the highest level and in the next realm of *Beriya*, beyond physical reality. Within physical reality, the realm of *Assiya*, *Malchut* ("sovereignty") is the lowest of the *Sefirot*. *Malchut* completes the cycle as the last in a chain of events beginning with the moment of inspiration that finds its actualization in the finite, real world. We would possess thoughts and feelings but never actualize them without *Malchut*, an unusual *Sefira* in that it has two spiritual characters. It is feminine in nature, carrying all of the other *Sefirot* to full term, as in a womb, nurturing the upper nine *Sefirot* until their manifestation as words or behaviors. It is the end, and yet it appears in the beginning. Kabbalah refers to *Malchut* as the *Imma Tat'ah*—"the Lower Mother." *Malchut* is derived from the Hebrew word for "king"— *melech*. It gives us a glimpse of divine sovereignty in the world, which can be expressed through our speech and behavior. Who we are is revealed in what we do.

Rabbi Wolf notes that Mahatma Gandhi said, "Be the change you want to see in the world,"[24] suggesting that each of us, as Kabbalah teaches, is a reflection of the whole cosmos. We begin our life from where we end, we end where we start. *Malchut* is the end and the beginning. The lowest *Sefira* in *Atzilut* is *Malchut*, but *Malchut* becomes *Keter*.

An Ancient Knowledge of the Cosmos

For further understanding of the origin of the kabbalistic Tree of Life we turn now to a recent analysis by the late Leonora Leet, who was a professor of English at St. John's University. It is Leet's thesis that the tradition associated with kabbalistic thought originated in the ancient Hebraic priesthood. Her thesis is based on two components of this hidden knowledge: 1) a cosmology linking the processes of cosmic creation to its goal in human transformation, and 2) the sacred science that could demonstrate this purpose. She finds that those two ideas were derived from, or informed, the two major functions of the Hebraic priesthood: the ritual of animal sacrifice and the consecrating of sacred space. It is the second of these two priestly functions that concerns us here, and so it is on Leet's "new experimental pathways to the sources of this knowledge,"[25] as it regards the form of the Tree of Life, that we will now focus.

Leet's argument begins with the finding that the traditional body of esoteric learning implies that the institutions built to conserve those teachings would have been the ancient priesthood and the temple, historical features having an inherent connection with geometrically based esoteric science. This elite group would have been entrusted with study and transmission of ancient geometric knowledge, bringing together science and cosmology and erecting temples employing an architectural form that demonstrates an understanding of sacred space. Erection of a sacred space requires knowledge of geometric proportions and selection of ideal forms for expressing ideology, exemplified by the cubic symmetry of the Holy of Holies in the Mosaic Sanctuary. Leet suggests that construction of a temple implies an imaginative grip of geometry, and therefore both the structure and the priests would have been dedicated to further study and development of this hidden knowledge.

With this as her foundation, she postulates that a continuous esoteric study of geometry was a priority during three periods that demonstrate enthusiastic interest in temple construction: construction of the Mosaically inspired desert sanctuary (discussed in the Book of Exodus); construction of the First Temple by Solomon, noted for his esoteric wisdom and use of the hexagram as his ring seal; and reconstruction of Solomon's Temple (inspired by the prophecies of Ezekiel) after the Babylonian captivity.

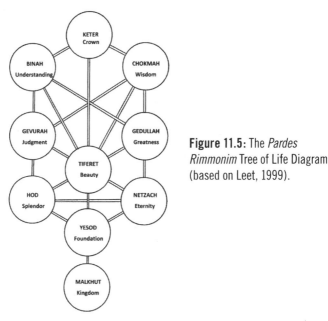

Figure 11.5: The *Pardes Rimmonim* Tree of Life Diagram (based on Leet, 1999).

In the Zohar, a major text of Kabbalah, there is a synthesis between geometrically based cosmology and a midrashic mode of interpretation that characterizes kabbalistic thought. But a differentiation exists between kabbalistic and Talmudic thought, in that the Talmud is modeled from biblical texts, while Kabbalah applies geometry as its basis. "In every text of true Kabbalah," Leet writes, "the eye of its authors is ever set upon the diagram called the Tree of Life. Whatever other geometric processes and forms are part of their cosmological thinking and illustrations, it is the Tree of Life Diagram that all texts included within the narrowest definition of the Kabbalah are concerned to explicate."[26]

Importantly, Kabbalah shares an ancient wisdom with neo–Platonism and the hermetic tradition, the former derived from Pythagorean sources and the latter from Egyptian studies devoted to geometry. While there might be a common Egyptian source for all of Western sacred geometry, it is certain that there have been associations between native Judaic esoteric traditions and Pythagoreanism, the differentiation of geometric knowledge being difficult to discern. However, there is a certain Hebraic interest in the hexagram that features in Indian and Far Eastern symbolism, including geometric diagrams such as *yantras* and mandalas. Leet writes,

> Whatever may be the truth of the cultural contacts between
> the ancient Hebrews and Indians regarding the dissemination of

the hexagram as the paramount symbol of creation . . . it was the preeminent geometric symbol throughout the long history of the Jewish esoteric tradition, though rarely surfacing . . . until it is finally embraced by the whole Jewish people as its symbol. Equally obscure is the origin of the distinctively kabbalistic symbol of the Tree of Life Diagram . . .[27]

Leet draws a similarity between the Tree of Life diagram and the menorah, suggesting their possible shared origin with Moses, but regardless, she concludes that it is likely that the geometric complex, including those forms, was part of a hidden tradition transmitted by the priesthood.

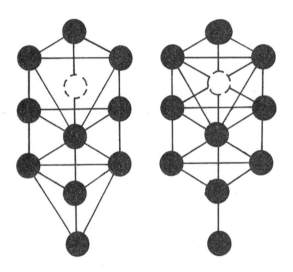

Figure 11.6: The Tree of Life according to Moses Cordovero (left) and Issac Luria (right).

The largest Jewish communities after the fall of the temple at Jerusalem in 70 AD were located at the cities of Alexandria in Egypt, and Babylon in Mesopotamia. Only a few small Jewish communities remained in Palestine. The spiritual community of Safed became a stronghold of mystical Kabbalah, and it was in the sixteenth century that kabbalists in Palestine defined the names and diagrammatic positions of the *Sefirot* on the Tree of Life, with particular importance placed on the *Pardes* because of its combination of written commentary concerning the *Sefirot* and the paths that connect them as depicted in Figure 11.5. However, Leet notes that the diagram and commentary raise concern as to the true form of the tree, in that the diagram and commentary (of which there are differing versions) do not agree in the placement of all of the paths. Leet notes that,

> one obvious error in this version of the tree is that it only has twenty paths or channels between the Sefirot rather than the twenty-two specified in the text, the missing ones being the horizontal paths between Chokhmah and Binah and between Chesed and Gevurah . . . [while] the summary commentary reveals that there are other differences and problems being caused by textual differences between the standard version . . . and a different version . . . closer to the drawn version of the diagram than is what seems to be in Moses Cordovero's Pardes Rimmonium (Garden of Pomegranates) original commentary written in 1548 CE.[28]

Cordovero, considered to represent the culmination of the earlier Zoharic tradition, was most concerned with the cosmic process of *emanation*, while Issac Luria, who provided a later continuous history of Jewish mystical understanding and practice, was concerned with the cosmic process of *return*. Both Cordovero and Luria were leading sixteenth-century Jewish mystics. Leet refers to the Cordovero diagram as the *Tree of Emanation* and the Luria diagram as the *Tree of Return* (Figure 11.6). Both structures include the same names for the *Sefirot*: *Keter* ("crown"), *Chokhmah* ("wisdom"), *Binah* ("understanding"), *Chesed* ("mercy"), *Gevurah* ("judgment"), *Tiferet* ("beauty"), *Netzach* ("eternity"), *Hod* ("splendor"), *Yesod* ("foundation"), and *Malkhut* ("kingdom"). The difference between them is in the positions of some of the paths. Cordovero illustrates paths from *Malkhut* to *Netzach* and *Hod*, while those connections are not shown in the Luria diagram. Also, Luria's diagram has paths from *Chokhmah* to *Gevurah* and from *Binah* to *Chesed* that are not included in Cordovero's structure. However, both trees include the same three horizontal paths, seven vertical paths, and twelve diagonal paths that correlate "with the similar divisions of the letters of the Sefer Yetzirah, having shown how the letters are so distributed in the form of the diagram attributed to Luria."[29]

The Hexagram Theory

R. Buckminster Fuller demonstrated that the fundamental structure in two-dimensional space is the triangle. We can see an indication of this fact by observing the numerous triangles formed by the twenty-two *Nativoth* pathways that intersect across the Tree of Life. A more basic geometrical form comprised of triangles, the hexagram is constructed with two oppositely pointed equilateral triangles whose apexes are defined by the radial arc points of a circle (Figure 11.7). This structure goes by several names, including the Shield of David, the Star of David, and Solomon's Seal. Today it is the symbol most identified with the state of Israel and the Jewish people. As an esoteric symbol, the star with six points is associated with the heart chakra of tantric yoga and has been discovered in the ground plan of Stonehenge; and so it may be assumed to have been derived by many cultures across the world for millennia. Leet provides grounds for the possibility that there was an early association of the hexagram with Jewish esoteric understanding and practice reaching as far back as the biblical period.

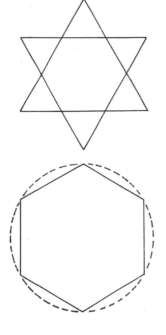

Figure 11.7: The hexagram, constructed of two oppositely pointed equilateral triangles with apexes defined by six radial arc points of a circle.

Figure 11.8: The hexagon shown within a circle.

Leet details the geometric construction of the hexagon and hexa-gram using the limitations of the compass and straightedge as was taught in ancient Greece and remains a fundamental part of high-school geom-etry today (Figure 11.8).

> [T]he measuring of the radial arcs on the circumference by moving the fixed foot to the circumference, results in the produc-tion of six and only six points, the multiplication of 3×2 equaling 6. . . . The switch from the geometry of the circle to that of straightedged surfaces, from what might be considered the unmanifest to the manifest aspects of creation, is required for precision because the circle and any curved line is, in every sense of the word, immeasurable. The world of the determinate is the world of straight lines, and such straight lines can be found within the general curvature of nature, as in the crystalline form of all chemical substances and in the lines that can be drawn between the centers of circular cells in organic matter to reveal its determining geometric structure.[30]

Leet's detailed discussion of the basic geometric construction of the hexagram at first seems rather laborious, but she is laying the foundation for her explanation of constructing the Tree of Life diagram in a previ-ously unrealized way, shedding light on not only kabbalistic mysticism, but also orthodox Judaic teachings in general. So, let's follow her discus-sion as she explains the importance of the hexagram and the relationship between this simple geometric symbol and the more esoteric Tree of Life diagram. There are several fundamental ideas expressed in the text concerning the hexagram. Leet describes a process overlapping two equi-lateral triangles to form an inner hexagon and then constructing ever-larger hexagrams such that "a key to the geometric derivation of the Tree of Life Diagram can be found."[31] She argues that drawing a hexagram using the six radial arc points on a circle is symbolic of the initial act of creation, and therefore subsequent acts of creation may be likened to construction ever-more complex patterns of hexagrams upon the first. Then,

> if we join the six points that the radius defines on the circum-ference of a circle not in a hexagram, as previously, but in a hexagon, and build other hexagons of the same size on each of its sides, we will find that these six close-packed hexagons define the outline of the central hexagon: in a two-dimensional drawing, the

six outer hexagons can be thought of as producing the inner seventh . . .[32]

On this basis Leet proceeds to construct geometrical configurations called the Seed of Life, Flower of Life, and the Sabbath Star diagram, building a framework of hexagons that ultimately produces a structure upon which the Tree of Life diagram and numerous other paths of unknown purpose can be seen. This progression is illustrated in Figure 11.9 with six hexagons surrounding a central seventh hexagon, and Figure 11.10 with completion of the Sabbath Star diagram. The method is one that applies basic two-dimensional geometric principles with a heavy reliance on the hexagon as the module to create more complex forms. Leet goes so far as to claim that "the main basis for the claimed superiority of the Sabbath Star diagram as the ultimate source of kabbalistic geometry is that the 'diagonals' by which the cubic Sefer Yetzirah tree can be infinitely expanded are actually the hexagram components of this diagram and that such expansion is thus based upon the principles of hexagram expansion."[33]

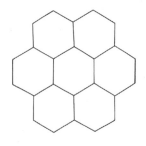

Figure 11.9: Six hexagons surround a seventh.

Figure 11.10: The Sabbath Star Diagram, building a framework of hexagons that produces a structure including the Tree of Life Diagram. Beyond the twenty-two paths of the Tree of Life, numerous other paths of unknown purpose are illustrated in the Sabbath Star Diagram (based on Leet, 1999).

In an early stage of this construction she finds common ground between Giorgio de Santillana and Heretha von Dechend's review of comparative mythology and Albert Einstein's theory of relativity, each explaining that space and time are not to be separated: She proclaims, "archaic time is the universe, like it circular and definite."[34] Leet likens the cosmic frame implied in the Sefer Yetzirah to the basic paradigm that de Santillana and von Dechend "discovered to inform the mythology of all the preliterate cultures worldwide," that being, "the temporal revolutions of the heavenly bodies to the ecliptic pole and whose mythological image is the *whirlpool*."[35] Similarly, she notes references to *Teli* and *Gilgal* in the Sefer Yetzirah indicative of,

> this most ancient understanding both of cosmic periodicity and of the means by which it can be transcended, the former defined as the "ecliptical world" and the latter as the nodes where its two great circles meet. Whether the later kabbalistic tradition is correct in its association of the Teli with the lunar nodes, or de Santillana and von Dechend's new analysis is accepted that understands the mythological references to involve the points at which the ecliptic path of the sun intersects the plane not of the moon but of the galaxy, it is through its harmonization of two such great circles of space-time, a balancing performed by its heart, that the world-soul of Adam Kadmon can maintain or regain its connection with the eternal.[36]

By exploring qualities of kabbalistic geometry that are unique and encoded in major texts including the Bible and later Jewish esoteric traditions, Leet finds that only one major form of this geometry has emerged from that tradition—the Tree of Life diagram, the structure she calls "the most famous cosmological diagram in the Western esoteric tradition."[37] It is from her building of the larger geometric complex based on the hexagram that the Star of David is shown to have been "an intrinsic element of Hebraic sacred science from its biblical origins."[38] Her diagram of expanding hexagrams is used as a model for the emanation of the four cosmic worlds of the Kabbalah.

Of particular interest is the decoding of the central mystery hidden in texts and diagrams of Kabbalah. By applying an interpretive strategy linking form to text, as is the hallmark of kabbalistic geometry, Leet is able to decode the geometric secrets embedded in major Hebraic cosmological texts such as Genesis, the Sefer Yetzirah, the Bahir, the Zohar, and the Otz Chiim. The science that the Hebraic priesthood shared with other cultures was most concerned with the interrelationships between sound, geometry, and number, becoming the central gnosis of Western sacred science. As an example, Leet notes that an interpretation of the human soul referenced in the Sefer Yetzirah yields truth on a secondary level of analogy on a cosmic level, in that references to bodily organs of the *nefesh* refer not to the cosmic body of planets and constellations but to an intermediate spiritual or astral body. And so the cosmos extends across space and time, informed by the soul. "But these two coordinates," she writes, "between whose numerical proportions the heart and soul body is the mediator, are not to be understood in modern Cartesian terms."[39] And this is where we find both the geometrical and cosmological aspects of the Tree of Life diagram taking hold, becoming one. She continues, "It is through geometry that the mind can embrace with rational precision and clarity those universal principles that otherwise appear so paradoxical to reason, such mystical paradoxes as that expressed in the hermetic formula, 'as above, so below,' the laws of universal correspondence in an all-embracing and unified cosmos that is the secret gnosis of esoteric spirituality."[40]

A Problem with the Design

At the beginning of the Zohar, there is mention of a diagram that underlies all of creation. The first reference to the diagram is said to have been by Elijah through direct revelation.

> This mystery remained sealed until one day, whilst I was on the seashore, Elijah came and said to me . . . "When the most mysterious wished to reveal Himself, He first produced a single point which was transmuted into thought, and in this He . . . graved within the sacred and mystic lamp a mystic and holy design, which was a wondrous edifice issuing from the midst of thought. This . . . was the unknowable by name. . . . And upon this secret the world is built."[41]

Leet concludes that the famous Tree of Life diagram could not be the "most holy design," since the edifice had an "unknowable" name, and the author of the Zohar in the thirteenth century presumably would have known the name of the Tree of Life.[42] Also, the Zohar provides us with a further description of the form of the structure:

> R. Simeon proceeded: "See now, it was by means of the Torah that the Holy One created the world. . . . He looked at the

Torah once, twice, thrice, and a fourth time. . . . indicated by the verse, 'Then did he see, and declare it; he established it, yea, and searched it out.' (Job. XXVIII, 27). Seeing, declaring, establishing and searching out correspond to these four operations which the Holy One, blessed be He, went through before entering on the work of creation. Hence the account of the creation commences with the four words Bereshith Bara Aelohim Aith ('In-the-beginning created God the'), before mentioning 'the heavens,' thus signifying the four times which the Holy One, blessed be he, looked into the Torah before he performed His work."[43]

For Leet, these quotations from the Zohar suggest that: 1) there was a geometric model that the kabbalist correlated with the process of creation, 2) the "mystic and most holy design" within the supernal mind was a map of the processes of emanation and material creation and their meaning, and 3) the divine geometry was discoverable at all levels of the cosmos to reveal its mystic secrets to the knowing. In the end, it is possible that the Sabbath Star diagram is associated with the basic hexadic geometry at the heart of kabbalistic tradition, expressed in the two geometric symbols popularized by Kabbalah since the 1600s—the hexagram and the Tree of Life diagram—and Leet ponders whether it might help explain the covert nature of their association.

The Sefer Yetzirah is understood to allude to a diagram with what is referred to as "cubic" symmetry, with the first chapter providing a definition of the six directions of space, like the six directions that point away from the six faces of a cube. Leet states that construction of the figure has been impossible because the definition addresses twelve diagonals, rather than twelve edges, of this cube, finding that the significance of the cube in this context is with regard for, "the subtle container of the higher cosmic worlds and defines both the process of divine emanation and or spiritual return. But since this is the Book of Yetzirah, the third kabbalistic world associated with the second dimension, the cube to be thus constructed would not be an actual solid, rather a two-dimensional representation of such a cube."[44] Leet believes that the six points are inscribed on the circumference of a circle in the regular geometric progression of hexagram growth, and the cube that appears within an orthographic projection forms the basis of the kabbalistic references to the six supernal directions, this being of the highest interest for construction of the Sefer Yetzirah diagram.

Solving the Problem

In later chapters we will discuss in detail the geometric characteristics of the five platonic solids that are of prime importance to sacred geometry. However, Leet provides interesting commentary on the value of the platonic solids to the two-dimensional world of Yetzirah, and she notes the following:

- The grid enlarging hexagrams she describes accommodates the accurate orthogonal projection of four of the Platonic solids: tetrahedron, octahedron, cube, and icosahedrons. However, it cannot provide the same accommodation for the dodecahedron because it necessarily alters some segments of the hexagram grid.

- Since the two-dimensional hexagram grid cannot be perfected in two dimensions to accommodate the dodecahedron, then as a model of creation the true expression of the form must include the third dimension that is inherent in Earth and the cosmos.

- Understanding of the relationship between the subtle structure of Earth and the dodecahedron has been a part of spiritual philosophy since Plato, and there is recent evidence of greater complexity of Earth's structure. Leets finds that ". . . if we can accept this evidence, then the dodecahedron becomes more than just a geometric symbol of Mother Earth; it conveys the true essence of this planet and, it would also seem, of the divine personality immanent within it."[45]

- She also concludes that the final dodecahedral form can be drawn such that its only true orthogonal projection is geometrically allowed with respect to the Sefer Yetzirah tree, shown in Figure 11.11 based on her analysis.

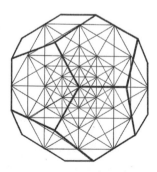

Figure 11.11: The only orthogonal projection of the dodecahedron geometrically allowed in relation to the Tree of Life (based on Leet, 1999).

Leet's study suggests that the Tree of Life Diagram might be the key to discovering a more fundamental form of creation. The center of the Tree of Life consists of an underlying hexagram diagram where the non-Sefirah Da'at (Knowledge) is located. If this is the ultimate knowledge—gnosis—to be recognized from its structure, then the cubic diagram that emerges has profound cosmological implications. It may represent both the Tree of Life and the Tree of Knowledge with a sixfold rotation centered about Da'at (Knowledge), and if so, then its "knowledge" might be the central gnosis of Hebraic sacred science.[46] The importance of the cube to Hebraic tradition is readily apparent in Exodus, where instructions are given for the construction of the Holy of the Holies to house the Ark of the Covenant. In fact, in all commentaries concerning construction of the temple and the Holy of Holies, the structure was to be in the form of a cube.

The first text and the foremost diagram of the Kabbalah could be referencing not only the mystery of the cube, but the archetype of creation as well. Leet recognizes that the cube, as a solid polyhedron symbolizing earth as one of the five elements in esoteric geometric tradition, can be viewed with the three lines crossing at the center, and in two dimensions that view transforms the diagram of a hexagram within its single-point hexagon back into a cube.[47] Therefore, the three intersecting lines are symbols of the three-dimensional solid world. This is illustrated in Figure 11.12. However, this is as far as Leet takes us with regard to the possible three-dimensional aspects of the Tree of Life as it is so often represented, with the twenty-two pathways crossing over and under each other between the *Sefirot*. As to the symmetry of the cube, we read in the Zohar:

> From this point onwards *bara shith*, "he created six," from the end of heaven to the end, thereof, six sides which extend from the supernal mystic essence, through the expansion of creative force from a primal point. . . .[48]

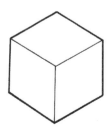

Figure 11.12: The cube viewed directly over one of its corners (vertices).

The fact that geometry is particularly expressive of cosmological meaning was understood by the Greeks as well as the Hebrews. Plato developed his cosmology in *Timaeus* on the basis of the earlier geometric insights of Pythagoras, and a similar core of geometrically encoded knowledge is to be found in the Western esoteric tradition, including the Kabbalah. For it is geometry that can provide the meaningful structure to cosmological and related areas of speculation that seems to partake of absolute truth, the truth inherent in the processes and forms of geometric construction. If geometry holds a central position in Western esoteric thought, it is because of its unique ability to supply an explanation, a blueprint, for those laws of existence that seem to apply to the processes of immaterial thought and to the material objects of such thought, an explanation of how consciousness and the world of material solids can be but different aspects of a coherent cosmos ordered by intelligible functions.

If all esoteric traditions have held geometry to be sacred, it is because their adepts understood it to encode the basic principles on which the cosmos is founded, and their geometric practice with the compass and ruler instructed them in its meaning. Leet suggests that this dual practice of construction and interpretation provides a technique for discovering cosmic principles that would seem fitting to name the science of expressive form. It can be considered a science because its basis is not arbitrary, but rather determined by discoverable geometric laws involved in geometric construction. The forms produced by such construction are also expressive of meanings that go beyond their purely mathematical properties and whose interpretation can be considered an art. The science of expressive form, of sacred geometry, has always required, then, a synthesis of analytic and intuitive capacities.

The Tree of Life diagram is not simply an arrangement of circles and the lines connecting them. Its parts are named, and their spatial relationships are given an associated mythological interpretation. Thus kabbalistic geometry, though having certain historical links with Pythagorean geometry, can also be distinguished from it in its quasi-Talmudic approach to geometric form as something very much like a literary text, one whose meanings must be grasped not through logical demonstration, but through mythological association.

Back to the Cosmos

As we noted earlier, one of the components of Leet's thesis linking the origin of kabbalistic thought with the ancient Hebraic priesthood is a cosmology relating the processes of cosmic creation to its goal in human transformation. After considering the historical relationship of Hebraic sacred science to the analogous traditions found in Egyptian, Greek, and Indian esotericism, she emphasizes the foremost cosmological principle demonstrated by these sacred sciences—"the genesis of the divine son."[49] For the Hebrews, sacred science focuses on the secret doctrine of the son being first announced by God when He said, "Israel is my son, even my firstborn" (Exod. 4:22). Analogues to this doctrine are found in all of the world's ancient religions, including the relationship between the Egyptian figures of the sun god Horus and the pharaoh, his earthly avatar, or the figures of the Hindu Krishna and Buddhist Amitabha, or the Christian Jesus as derived from Hebraic tradition.

Does the universal appearance of this understanding of the oneness of cosmic and human destinies point to a singular source derived from a common Neolithic culture, more direct influences, cross-fertilizations, or similar prophetic intuitions? While the answer to this question remains unknown, Leet concludes that each tradition of the world's major religions "developed its own authentic formulation of this savific belief and its attendant rituals, practices, and proofs. But from the Bible, through Merkabah mysticism and the Kabbalah, to Hasidism, the central mystery of the son, as of the cosmic process, has ever had but one meaning, *that ultimate unification of the human and divine that could affect both the personalization of the divine and complementary divinization of perfected humans.*"[50]

Associating the Tree with the Sphere

After this extensive review of Leet's thesis for the origin of the Tree of Life diagram, it is worth noting Jay Weidner and Vincent Bridges' concept of a similar "geometric pattern that emerges from the intersecting lines of the Tree [of Life] like a moiré pattern in a holographic projection."[51] In fact, as an alternative to Leet's detailed path through hexagonal expansion to arrive at the Tree of Life, the intersection of four circles stacked upon each other will also produce the locations of the ten *Sefirah* (Figure 11.13).

Weidner and Bridges suggest that this diagram illustrates reality as the points of intersection of the *Olamot*, the four great realms of abstraction—Atzilut, Beriya, Yetzira, and Assiya. Note that the three uppermost *Sefirah* (*Keter*, *Chokmah*, and *Binah*) form the vertices of an upward-pointing triangle. This first triangle is followed beneath by an inverted triangle of *Chesed*, *Gevurah*, and *Tiferet*, in turn overlaying a third triangle including *Netzach*, *Hod*, and *Yesod*. Weidner and Bridges find the whole pattern, "resolved by, and enfolded into, the last Sefirah, Malkuth. . . . The Bahir adds that portions of the celestial sphere can be equated with the spheres of each Sefirah, or globe, on the Tree of Life."[52]

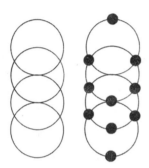

Figure 11.13: The intersection of four circles stacked upon each other to produce the locations of the ten Sefirah.

Of particular interest for the astronomical gnosis hidden in the Bahir and the Tree of Life, Weidner and Bridges refer to the first verse of the sixth chapter of the Sefer Yetzirah, in which proof of the existence of the Tree of Life is provided:

> These are Three Mothers . . . And seven planets and their hosts, and twelve diagonal boundaries a proof of this true witness in the Universe, Year, Soul and a rule of twelve and seven and three: He set them in the Teli, the cycle and the heart.[53]

With *teli* derived from the root of the Hebrew word *talah*, "to hang," Weidner and Bridges find that the mystery hidden in the Bahir concerns the axis about which the celestial sphere rotates (Figure 11.14). They also find reference to three Leviathans (dragons) that are, in fact:

1.   the constellation Draco (Great Dragon) which is coiled and rotates around the north celestial pole (currently represented by Polaris, the north star), but is never within the confines of the circular pole of the ecliptic; Draco, then, is the Teli from which the stars hang over the northern sky, "like a king on his

throne"; this Teli is perceived to extend through the earth from the north ecliptic pole to the south ecliptic pole;[54]

2. the Coiled Serpent defined by the two points located at the solar solstice standstills, where the celestial equator intersects the plane of the ecliptic; currently these two points are generally located along a line extending through the earth between the constellations of Leo/Virgo and Aquarius/Pisces; and,

3. the dragon axis defined by the two points where the galactic plane intersects the ecliptic plane; currently these two points are generally located along a line extending through the earth between the constellations of Scorpio/Sagittarius and Taurus/Gemini.

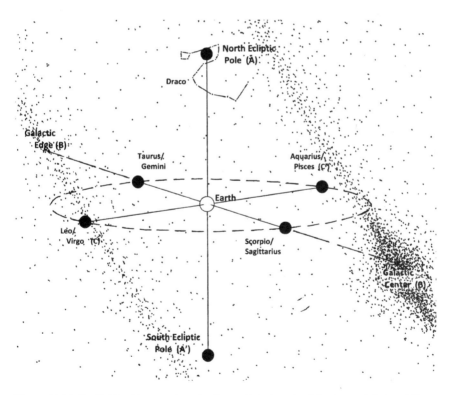

Figure 11.14: The Bahir addresses the axis about which the celestial sphere rotates, and the three serpents of the cosmos (based on Weidner and Bridges, 2003).

Each of these "dragons" results from precession. Draco encircles the confines of the temporal north pole as the four points defining the other two dragons rotate through the twelve signs of the zodiac, each of these rotations defined by the rate of precession, about 1 degree of arc per seventy-two years. Knowledge of this backward spin of precession, caused by the 23.4-degree tilt of the earth in relation to the ecliptic plane, is, of course, the basis for much of the cosmology that developed over the millennia across the world, as evidenced by the traditions outlined herein.

Returning, then, to Ezekiel's vision of the Merkaveh we find greater meaning to the metaphors that fill the biblical text. Verse 96 of the Bahir states, "What is the earth from which the heavens were graven [created]? It is the throne of the Blessed Holy one. It is the Precious Stone and the Sea of Wisdom." Plainly, Ezekiel's sapphire-blue throne is the sky arching over the earth, the beryl green wheels represent the churning seas that extend beyond the limits of land in each of the four cardinal directions, the likeness of man is within the firmament between sky and sea, and the sun shines with the luster of electrum as precession proceeds from the energy of the north wind—Teli—extending down from the pole and described in verse 106 of the Bahir as "the likeness before the Blessed One," the face of God.[53] From this picture it becomes apparent that the cubic symmetry of the three-dragon axis, as described in Ezekiel's vision, recalls the symmetry expressed in the Tree of Life. Centered within the vision is the "terrible ice" and the bright rainbow brightness surrounding the chariot. The Tree of Life, then, is the multifaceted clear stone, the ice, that receives the light refracted onto the celestial sphere about us as our chariot, Earth, continues on its journey through the heavens.

Before proceeding further, let's summarize the geometry built into this conception of Earth's movement through space relative to the polar, ecliptic, and galactic markers noted above.

1. The three lines extending through the north and south celestial poles, and four points defined by the intersection of the ecliptic plane with the celestial equator and the galactic plane with the ecliptic, define six directions uniformly distributed about the earth, such that the six planes oriented orthogonally to those lines would produce a cube.

2. This cubic symmetry contains the symmetries of the five Platonic solids.

3. The same symmetry is expressed in the Tree of Life.

4. Earth's rotation is tilted at an angle of 23.4 degrees from the plane of the ecliptic.

5. The galactic plane is oriented at an angle of approximately 60 degrees from the celestial equator.

This geometry is summarized in Figure 11.15. We have seen this cubic symmetry time and again in the sacred traditions of people around the world. However, these geometric relationships are also expressed in the more mundane, and for some, troubling aspects of secular thought. The foundations of modern Freemasonry and kabbalistic magic are similarly derived from this geometry, and as we shall now show, for good reason.

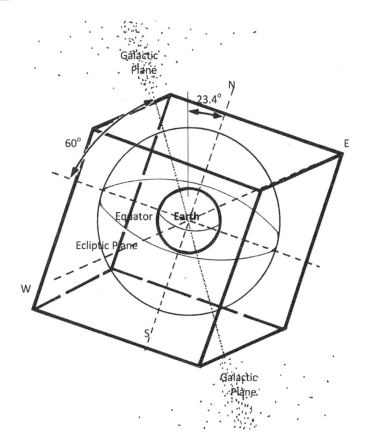

Figure 11.15: Cubic geometry built into a conception of Earth's movement relative to polar, ecliptic, and galactic markers.

For brief consideration here I look at symbols as instruments of expression, of communication, of knowledge and of control. . . . symbols to a supreme degree tools of the artist . . .
clearly expressing values regarded emotionally and intellectually as important by the people who assert them . . . is especially clear with political and religious symbols.

RAYMOND WILLIAM FIRTH
Symbols: Public and Private

Chapter 12
Magic and Freemasons

Each separate symbol and sigil born by the God on some part of his person was a clue to his inherent nature. The myths and legends passed down to posterity by the Egyptian priests concerning the Gods . . . in each one of these legends and pictorial descriptions of the Gods is concealed a wealth of transcendental knowledge for whomsoever has the ability to perceive it. . . .

The evolution and development of the Cosmos, spiritual and physical, were first recorded by philosophers in geometrical changes of form. Every esoteric cosmology used a circle, a point, a triangle and a cube and so on. These later were incorporated into a simple geometrical form such as is called in the Qabalah "the Tree of Life."

ISRAEL REGARDIE
THE TREE OF LIFE

Intent in Flux

Having described in some detail relationships between various ancient and contemporary cultures and mythologies and their associated religious symbolism and geometrical constructs, topics such as magic and fundamental ideas of contemporary freemasonry may seem rather tangential to our discussion. This is far from the case. In this chapter I show that rituals and ceremonies found in magic (magick) and freemasonry are filled with lines, pillars, curves, circles, and spheres, symbols that evolved in the antiquity of ancient Egypt, orthodox Judaism, and mystical Kabbalah. Those symbols appear within a matrix of cultural and religious traditions that have shaped Western thought throughout the ages.

The ideas coalesced to produce rituals practiced by modern magicians of the occult and social organizations such as the Freemasons. However, the individual symbols take on their own meanings and continue in a state of flux. The intent in reviewing those symbols is to illustrate this continuing fluctuation in symbolism while demonstrating that its underlying intent generally remains in accord with the ancient understandings explored earlier.

Eliphas Levi

The year 1810 was another in a series of eventful years in France. The Treaty of Paris ended the war between France and Sweden, a conflict that had also pitted Russia and Denmark–Norway against the Swedes (thanks to some prompting by the French). The marriage of Napoleon and Josephine de Beauharnaiss was annulled in January, and Napoleon married Marie-Louise of Austria in March. By July he had engaged France in the Peninsular War against Spain and Portugal and was annexing the Kingdom of Holland. In August the French prevailed over the British fleet at Grand Port, Mauritius, and Jean-Baptiste Bernadotte, one of Napoleon's marshals, was elected crown prince of Sweden.

In February of that year, a baby boy by the name of Alphonse-Louis Constant was born to a shoemaker in Paris. The boy eventually attended a Roman Catholic seminary in anticipation of entering the priesthood. But his plans fell by the wayside when he fell in love while at seminary.

He left his studies in 1836, never to be ordained. Nonetheless, his interest in theology remained with him. In the following years he studied the occult, and in 1845 he changed his name to Eliphas Levi, a moniker that he felt sounded a Judaic tone more in line with his newfound esoteric interests—the study of magic and mystical Kabbalah.

In his book *The Tree of Life,* the twentieth century magician Israel Regardie defines magic and mysticism as follows, and it is important to keep in mind these definitions in this chapter, as they do not necessarily correspond with ancient understanding and application thereof.

> **Magic:** The art of causing change to occur in one's environment and one's consciousness. Derived from the Greek *mageia* the science and religion of the priests of Zoroaster [also known as Zarathustra, a philosopher and prophet who lived in Persia, possibly modern Iran or Afghanistan; current estimates of the date of his compositions range from 6000 BC to 1000 BC]. Willpower, imagination, intention, and the use of symbols and correspondences play a major role in this art.[1]

> **Mysticism:** The belief in intuitive spiritual revelation, the objective of which is union with the divine. This mystical union is achieved through intuition, faith, ecstasy, or sudden insight rather than through rational thought. Mysticism is sometimes considered a more passive approach to the divine, as opposed to magic, which is more active. The true theurgist [someone inducing supernatural beings to behave on the magician's behalf] must be both mystic and magician.[2]

Mystics dedicate themselves to exploring, understanding, and expressing the spiritual and divine nature of the world. They comprehend forces that are transparent and yet still exist within our material reality. Although those forces are structured in a manner not readily perceived by our senses of sight, hearing, smell, taste, and touch, the mystic seeks greater understanding of reality through exploration of the physical and metaphysical nature of these forces. Dedopulos identifies three broad categories of mystical thought and practice.

- Animists recognize that every object and phenomenon in the universe has a unique, sentient spirit that generally is invisible, yet a part of the natural world.

- Shamans understand that there is a spirit world inhabited by archetypal spirits (for example, all snakes are represented by the spirit Snake). The spirit world also contains the souls of the dead.

- Deists believe in a divine force that underlies the world. Deism has been expressed by three separate links of intention:

  - Theosophists and Gnostics, who seek ever deeper knowledge and understanding of the divine, wanting to improve their lives and make sense of the world.

  - Theurgists and magicians pursue control and channeling of divine energies, the goal being to affect the world in their favor.

  - Ecstatic mystics explore direct contact with the divine force, anticipating a deep sense of joy achieved through such contact.[3]

Kabbalah serves as a vehicle for Deist mystics to increase their understanding of God's purpose and intentions, the design and mechanics of the universe, and relationships between self, God, and world. Its widespread appeal—from serious Talmudic scholars to English ritual magicians—results from this broad scope of application.[4] This very reason may explain why Eliphas Levi began his study of mystical Kabbalah. It is also possible that he discovered some measure of comfort in the study and application of magic and mysticism after experiencing conflict between traditional organized religion and the reality of life and love. As Dedopulos explains, "Traditional religion aims to explain the trials and tribulations of life, to give meaning and purpose to the pain and hardship of existence. Organized religion sets out to offer a set of structures and purposes that can help people make sense of the world around them. Mystics, on the other hand, have little interest in the daily grind."[5]

Levi did not choose a life of seclusion as he sorted out the secular conflicts in his life. Rather, he was inspired to share his insights concerning the occult, sparking new interest across the West during the late nineteenth and early twentieth centuries. That interest bloomed into a revival in the practice of magic and mysticism informed by teachings from around the world, and produced the Hermetic Order of the Golden Dawn, a society founded by Levi that practiced a multifaceted magical tradition that has a continuing impact on Western occult practices. In

1861 Levi was initiated as a Freemason in the Lodge Rose of Perfect Silence. No doubt his fellow masons were taken back by Levi's reception speech, in which he stated, "I come to bring to you your lost traditions, the exact knowledge of your signs and symbols, and consequently to show you the goal for which your association was constituted."[6] His brethren would not accept his belief that Kabbalah was the source of Masonic symbolism, yet Levi obtained the degree of Master Mason later that year. Troubles persisted at the lodge when Levi was the keynote speaker on the topic of the Mysteries of Initiation.[7] A fellow mason attempted to comment on the points made in the address. Levi did not welcome the harangue and walked out of the meeting, never to return.[8]

While Kabbalah was the common thread that bound the ritual and ceremonial magic of the Golden Dawn, the society developed its philosophy from a variety of theologies. This philosophy included an understanding that there was one god who was the source of all gods that had been perceived by mankind since time immemorial. The various deities were simply expressions of the various aspects of God. John Symonds explains, "From there, it was obvious that across the world, certain deities—and qualities, and colors, and planets, and so on—were equivalent."[9]

The questions of whether there is but one god or many, if possession of a soul is limited to human beings or granted to all things, animate or inanimate, and whether we live in a world of illusion or reality—duality or a unification of all that exists—remain without universal consensus. Cultural perceptions and traditions cause confusion between societies and conflicts between nations. But some Deists have found in the Kabbalah a philosophy that can resolve this confusion. Levi's comparison of the tarot (cornerstone of philosophy of the Golden Dawn) to the ten Sephiroth and the twenty-two Nativoth allowed Hermetic researchers to relate the Tree of Life to gods worshipped from the cultures of ancient Greece and Egypt. They extended a pantheistic view of the universe and explored implications for such worldly materials as gemstones, vegetation, animals, and so forth. Ultimately, this bringing together of "lost traditions, the exact knowledge of . . . signs and symbol" proved to be the concept that unified the Golden Dawn and served as the "foundation on which they could build their rituals and magical practices, and gave the Society an internal coherence that it would otherwise have lacked."[10]

Israel Regardie

Israel Regardie, born Francis Israel Regudy in London in 1907, followed in Levi's footsteps to further the legacy of the Hermetic Order of the Golden Dawn by encouraging popular support of the occult. His father was a cigarette maker, his mother an orthodox Jewish immigrant from Russia. The family immigrated to the United States in 1921. There Regardie studied art and learned Hebrew. He educated himself in Hindu philosophy, yoga, and theosophy. After associating with members of the Golden Dawn and living in England for several years, he published *The Tree of Life* and *A Garden of Pomegranates*, which address his interests in magick and Kabbalah based on work with Aleister Crowley. Crowley was an English mystic, occultist, magician, and member of the Golden Dawn who espoused the rule "Do What Thou Wilt."[11] He was a hedonist, a rebel, and drug user, and his ideas regarding magic, mysticism, philosophy, and ethics continue to produce significant debate to this day.

Regardie wrote a biography of Crowley *(The Eye in the Triangle)* and republished several of Crowley's works into the 1970s. He joined the Stella Matutina Temple of the Order of the Golden Dawn in 1934. His book *The Golden Dawn* is based on documents he acquired after the order dissolved. The compilation only served to increase the popularity of the occult in the West, an expressed purpose of Regardie's writings: "it is essential," he wrote, "that the whole system should be publicly exhibited so that it may not be lost to mankind. For it is the heritage of every man and woman—their spiritual birthright."[12] His other works as an initiate at the Stella Matutina Temple included *The Middle Pillar*, *The Art of True Healing*, and preparations for *The Philosopher's Stone*.

Regardie believed magic was a subset of psychology with therapeutic benefit to patients within a clinical setting. He presents his case in *The Middle Pillar*, describing an exercise of the same name that became the foundation for one of the most fundamental rituals in contemporary Western magic. But it is in *The Tree of Life* that he provides a detailed survey of the Western magical tradition, including the theory upon which those practices are based. He is in agreement with Levi that magic was "the traditional science of the secrets of nature which has been transmitted to us from the Magi."[13] As Regardie explains, magic is far more noble in its purpose and holy in its results, relating the self to the universe:

From academic sources Magic is defined as "the art of applying natural causes to produce surprising effects." ... The result, which the magician above all else desires to accomplish, is a spiritual reconstruction of his own conscious universe and incidentally that of all mankind, the greatest of all conceivable changes ... In the ... case [of Yoga], the spiritual axe is laid to the root of the tree, and the effort made consciously to undermine the whole structure of consciousness in order to reveal the soul below. The magical method, as opposed to this, endeavors to rise altogether beyond the plane where trees and roots and axes exist. The result in both cases—ecstasy and a marvelous outpouring of gladness, wildly rapturous and incomparably holy—is identical.[14]

The Hermetic Order of the Golden Dawn understood that there was one god, the source of all gods perceived by men, and that all other deities were expressions of the aspects of God. This idea is surprisingly reminiscent of the Lakota expression *Mitakuye Oyasin,* meaning "We are all related" or "All our relations." While the Lakota perceive a variety of divine life forms that created the world, the stars, the animals and plant life, and humans, they in fact recognize that the entire universe including the creator are unity—parts of the whole, one and the same. But Regardie recognized an inherent problem in Western society. This problem pertained to a separation, a classification, an attempt to cloak our eyes from the truth and the vision we must have if we are to improve our relationship with each other, the universe, and the creator. He wrote,

> ... where there is no vision the people perish." [Humankind] ... has lost in some incomprehensible way its spiritual vision. An heretical barrier has been erected separating itself from that current of life and vitality which even now, despite willful impediment and obstacle, pulses and vibrates passionately in the blood, pervading the whole of universal form and structure.[15]

Regardie understood the traditional philosophy of magicians, that each "man is a unique autonomous center of individual consciousness, energy, and will—a soul, in a word."[16] He likened every man to a star shining with its own inner light, solitary and moving through the cosmos without interference other than the gravity of other stars, whether proximal or distal to its course. And so he finds human lives analogous to the mechanics of the universe, and he applies the philosophical terms of Kabbalah as he weaves a blanket of understanding "around the central structure of the Tree of Life."[17] The source of this arboretum of philosophy is ancient Egypt:

> The principal ingredients of the magical system are the source of reference which is the Tree of Life of the Qabalists, and the hieratic religion of the sacerdotal caste of Egypt . . . Other people firmly maintain that if ever such a person as Moses existed historically, and if the Qabalah and its corollaries emanated from him, then he obtained it from the Egyptian priests with whom he indubitably studied in the Nile temples.[18]

The Golden Dawn believed that Tree of Life developed from a philosophy first developed by the ancient Egyptians. In this light Kabbalah was a philosophy of evolution in which everything in the universe emanated from a primeval substance or principle. That entity is infinite source of *Ein Sof*—Without End—that we encountered in chapter 11. Regardie states, "In variegated forms, the archetypal ideas find their particular representation below; stones, jewels, perfumes, and geometrical forms all being peculiarly indicative in the mundane sphere of a celestial idea."[19] He then provides further information on the meaning of various geometrical constructs, illuminations, and vibratory frequencies used in magic, the fundamental geometrical structure being the hexagram (Figure 12.1). We've already seen this structure as the symbol of the principle "as above, so below." Finding an equivalence with the hexagram, Regardie suggests that, "the Astral Light contains the builder's plan or model . . . on which the external world is constructed, and with whose essence lies latent the potentiality of all growth and development . . . It is at once substance and motion, the movement being one which is "simultaneous and perpetual in spiral lines of opposite motion."[20]

Figure 12.1: The hexagram is used as a fundamental geometrical structure in magic.

Again we see in these statements a significant similarity with an array of mythologies and traditions previously described in this volume. Regardie refers to Jungian psychology and the term *collective unconscious*, which describes the "mental patterns and primordial images that are shared by all of humanity."[21] Regardie further states that "One of the fundamental postulates of magic is that Man is an exact image in miniature of the universe, both considered objectively, and that what man perceives to be existent without is also in some way represented within."[22] In this we see a conceptualized relationship between man and God, a spiritual power within man, and an exertion of that power by the magician. Regardie makes this most plain when he concludes that the purpose of religion is to develop humanity toward a state of perfection via spiritual processes and a union with Ein Sof.[23] This spiritual power and human objective appear very similar to the power and objective of the shaman, but they are received and applied through other means. Regardless, the power is derived from the creator, whom Regardie relates to the Egyptian god Ptah whose first appearance initiated, or opened, a cyclical manifestation of the cosmos.[24]

Perceiving Ptah as the essence from which the universe was created upon a potter's wheel is analogous to the magician working within his magical circle—a protective circle or sacred space barring all outside or negative forces—drawn on the floor or envisioned on the astral plane.[25] This is a very direct correspondence between ancient mythology and the magician's traditional practice. Observing the various cultural traditions compiled by the Golden Dawn to produce its own form of mysticism, Regardie noted that "All the legends and myths of the ancient peoples in association with the Gods disclose a valuable account of their true nature, if but a little discrimination be employed with an understanding of the fundamentals which form the basis of the Qabalah."[26] And it is the Kabbalah, and more specifically the Tree of Life, that was held in high regard in the Golden Dawn's philosophy, one in which various geometries continue to be applied in ritual and ceremony. These sacred geometries are also related to numerical values common to mythologies associated with an understanding of precession.

Regardie observes that while in ceremony a Theurgist stands inside of an octagon with names suggestive of Mercury and Hermes around a circle. This "Magical Circle" separates inside from outside, imposing a spatial confinement to the magician's activities such that illusion and the constant state of flux perceived in the universe are removed from thought. It is not only a "symbol of the infinite," but "also typifies the astral sphere . . . The Circle in which the magician is enclosed represents his particular cosmos; the conquest, self-inaugurated, of that universe is part of the process to attain complete self-consciousness."[27] Regardie also notes *The Lesser Key of Solomon the King* or *The Goetia* that refer to seventy-two spirits invoked by Solomon. He discusses qualities (colors, dimensions, inscriptions, and names of the divine) to be painted around the Magical Circle and a Triangle used in ceremony, their meaning associated with "the dominion of the Four Greater rulers or Elemental kings of the Cardinal Points are these hierarchies of seventy-two spirits."[28]

Dedopulos concludes that "Undoubtedly, the original architects of the Kabbalah would have been deeply dismayed by the use that the Golden Dawn made of their work."[29] At the same time, he finds that as the underlying structure of the creator's universe, "the Kabbalah should manifest itself within all creatures and belief systems."[30] Dedopulos further notes that the biblical doctrine of the Sundering is an expression of reality being an aspect of God's divided self, but in seeking a reunification with divinity the universe at all levels is connected to itself. We encountered this same idea expressed in the Upanishads, Native science, ancient Egyptian philosophy, and so forth. Dedopulos finds in these ancient and contemporary understandings a remarkable correspondence with the theory of quantum mechanics, in that each particle of matter, from subatomic to macrocosmic contributes to what may be envisioned as a holographic universe.[31] To paraphrase d'Artagnan's famous exclamation *"tous pour un, un pour tous,"*[32] science and religion have found common ground . . . all is one, one is all.

How reasonable is it to accept the interpretation of the Tree of Life as perceived and applied by magicians? There is really no way to know for sure. The search for universal truth and meaning is a journey meant for each of us to undertake individually. Dedopulos reminds us that various interpretations and applications of the Tree of Life have merit, as it represents a map that is versatile and reflective of other philosophies that are structured to help us discover ourselves and the relationships we share with all matter and energy in the universe: "A profoundly versatile map of symbolism and linkage, it reflects . . . God's creation of the universe, and humanity's way of purification and enlightenment—but also all other structures. The art of interpreting the tree in the light of other systems is known as correspondence.[33]

Completing the Picture

Harold Bloom, Sterling Professor of Humanities at Yale and Berg Professor of English at New York University, is a self-described Jewish Gnostic. His book *Omens of the Millennium: The Gnosis of Angels, Dreams, and Resurrection* reviews the intensifying interest in angels and other spiritual contacts during the latter years of the twentieth century, and he concludes that those contacts (through conscious or unconscious means) are associated with a pervasively Gnostic outlook on the part of Western culture.[34] In other words, there is a popular realization that God is not external to our world, but is found within each of us and interacting with us through supernatural beings, or what we previously suggested to be various aspects or facets of God. This expression of Gnosticism might not bode well for institutionalized religion, but it does suggest that Western society takes spirituality very seriously.

As if discovering unity within the diverse philosophies of ancient traditions, Bloom identifies the foundation for this contemporary awakening of spirituality in an amalgamation of Jewish Kabbalah, Christian Gnosticism and Muslim Shi'ite Sufism. The resounding interest in this Gnostic outlook is not a realization of the ending of an age, or the coming of a New Age. It isn't a popular attempt at using magic or the occult (hidden knowledge) for self-interest or personal gain, to force harmony among people, or just to feel good about one's self.

Bloom contends that, rather than having faith in God or a belief in a creator who encourages universal peace and understanding, people are receiving and acting on knowledge, a gnosis, that they are one with the universe. The action of one person sends a ripple outward in all directions, affecting the entire universe.

This, of course, is the schema upon which Freemasonry is founded—that the religious dogma with which each member identifies is not of import to the organization. Freemasonry is not a religion. It is not a substitute for religion. But there are general requirements that each candidate must meet to be considered for initiation as a regular Freemason, none of which include that the candidate be employed as a mason. An initiate may be Jewish, Christian, Muslim, Hindu, Buddhist, Sikh, Mayan, or any other organized or unorganized religion. However, aside from age requirements and anachronistic concerns related to free birth and having sound mind and body, the candidate must: come of his own

free will; believe in a Supreme Being; have good morals and a good repu-
tation; and provide character references. These requirements help ensure
that each candidate will behave in accordance in the setting of lodge ceremonies, rituals, and social functions, and that each member of the
organization is understanding and accepting of a higher power in this
universe. I can only assume that the last of the four requirements, that of
providing character references, is a matter of "trust but verify."

Variations in Unity

Don Karr is a well-known artist and linguist. His study of Kabbalah
produced scholarly discussions of the history and significance of Kabbalah
as it relates to ancient mythology. His 2010 essay, "Approaching the
Kabbalah of Maat: Altered Trees and the Precession of the Æons,"
provides significant insight into the forms and uses of the Tree of Life
and the precession of the eons—a schema of three evolutionary steps of
mankind outlined as follows by Aleister Crowley in *The Book of the Law*:

1. Worship of the Mother, continually breeding by her own virtue

2. Worship of the Son, reproducing himself by virtue of voluntary
    death and resurrection

3. Worship of the Crowned and Conquering Child[35]

Noting that there was a tremendous outpouring of literature
concerning Kabbalah during the latter part of the twentieth century,
Karr argues that the various forms and uses of the Tree of Life and the
precession of the eons were not independent matters. He describes the
Tree of Life as a "western mandala," used in meditation, but also serving
as an icon of occult knowledge that can form the basis of "Ritual Magick."
He writes, "These may be called the subjective and objective ways by
which to use the Tree of Life. The Tree comprehends and synthesizes all
forces, forms, and concepts of the Universe, and it embraces all essential
keys to attain true union with the Divine. It is a most excellent model by
which to view your entire Universe."[36]

I began this chapter by noting that a breadth of cultural and religious
traditions provided symbols that have shaped Western thought for thousands of years. At this point those symbolic geometries are quite familiar
to us. The degree to which ancient symbolism affected the philosophy of
the Golden Dawn is made plain in Kerr's commentary regarding

Crowley's book *Liber 777*, of which Kerr notes nearly two hundred columns of correspondences associated with the Tree of Life. He notes,

> Among the sources which Crowley's introduction acknowledges are *Kabbala denudata*, "the lost symbolism of the Vault in which Christian Rosenkreutz is said to have been buried," John Dee, H. C. Agrippa, Pietri di Abano, the "Art" of Ramon Llull, Pietro di Abano, Eliphas Levi, the Hermetic Order of the Golden Dawn, "Swami Vivekananda, the Hindu, Buddhist, and Chinese Classics, the Quran and its commentators, the Book of the Dead, and, in particular, original research." The preface of *777* goes on to say, "The Chinese, Hindu, Buddhist, Moslem, and Egyptian systems have never before been brought into line with the Qabalah; the Tarot has never been made public."[37]

Of particular importance here is the reiteration of specific geometrical relationships expressed in the various forms of the Tree of Life and associated diagrams that the Golden Dawn applied to illustrate fundamental philosophical concepts relating the human being to God, the microcosm to the macrocosm. As Regardie writes, "The Tree of Life establishes the template for the organization of *everything*: the universe; the body, mind and soul; the initiated grades; courses of meditation; etc. The various systems (structures or pantheons) arrayed on the *sefirot* and paths are supposedly helpful in understanding of the parts *and* the whole of the Tree."[38] Figure 12.2 illustrates various versions and attributes of the Tree of Life that have been applied in magic.

In 1923 Frater Achad (Charles Stansfield Jones), a prodigy of Crowley's, reconstructed the Tree of Life in a manner that would have caused a considerable concern even to Crowley and other members of the Golden Dawn, let alone scholars of mystical Jewish orthodoxy. Essentially, Achad reassigned Hebraic letters and the associated tarot trumps to the paths in reverse order—overturning of the Tree's structure by one hundred eighty degrees—ascending through the Tree, by way of what the Golden Dawn referred to as the Path of the Serpent of Wisdom. The rationale for this turn of events we can forego in this discussion. For us the important point is that the geometrical construct of the Tree and the philosophical interpretations thereof, are many and varied. Yet the general form of the symbol remains, regardless of the minor variations in pathways or orientation of the construct as a whole.

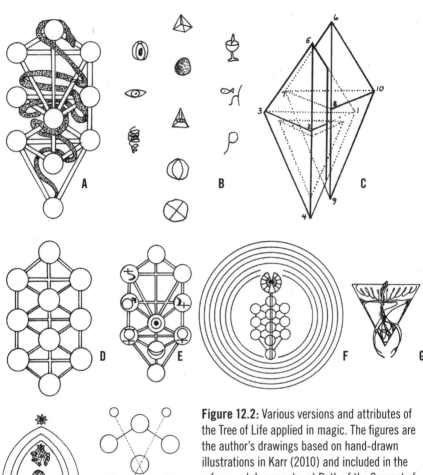

Figure 12.2: Various versions and attributes of the Tree of Life applied in magic. The figures are the author's drawings based on hand-drawn illustrations in Karr (2010) and included in the referenced documents. a) Path of the Serpent of Wisdom on the Tree of Life by Frater Achad; b) Tree of Life from *Liber Magnus Conjunctiones* prepared by Kenneth Grant; c) Variation based on Figure 5, page 15: Bi-pyramids, prepared by Ordo Adeptorum Invisiblum [OAI]; d) The Perfected Tree prepared by 416; e) Golden Dawn Tree of Life—top sefirot is "Primum Mobile," upper right "Zodiac," and bottom is Elements; letter attributions of paths not shown for clarity; f) Golden Dawn "Perfected tree" from Figure 16 in Rosenroth's Kabbala denudate, Tome I (Sulzbach: 1677); g) drawing based on Axil/Aion Tarot, Card 12, by Aion 131; h) Processes of Initiation, from The Process of Initiation by RA Oh 1043; i) Passage of Influences on the Perfected Tree on page 30 of Karr, 2010.

History has demonstrated a continuity of thought and expression that is only now becoming clear to us in an amalgam of human traditions. In this and previous chapters we've looked at the cultural and religious traditions of both ancient and contemporary societies and noted geometries that pervade symbolism expressing relationships between microcosm and macrocosm, man and creator, and all things in the universe. It is readily apparent that the same geometry has been applied time and again in different cultures. While some specifics of the symbolic constructs have varied over time, there is a general theme to the geometry of these symbols.

The geometric constant is the circle. As we have seen, the circle has been used for thousands of years as a two-dimensional symbol representing the sphere. This theme continues with us for the remainder of the book. That the circle symbolizes the sphere as an expression of universally understood spiritual and secular relationships is a concept surprisingly unrealized in today's world. In section III I identify the components of this geometry, a reverse engineering of sorts, and then bring all the pieces together again to view this theme in its entirety.

Part 3

The Third Dimension

Chapter 13

The Disdyakis Dodecahedron

Moyers: Now what do you make of that—that in very different cultures, separated by time and space, the same imagery emerges? Campbell: This speaks for certain powers in the psyche that are common to all mankind. Otherwise you couldn't have such detailed correspondences . . . . Schopenhauer's answer is that such a psychological crisis represents the breakthrough of a metaphysical realization, which is that you and that other one, that you are two aspects of the one life, and that your apparent separateness is but an effect of the way we experience forms under the conditions of space and time. Our true reality is in our identity and unity with all life. This is a metaphysical truth which may become spontaneously realized under circumstances of crisis. For it is, according to Schopenhauer, the truth of your life.

JOSEPH CAMPBELL
THE POWER OF MYTH

The Same Imagery Emerges

In chapter 4 we investigated the possibility that the ark built by Utanapishtim, king of Shuruppak, was constructed in a spherical shape. Named *Preserver of Life*, the ark had dimensions equal in width and length: "[t]he length, width, and height each measured 120 cubits." In another Sumerian myth about the flood, Atra-Hasis is told, "Destroy your house, build a boat; despise possessions And save life! Draw out the boat that you will build with a circular design; Let its length and breadth be the same." Irving Finkel concluded that Atra-Hasis's vessel was circular in shape. However, by reviewing each bit of information provided to us in the myths, we are able to conclude that the ark was not only circular, but spherical. This shape was far more advantageous than the common bow and stern ark form we think of for surviving the Great Flood. Recall, too, that the *Preserver of Life* included floor space measured as one field, an interior divided into seven levels, and levels each divided into nine sections.

Our spherical design satisfies each of the design requirements listed in the myths without adding further design, construction, and operational complications.

In reviewing other world myths we encountered geometrical information in oral traditions that describe the origin of the world and the creation, destruction, and re-creation of our world and of life itself. There are a number of geometrical relationships common to the two-dimensional symbols presented in chapter 2, and numerous descriptions of two-dimensional and three-dimensional geometries are exemplified in chapters 3 through 11. This geometry shows itself again and again, proving its relevance to sacred relationships even today.

We can see this same geometry in the Cloud Peak medicine wheel. Note how the configuration of a subgroup 6 medicine wheel—and particularly the Cloud Peak medicine wheel, with its central cairn, stone ring, and eight stone lines radiating from the cairn to the ring—conforms with the configuration of rocks placed to form a traditional Lakota altar as described by Black Elk (Figure 13.1):

> One of the rocks . . . is placed at the center of the round altar; the first rock is . . . at the center of everything . . . The second rock is placed at the west . . . the next at the north, then one for the east, one for the south, one for earth, and finally the

hole is filled up with the rest of the rocks, and all these together represent everything that there is in the universe.[1]

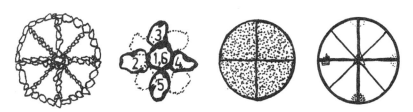

Figure 13.1: Depictions of Cloud Peak medicine wheel, Black Elk's grouping of stones representing everything in the universe, a stick-drawn diagram of the Sacred Hoop, and Wohpe's disk painting.

Again, in use of earth (soil) and lines drawn on the ground surface:

A pinch of the purified earth was offered above and to the ground and was then placed at the center of the sacred place. Another pinch of earth was . . . placed at the west of the circle . . . earth was placed at the other three directions, and then spread evenly around within the circle . . . He first took up a stick, pointed it to the six directions, and then, bringing it down, he made a small circle at the center; and this we understand to be the home of Wakan Tanka. Again, after pointing the stick to the six directions, . . . [he] made a mark starting from the west and leading to the edge of the circle. In the same manner he drew a line from the east to the edge of the circle, from the north to the circle, and from the south to the circle . . . everything leads into, or returns to, the center.[2]

And in "A Myth of the Tetons as It Is Told in Their Winter Camps," told by holy man George Sword,

Wohpe is instructed to take a disk and paint it green; paint a blue stripe around the edge of the disk; paint a broad red stripe across the disk, over its center; paint another broad red stripe across the disk over its center so as to divide the disk into four equal quadrants; and paint four narrow red stripes across the disk over its center and between the red lines so as to divide the disk into equal parts [eighths].[3]

The Geometry Defined

Viewing the Cloud Peak medicine wheel orthogonally from above, over the center of the cairn, the two-dimensional configuration of lines formed by the cairn, circle, and eight stone radii matches the configuration of lines of a disdyakis dodecahedron projected onto a plane (Figure 13.2).

 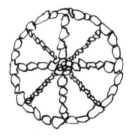

Figure 13.2: Cloud Peak medicine wheel reflects the geometrical form of the disdyakis dodecahedron projected onto a plane.

For brevity I will generally refer to the disdyakis dodecahedron (also called a hexakis octahedron) as DD. A DD is a polyhedron, a three-dimensional solid, just as the tetrahedron and cube are polyhedral solids. Drawn on paper in two dimensions, for example, it has many of the characteristics of the geometries listed above from mythology, as well as in historic and modern spiritual and religious practices. The exterior outline consists of an octagon with eight equivalent sides. Geometricians call each line segment an edge and each corner a vertex. Eight more lines extend from each exterior corner to the center of the drawing. Additional

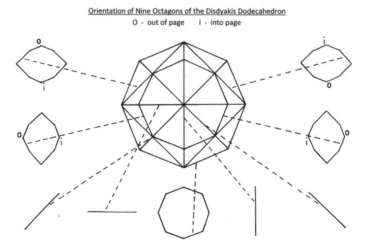

Figure 13.3: The DD is constructed from nine octagons.

line segments form more edges and vertices, and the result on paper appears as the front half of a solid DD, with twenty-four triangles separated by a certain number of edges and vertices. In fact, in three dimensions the DD is seen to be constructed from a total of nine octagons (Figure 13.3). Three of the octagons are regular, each consisting of eight sides of equal length. The other six octagons are irregular, having four sides of equal length and four sides of another length.

Each of the forty-eight triangles of the DD is constructed from three edges, each edge having a unique length. The result is that each of the triangles (or faces) is a scalene triangle (Figure 13.4). If the shortest edge has a length of unity (1.00000 . . .) then the other two edges have lengths of about 1.34 and 1.63. In other words, the ratios of their lengths are approximately 3:4:5. Here are some additional geometrical facts about the DD:

- The DD is a convex, three-dimensional polyhedron. It is one of thirteen Catalan solids named for Eugène Catalan, the Belgian mathematician who was the first to describe them in the mid-nineteenth century.

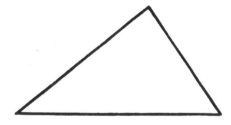

Figure 13.4: A scalene triangle has three sides of unique length.

- Its dual is the Archimedean truncated cuboctahedron (great rhombicuboctahedron). In dual polyhedra, the faces of one of the solids correspond to the vertices of the other.

- The DD has forty-eight faces, all equal in size, each consisting of a scalene triangle. It is face-uniform, with twenty-four of the faces identical (face transitive) and the mirror image of each of the other twenty-four faces (which are also face transitive). Let's call the first twenty-four faces A and the other twenty-four faces B. As a triangle, each of the faces adjoins three other triangles. Each A face adjoins three B triangles, and each B face adjoins three A triangles.

- It has seventy-two edges: nine octagons times eight sides per octagon equals seventy-two sides (edges).

- It has twenty-six vertices, where the corners of three triangles meet. It has a face configuration of V4.6.8, indicating that alternating vertices of the triangular faces are associated with 4, 6, or 8 faces each.

- Based on the number of faces, edges, and vertices of the DD, the Euler characteristic (or Euler–Poincaré characteristic), defined as $x = V - E + F$ where V is the number of vertices, E is the number of edges, and F is the number of faces, is 26 - 72 + 48 = 2. The Euler characteristic is a topological invariant describing the shape or structure of a topological space. Topology is area of mathematics addressing the properties of space. The value of 2 is equivalent to the Euler characteristic of the sphere (2), applied identically with spherical polyhedra.

- The dihedral angle (the angle between faces that share a common edge) is 155° 4' 56."

- It has full (or achiral) octahedral symmetry (Oh or *432) of order 48. The Oh group has the same rotation axes as chiral octahedral symmetry, but with the addition of mirror planes. This group is isomorphic and the full symmetry group of the cube and octahedron. It includes six orthogonal vertices (perpendicular to the six faces of a cube). Included in the isometric crystal form, the dodecahedron has one of the highest degrees of symmetry. The isometric crystal system includes forms approaching sphericity. It is the hyperoctahedral group for n = 3. It has the following conjugacy classes (see Figure 13.5):

  - identity
  - 6 × rotation by 90°
  - 8 × rotation by 120°
  - 3 × rotation by 180° about a 4-fold axis
  - 6 × rotation by 180° about a 2-fold axis

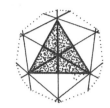

Figure 13.5: Conjugacy classes of the DD.

- The faces being triangles, if the shortest of the three sides has a length of 1 unit, then the three edges of each face have lengths as follows (see Figure 13.6):

    - $S_1 = 1.0000000$ ....
    - $S_2 = 1.3377087$ ....
    - $S_3 = 1.6306010$ ....

and

- The inradius (radius of the largest circle inscribed in a face) is 0.3366960 ....

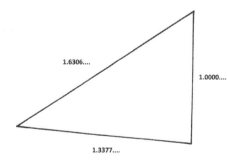

Figure 13.6: The scalene triangle of the DD. Each side is labeled with its relative length.

1.6306....

1.0000....

1.3377....

- Scaling $S_1 = 1$ gives a solid with surface area (S) and volume (V):

    - $S = 21.4224$ ....
    - $V = 4.0$ ....
    - The surface area to volume ratio is $S/V = 5.35$

Implications

Included in the isometric crystal form, the dodecahedron has one of the highest degrees of symmetry. The sphere is considered to have perfect symmetry, with infinite planes of symmetry passing through its center, the presence of infinite rotational axes, and yielding the same apparent form under any given amount of rotation. The isometric crystal system includes forms approaching sphericity, or what the Lakota of the nineteenth century might have referred to as *roundness*. Minerals that exhibit an isometric crystal form like the symmetry of the DD include diamond (carbon), the hardest mineral on Earth (Figure 13.7).

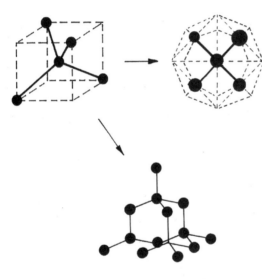

Figure 13.7: Diamond exhibits an isometric crystalline structure like the symmetry of the DD.

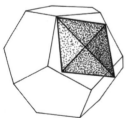

Figure 13.8: By replacing each pentagon of the dodecahedron with four triangles (forming an inflated rhombus), the result is the disdyakis dodecahedron.

Recall that the regular dodecahedron is a regular polygon and one of the five Platonic solids. It consists of twelve pentagons. If we replace each pentagon with a rhombus (a parallelogram with four equal sides) then the structure becomes a rhombic dodecahedron. If each rhombus is then subdivided into four triangles and inflated a bit, the resulting polyhedron is the disdyakis dodecahedron (Figure 13.8).

Another way of looking at the structure of the DD is to recognize that it consists of six sets of eight triangles (Figure 13.9a). Each of these octets of triangles covers one-sixth of the total surface area. Each octet is centered on a unique vertex. If the pole of one of those vertices is oriented straight up toward the sky, then there is another octet with the pole of its central vertex pointing toward Earth. If the pole of one of the remaining four octets is then pointed due west, then the poles of the other three octets will point due north, east, and south. This indicates the cubic symmetry of the DD.

We can perform a similar exercise when we observe eight sets of six triangles, with four sets in the upper half, four in the lower half, and a regular octagon serving as the equator separating the two halves (Figure 13.9b). If we align two of these triangular sextets, one in the northern hemisphere and one in the southern hemisphere, toward the north, then the pairs of similar sextets about the polyhedron will be oriented toward the east, south, and west. The pole of the central vertex in each sextet in the northern hemisphere will point upward at an angle of 30 degrees from horizontal, while the pole of the central vertex in each sextet in the southern hemisphere will point downward at an angle of 30 degrees below horizontal. This implies the octahedral symmetry of the DD.

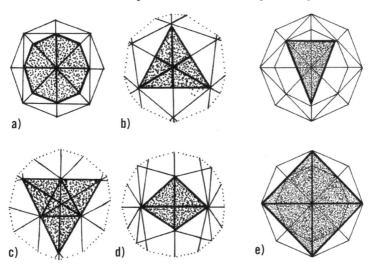

Figure 13.9: The DD can be seen consisting of a) six sets of eight triangles, b) eight sets of six triangles, c) four sets of twelve triangles, d) twelve sets of four triangles, e) other combinations.

We can also see four sets of twelve triangles that form four large equilateral triangles over the surface of the DD (Figure 13.9c). From this geometry, of course, we find the symmetry of the tetrahedron. And after recognizing that the symmetries of four of the Platonic solids are found in the DD, we should not be surprised that the symmetry of the icosahedron, the fifth Platonic solid and a regular polyhedron with twenty identical, equilateral triangular faces, thirty edges, and twelve vertices, is also contained in the DD. This becomes quite apparent once we realize that the dual of the icosahedron is the dodecahedron itself.

With this background information, and the assistance of Buckminster Fuller's career-long study of geopolyhedra, we have the tools to look at the specific case of the disdyakis dodecahedron and its projection onto a sphere. Recall that the faces of the DD are forty-eight equivalent scalene triangles. There are twenty-six vertices and seventy-two edges. Euler's formula says that the number of vertices plus the number of faces in every system will always equal the number of edges plus two (V + F = E + 2). Applying this formula to the disdyakis dodecahedron we find, of course:

$$26 + 48 = 72 + 2 = 74$$

Nothing unusual here. Next, let's look at the number of great circles that result from the polyhedron's symmetry. The objective is to confirm Fuller's method for intertransformability between the disdyakis dodecahedron (DD) and its spherical cousin. Thirteen great circles result from the thirteen axes defined by the DD's twenty-six vertices. Projected onto a sphere, these axes define thirteen great circles that divide the sphere's surface into forty-eight convex, triangular areas. Those thirteen great circles form the edges of the spherical DD. I call this structure a *disdyakis dodecasphere*. The forty-eight triangular faces result in arcs producing the most symmetrical arrangement of nine great circles. The DD's seventy-two edges include thirty-six opposing pairs, and the midpoint of each edge of a pair can be connected by thirty-six intersecting axes. Those axes then generate thirty-six great circles. Finally, the axes of symmetry connecting the face centers of the DD create twenty-four great circles. Shown in Figure 13.10, the total (13 + 36 + 24) is seventy-two great circles. But, there are some surprising relationships between a number of these great circles and those produced from polyhedral projections we have seen before.

Platonic Relationships

Recall that the forty-eight faces of the DD can be envisioned as eight groups of six, forming the eight faces of the octahedron, but with each of the octahedron's triangular faces divided into six equivalent triangles, illustrated in Figure 13.10. By this fact we find that the DD replicates what Fuller described as "unique topological aspects" of the octahedron, the thirteen great circles corresponding to the octahedron's symmetry! Since Fuller also found that the resulting great circle pattern of the cube is identical to that of the octahedron, it follows that the equivalents of the thirteen great circles of the cube are found in those of the DD.

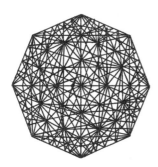

Figure 13.10: Seventy-two great circles are formed from circles derived from the DD's vertices, edges, and face centers.

Let's back up for a moment and look at the tetrahedron—four faces, four vertices, six edges. The forty-eight faces of the DD can be grouped into four systems of twelve triangles each. Those groupings are geometrically equivalent to each other, and the four groups are oriented around the center of the polyhedron such that they are equivalent to the four faces of the tetrahedron. Therefore, we have equivalence between the great circles resulting from the spherical tetrahedron and a portion of the great circles of the spherical DD. Applying Fuller's intertransformability analysis to the tetrahedron, we find that four great circles result from the four axes defined by a line extending from a vertex through the opposite face. Three additional great circles result from the tetrahedron's six edges, forming three opposing pairs. And finally, we find that the axes of symmetry connecting the face centers with the opposite vertex have already been accounted for, this adding no additional great circles. So, our total is $(4 + 3 + 0 = 7)$ great circles for the tetrahedron. Thus, these seven great circles are equivalent to seven of the great circles of the spherical DD.

In fact, by applying the same method of analysis to the icosahedron and the cuboctahedron (Fuller's vector equilibrium or VE), we discover the same result as we found with regard to the tetrahedron, cube, and octahedron: Each of the great circles associated with the spherical projection of those polyhedra is found within the various great circles of the spherical DD. We also saw from Fuller's analysis that the thirty-one great circles of the spherical icosahedron are equivalent to the spherical edges of the pentagonal dodecahedron (as well as the octahedron). In summary, the great circles associated with the five Platonic polyhedra as well as the VE are all equivalent to great circles associated with the spherical DD. In other words, the great circles resulting from each of these polyhedra are a subset, a subsystem, within those of the spherical DD!

With this discovery, we can conclude that the geometry of the DD contains all information that is found in each of the five Platonic solids, as well as the VE. The importance of this conclusion can be understood by looking back at what history has to say about the symbolism of each of those structures. In other words, at least part of the symbolism, the metaphor of the DD, can be understood by what history says about the symbolism of its subsystems. Let's take a look at several of the important ideas that developed from the geometry of the Platonic solids.

The five Platonic solids had been carved out of rock in Scotland during the late Neolithic more than a thousand years before Plato began his studies of them, and many speculate that the structures were known long before then.[4] But it is the ancient Greeks who are credited with their discovery via their documentation of the mathematical structure of the tetrahedron, cube, octahedron, icosahedron, and dodecahedron (Figure 13.12).

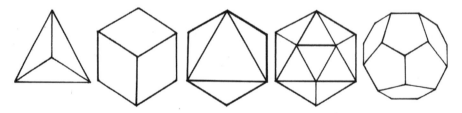

Figure 13.12: The Platonic solids: tetrahedron, cube, octahedron, icosahedrons, and dodecahedron.

Theaetetus, a friend of Plato's, appears to have been the first person to have written an account of these five regular solids—so states Euclid, who included discussion of the Platonic solids in the thirteenth and final book of his *Elements*. The importance that Theaetetus recognized in these polyhedrons was associated with their symmetry—each face of a regular polyhedron is a regular polygon (identical and equilateral), and all vertices of each solid lie on a sphere (the distance from center of the solid to each vertex is a constant). His study of how many regular solids there are began with determining how many solids could be constructed with only equilateral triangle faces. The tetrahedron was the first regular solid to be put on the list, with the other four following as the number of sides per face increased from three (tetrahedron, octahedron, and icosahedron) to four (cube) and then five (dodecahedron). But with six sides to a face (the hexagon), the tiling of polygons resulted in a plane, without the curvature needed to produce a solid, and so the accounting ended with a total of five regular solids. Theaetetus' impressive analysis included details such as calculation of the ratios of the edge lengths of each solid to the diameters of their circumscribing spheres.

It was Theaetetus' accounting of regular solids that likely inspired Plato toward further analysis. He explained his "theory of everything" in *Timaeus*, written in about 3650 BC in the form of a dialogue. The work is devoted to the study of those five solids. Interestingly, Plato derived each of the solids from right triangles that he proposed formed the basic structure of all matter in his theory of everything. We can immediately see the similarity between Plato's triangular particles and Fuller's study of triangles as the elementary structure of space. However, Plato's treatise heads off on a tangent as he associates four of the regular solids with the four classical elements (earth, air, water, and fire). Plato states that the fifth solid, the dodecahedron, is ". . . the god used for embroidering the constellations on the whole heaven."[5]

Plato's objective in *Timaeus*, then, is to describe the mechanics and origin of the universe through deeper understanding and application of the five regular solids. But there was a problem. While the gods might have used the dodecahedron as the template for applying constellations to the celestial sphere, Plato did not identify an element fundamental to the dodecahedron as he had by relating the other four regular solids to the four fundamental elements. Aristotle later identified the "aether" of the heavenly bodies, a substance believed to be inherent in all the universe, to be the fifth element. Was it possible that Plato missed this fifth element, this quintessential material that provides perpetual motion and change in the universe? If we follow this train of thought, we might very well find ourselves on the very same tangent as Plato.

What is of greater interest to us is that Plato's theory states that, except for the dodecahedron, four of the regular solids can be constructed using two types of right triangles—an isosceles triangle with angles of 45, 45, and 90 degrees, and the 30-60-90 degree triangle with the infamous 1:√3:2 relationship for the three legs. Plato then applies these geometrical properties of the polyhedra to a discussion of chemical principles that relates the four basic elements to each other. While Plato misses the mark in the arena of chemistry, his insight regarding the importance of symmetry and mathematics to understanding the fundamental building blocks of the physical universe is spot on. Buckminster Fuller built his career on just this idea.

Almost two thousand years later, the German astronomer Johannes Kepler discovered the relationship between five of our sister planets and the five Platonic solids. Kepler's *Mysterium Cosmographicum* (1596) proposes a model of our solar system with the five regular solids placed inside each another (Figure 13.13). The outermost solid is the cube, which contains the octahedron, which contains the icosahedron, which contains the dodecahedron containing the tetrahedron. Six inscribed and circumscribed spheres are situated between the polyhedra, representing Mercury, Venus, Earth, Mars, Jupiter, and Saturn. Kepler had calculated the distances between these planets and concluded that those distances corresponded to the difference in the radii that separated each pair of nested Platonic solids. Unfortunately he was wrong—another tangent to lead even the most learned astray. However, Kepler's astronomical studies did document that the planets orbit the sun in ellipses, and he developed other theories concerning the movements of the planets in our solar system, contributions that accelerated the further study and understanding of the cosmos.

Figure 13.13: Kepler's model of our solar system with the five regular solids placed inside one another (from Kepler's *Mysterium Cosmographicum*, 1596).

Exploring the Unknowable

Leet's geometrical analysis concerning the Tree of Life provides much of the background information included in chapter 11. That analysis includes consideration for the geometry of the Platonic solids. Recall that the Zohar alludes to a diagram that underlies all of creation and is said to have been disclosed through direct revelation by Elijah:

> This mystery remained sealed until one day, whilst I was on the seashore, Elijah came and said to me ... "When the most mysterious wished to reveal Himself, He first produced a single point which was transmuted into thought, and in this He ... graved within the sacred and mystic lamp a mystic and holy design, which was a wondrous edifice issuing from the midst of thought. This ... was the unknowable by name .... And upon this secret the world is built.[6]

Leet concludes that since the name of the "most holy design" was "unknowable," it could not the Tree of Life diagram of the Kabbalah.[7] Here, however, we must look at her solution to the enigma of the construction of the Sefer Yetzirah, the Hebrew Book of Formation. Recall that the Sefer Yetzirah, believed to have been authored by Abraham, describes how God created the world using "32 wondrous ways of wisdom": ten numbers, or *Sefirot*, and the twenty-two letters of the Hebrew alphabet. Note that the Sefer Yetzirah includes allusions to a cubic diagram (in the sense that it has the same symmetry as that of the cube) because it provides a definition for the six directions of space.

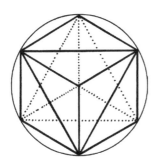

Figure 13.14: The geometry of the twelve elementals (based on Leet, 1999).

Figure 13.14 illustrates this geometry. Leet suggests that the solution to the enigma was indeed known and that a tradition existed of interpreting the Tree of Life diagram as contained within the Sefer Yetzirah cube. This solution is based on the following reference to the Twelve Elementals in the *Bahir* (*Book of Illumination*), the oldest and most important text of the Kabbalah:

> The blessed Holy One has a single Tree, and it has twelve
> diagonal boundaries:
> The northeast boundary, the southeast boundary;
> The upper east boundary, the lower east boundary;
> The south west boundary, the northwest boundary;
> The upper west boundary, the lower west boundary;
> The upper south boundary, the lower south boundary;
> The upper north boundary, lower north boundary;
> They continually spread forever and ever;
> They are the "arms of the world"
> On the inside of them is the Tree.[8]

Leet proposes that the compilers of the Sefer Yetzirah used the term *alakhson* (geometric diagonal) because, as practicing geometers, they recognized that the diagonal and the edge were different characteristics of a polyhedron, and irrational and rational measures needed to be differentiated as well. The cube holds mystical significance as a subtle container of higher cosmic planes and provides a means for emanation of the divine and a return to spiritual understanding. At the same time, the Book of Yetzirah represents the third kabbalistic world and has a geometrical associated with the second dimension, so the cube perceived by the mysticists "would not be an actual solid, rather a two–dimensional representation of such a cube."[9] Leet makes this distinction based on the following Zoharic passage:

> What is the meaning of Bereshith? It means "with Wisdom,"
> the Wisdom on which the world is based, and through this it
> introduces us to deep and recondite mysteries. In it, too, is the
> inscription of six chief supernal directions, out of which there go
> forth six sources of rivers which flow into the Great Sea. This is
> implied in the word BeReSHiTH, which can be analyzed into
> BaRa-SHiTH (He created six). And who created them? The
> Mysterious Unknown.[10]

The meaning of *Bereshith* as "He created six" is intended to suggest that the "totality of existence" was emitted from the "six chief supernal directions."[11] Since the supernal directions preceded existence, they can't be contemporaneous with the world of created solids, nor with the six directions perpendicular to the faces of a cube. Therefore, Leet concludes that these "six chief supernal directions" are those defined by the six points of the hexagram inscribed on a circle, as well as the cube that is perceived in an orthogonal projection. Applying "explicit suggestions within these two kabbalistic texts (*Sefer Yetzirah* and *Bahir*) and the normal process of geometric construction,"[12] Leet proceeds with a detailed presentation for constructing the Sefer Yetzirah Tree of Life diagram beginning with the twelve edges of a cube projected in two dimensions (as on paper), and similarly the twelve diagonals of the Sefer Yetzirah cube. This exercise leads to a relationship with the octahedron, finally arriving at the configuration of the Tree of Life diagram contained within the Sefer Yetzirah diagram through a method of hexagram enlargement "adapted to include the additional lines necessary for the two-dimensional projections of the interpenetrating octahedral and cubes, the final enlargement [containing] all the required lines of the Tree."[13] This process is illustrated in Figure 13.15.

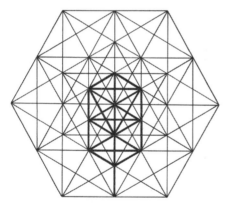

Figure 13.15: The Sefer Yetzirah Tree of Life Diagram (based on Leet, 1999).

Quintessence

This is a solid, methodical approach to constructing the Tree of Life, but rather complicated. In chapter 17 I propose a simplified method of constructing the Tree of Life, one that at the same time takes into account all of the sacred geometries we have seen, including that of the circle.

If there was an ancient recognition of the dual relationship between cube and octahedron, then was there also an understanding of the geometry of solids that would have necessitated knowledge of the remaining three regular polyhedra—the tetrahedron, icosahedron, and dodecahedron? Leet suggests there was, with a particular reinforcement of the association between the five Platonic solids and the five *Partzufim* ("divine personalities," a term from the Kabbalistic teachings of the sixteenth-century scholar Isaac Luria) arising from a similarity between them, specifically "that in the minimal Sefer Yetzirah Diagram the cube and octahedron are not only dualing but interpenetrating, a circumstance suggestive of the principal activity of the Lurianic Parzufim, that of sexual unification . . ."[14] Thus, Leet concludes that subsequent to the writing of the Sefer Yetzirah, Kabbalists could have used the cubic diagram of the Sefer Yetzirah in their own contemplations of the cosmos, and that they may have extended their interpretation of the diagram to "contribute to the mythological details of their later concepts."[15]

What is important is the uniformity of thought expressed in the ideas of Plato and his fellow Greeks, Kepler, the writers of the Sefer Yetzirah and the *Bahir*, Kabbalists, and Leonora Leet, an agreement that the five Platonic solids encode fundamental information, a geometrical key to understanding not only the form, but also the function of the universe. Hidden within geometry is vital information, ancient yet eternal. They are not metaphors. They are information unto themselves, there for us to discover.

Kepler's model of the solar system was wrong, and Plato did not identify a fundamental element for the dodecahedron. However, Aristotle suggested that the fifth element, ether, which pervaded the universe, was the quintessence of the dodecahedron.

Is this fifth element, Elijah's "single point transmuted into thought,"[16] the wisdom of the Zohar, the quintessential material producing motion and change in the universe? Is this the "Great Mystery" of the Lakota?

Is this the Sacred Sphere?

Chapter 19

The Disdyakis Dodecasphere

At first, the content of these documents was very basic, but as time went by improvements added layers of sophistication until around 800 BC when the Greeks created a full alphabetic writing system that finally separated consonants from vowels. The period immediately before these early records were left by the Sumerians and the Ancient Egyptians has become a virtual wall, separating what we call "history" from everything that happened before—which we label "pre-history." Everything that occurred before the advent of true writing is now considered to be myth and legend because every piece of human knowledge had to be transmitted by word of mouth from generation to generation.

CHRIS KNIGHT AND ALAN BUTLER
CIVILIZATION ONE

In Jung's view such, perhaps all, symbols are "natural" because they reach down to and express the unconscious in primitive fashion at the same time as they correspond to the highest intuitions of consciousness. But they can be seen as "natural" in another way, which Mary Douglas's view shares, of being related to the human body. "The symbols of the self rise in the depths of the body and they express its materiality every bit as much as the structure of the perceiving consciousness."

RAYMOND WILLIAM FIRTH
SYMBOLS: PUBLIC AND PRIVATE

What Is It?

It's certainly surprising to encounter descriptions of similar geometrical symbolism from traditions so spread out across time and space. However, the symbolism is everywhere for us to see. It is Native American, Babylonian, Hindu, Celtic, Norse, Egyptian, Mayan, Jewish, Christian, and many other ancient and modern religious traditions, as well as in magic and worldwide secular organizations such as Freemasonry.

The symbol is everywhere, but what is it exactly? And what does it mean? We can answer the first question by applying geometrical principles to the graphic symbolism presented to us. It will become much easier to construct the fundamental symbol if we look first at its meaning. At first this seems counterintuitive. After all, how can we understand what it means if we don't know what it is? However, it is important to remember that the origin of the symbolism is prehistoric. The symbol is a natural consequence of the human need to know where we came from, who we are, and how we relate to the cosmos. The Ancients and indigenous peoples paid attention—listened—and they discovered the meaning. They communicated its meaning through symbolism—pictorial, megalithic, and mythological. The geometry of the base symbol is common to each of the symbols we've looked at so far, as well as numerous other symbols that have appeared since humans first recognized the need to describe and preserve their understanding and beliefs about the world around them.

Our review of religious and life-way concepts, mythology, and shared beliefs began with a look at symbolism found in Lakota traditions. Our focus was on the Sacred Hoop, which is common to many Native American tribes, and in it we discover the *Sacred Sphere*. Recall that the configuration of a subgroup 6 medicine wheel—particularly Cloud Peak medicine wheel with its central cairn, stone ring, and eight stone lines radiating from the cairn to the ring—conforms with the configuration of rocks forming a traditional Lakota altar and the sacred use of earth [soil] and lines drawn on the ground as described by Black Elk, as well as Wohpe's painted disk as described in "A Myth of the Tetons as It Is Told in Their Winter Camps," told by holy man George Sword.

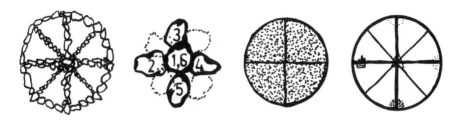

Figure 14.1: Cloud Peak medicine wheel, stone altar, earthen drawing, and Wohpe's disk painting, based on their respective descriptions.

Figure 14.1 shows renditions of the medicine wheel, altar, earthen drawing, and Wohpe's painting based on their respective descriptions. We see the same basic geometry expressed in Black Elk's description of an altar constructed from rocks, a stick drawing on the ground, instructions Wohpe received for painting the disk, and the configuration of stones at Cloud Peak medicine wheel. Each symbol includes a circle and four diameter lines equally spaced and dividing the interior of the circle into eight wedge shapes. The four diameter lines cross at the center of the circle. The medicine wheel includes a rock cairn at its center, as does the altar with its six stones. Bear Butte, the center of traditional Lakota territory, is represented symbolically in the middle of Wohpe's painting. Now, recall the two-dimensional symbols illustrated in chapter 2 and the various descriptions we've seen of symbols from cultures around the world. The geometry is the same: a circle bounding diameter lines ranging in number from two to six, but most commonly two or four.

The medicine wheel provides a key to unlocking the mystery. Black Elk's description of the rock altar is the confirmation. Keep in mind that the symbol of a circle containing two perpendicular lines is usually referred to as an earth cross or sun cross and that the configuration is usually meant to represent Earth, Sun, world, universe, and relationships in space and time. Every morning when we rise we immediately accept that we live in a three-dimensional world. In fact, all life functions quite well with this realization, or at least an innate acceptance of this perceived reality. Nature thrives in three dimensions. Matter consists of mass having volume. Energy can move in any direction. Whether particle or wave, matter and energy act across the three dimensions of space. We expect it. Our lives depend on it. However, humans have the ability to perceive a universe limited to two dimensions, and we are the only life form on Earth to do so. Every natural surface, from microscopic to macroscopic, includes a third dimension: water, soil particles, rocks, microbes, algae, grass, animals, trees, air, clouds, our planet and our solar system, the Milky Way, the universe. But people work hard to construct flat surfaces: lumber, steel, glass, plastic, paper, floors, walls, ceilings, boxes, roads. You name it, and we've built a flat surface on it. And we do so to our advantage. We build with it, draw on it, walk on it, paint it, drain water off of it, fill it, drive on it, cook with it, and on, and on. Not that there's anything wrong with that. It's what we do.

Perhaps most important of all, we tend to create nonverbal communications in two dimensions. We print letters and numbers—symbols—on paper and other surfaces to be read by others. We use a word processor to display words on a flat screen. Similarly, images of humans, animals, and the cosmos were drawn on cave walls tens of thousands of years ago. These various means use two-dimensional symbols to communicate concepts having three dimensions.

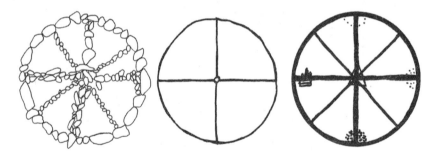

Figure 14.2: Cloud Peak medicine wheel, Black Elk's stick drawing, and Wohpe's painting.

With these ideas in mind, let's return to the medicine wheel, Black Elk's stick drawing, and Wohpe's painting (Figure 14.2). Again we see symbols given in two dimensions—the medicine wheel of rocks laid out across the surface of the earth, the drawing on soil, and the depiction of the Lakota world. Imagine the difficulty of constructing similar symbols in three dimensions. How easily could we build a 7-foot diameter dome of rocks or soil? What could we use to construct a sun sphere? How long would it take compared with just drawing a symbol of the sun? The point, of course, is that it is much easier for us to communicate by nonverbal means if we don't have to consider the third dimension in the means of communication. Why build a dome when a circle will do, and why make a sphere when an "o" on paper will suffice quite nicely?

In our example of the medicine wheel, the rock cairn at the hub of the circle gives a third dimension to the symbol. In Black Elk's description of the rock altar, six rocks are placed symmetrically in a pile and additional rocks are then placed over them. Thus the third dimension is included in the symbol, "and all these together represent everything that there is in the universe."[1] The symbolism of the circle is clear (Figure 14.3).

The two dimensional symbol of a circle is a graphic representation of the three-dimensional sphere.

Figure 14.3: The two-dimensional symbol of a circle is a graphic representation of the three-dimensional sphere.

Nine Circles

With this realization comes a far greater understanding and appreciation for the accomplishments—physical, intellectual, emotional, and spiritual—of our ancestors and of indigenous cultures around the world. Archeologists underappreciate the traditional, symbolic meaning of the circle. Its traditional interpretation as a simple circle will need reconsideration as the paradigm shifts to recognizing its deeper meaning. But before we delve into the greater meaning of the symbol, we shift the symbol itself from two dimensions to three.

Let's construct the symbol simply as a geometric structure. You will likely begin to see the relevance in the geometry to the mythologies and symbolism already discussed. You might also begin to recognize the relationship that the symbol supports between yourself and the world around you. The obvious starting point is a circle, the most common of all geometric symbols. It's time to draw.

- Draw a circle; it's the basis upon which the symbol is built.

- Draw a diameter from the top (north pole) to the bottom (south pole) of the circle.

- Draw another diameter from left (west) to right (east), equivalent to Earth's equator.

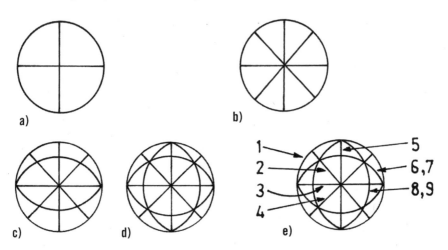

Figure 14.4: The two-dimensional construction of the nine great circles associated with sacred pictographic symbols. a) Three circles represented by a cross within a circle. b) Addition of two more circles represented by two more diameters within the circle. c) An ellipse represents two more circles. d) Another ellipse represents two additional circles. e) The result is a two-dimensional symbol of nine great circles.

We now have a circle containing two perpendicular lines (Figure 14.4a). Now imagine taking the crossing point at the center of the circle and pulling it out of the page while that point is similarly pulled under the page. Imagine the east–west diameter becoming a circle that extends out of the page to your left and back into the page on your right; it continues beneath the paper and back to its beginning point at the west end of the first circle. At the same time, the north–south line does the same, out of the page at the north pole and back into the page at the south pole, continuing beneath the paper and back to the paper at the north pole. We are envisioning three circles oriented orthogonally to each other. This is the spherical frame upon which additional circles are constructed.

- Draw two more diameters, from northeast to southwest and from northwest to southeast, subdividing the first circle into eighths (Figure 14.4b).

If we again imagine pulling the crossing point at the center of the circle out from the page to form our spherical frame, then the two additional diameters you've drawn are pulled out in similar fashion to create two additional circles, one oriented northeast–southwest and the other northwest–southeast. We now have a total of five circles. The next two steps may be a bit difficult to draw accurately, but you can refer to the accompanying figures if necessary.

- Draw the upper half of an ellipse (an oval) beginning at the west end of the first circle, curving up to just north of the midpoint of the line that extends between the north pole of the first circle and the center of that circle, and then curving back down to the east end of the first circle.

- Draw the lower half of the ellipse in a similar manner through the lower portion of the first circle, extending from the west end to the east end of the symbol.

The ellipse you have drawn represents a circle projecting out of the page along the equator of the first circle (Figure 14.4c). This sixth circle is inclined away from you at a 45-degree angle toward the north pole of the first circle, while also dipping toward you, but beneath the paper, at a 45-degree angle toward the south pole. In fact, as a two-dimensional ellipse also represents another circle, one that extends out of the page at a 45-degree angle but toward you, it continues at a 45-degree angle into the page along the equator of the first circle, and dips toward the north beneath the paper.

We have a total of seven circles. You may notice that so far we've drawn three orthogonal circles and then filled in the open spaces of the resulting sphere with two additional circles oriented at 45-degree angles in two of three dimensions. Two more applications of pencil to paper will complete our drawing, adding two necessary circles to the third dimension.

- Draw the left half of an ellipse beginning at the north end of the first circle, curving around to just west of the midpoint of the line that extends between the west pole of the first circle and the center of that circle, and then curving back down to the south end of the first circle.

- Draw the right half of the ellipse in a similar manner through the east portion of the first circle, extending from the north end to the south end of the symbol.

A Look Back

As before, this second ellipse actually represents two circles in the third dimension, one angled through the page down and through the north–south diameter toward the right, and the other angled through the page down and through the north–south diameter toward the left (Figure 14.4d). We have completed our construct with a total of nine circles. In fact, from a purely geometrical standpoint, what we have drawn are nine *great* circles, each of their centers being located at the very center of the sphere. Of course, on paper the center of the sphere is represented by the center of the first circle, which also happens to be the center of the symbol.

A quick review of the symbols illustrated in chapter 2 will show that each of the symbols shown have certain components of what we've drawn—the circle, diameter lines, and curves extending through the circle, although not necessarily through the center of the circle. And all of the lines on the page are not necessarily required for the symbolism to be recognized. Recall that the two ellipses each represented two circles. Rotating the symbol 90 degrees, 180 degrees, or 270 degrees, either clockwise or counterclockwise, will result in the very same symbol as the original. It looks the same whether viewed from the front of the paper or the back. From left or right, top or bottom, right side up or down. Qualitatively it appears to be very symmetrical about each of the two orthogonal axes within the plane of the paper and, indeed, about the three orthogonal axes associated with three spatial dimensions.

Recall from chapter 13 that, when viewed from above its central cairn of stones, the Cloud Peak medicine wheel appears as a two-dimensional configuration of lines formed by the cairn, circle, and eight stone radii, matching the configuration of lines of a *disdyakis dodecahedron* (DD) projected onto the plane of the earth's surface. The DD is a polyhedron, a three-dimensional solid which, when shown in only two dimensions (as on paper or on Earth's surface), has many of the characteristics of the geometries we saw in ancient mythologies and the various symbols illustrated in chapter 2. In three dimensions the DD is seen constructed from nine octagons.

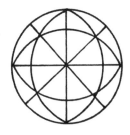

Figure 14.5: The nine octagons of the DD represent nine great circles encompassing a sphere.

In fact, the nine octagons of the DD represent nine great circles encircling a sphere (Figure 14.5). Each includes forty-eight faces (scalene triangles), seventy-two edges, and twenty-six vertices. The only differences are that the faces of a polyhedron are planar (flat), while each face on a sphere is curved; each edge of a polyhedron is a straight line segment bounded by two faces with a dihedral angle of 155° 4' 56," while each line on a sphere is curvilinear and the dihedral angle between adjoining faces is 0° 0' 0"; and each vertex of the DD is a point of maximum convexity where the corner of several triangular faces meet, while those same locations on a sphere are simply the locations where the corners of several curved faces meet—in other words, the curvature at those locations is no different than at any other location on the surface of the sphere.

Projecting the seventy-two edges of the DD onto the surface of a sphere produces the image of the nine great circles extending across the sphere that we just drew in two-dimensional form! I refer to this spherical construction as a *disdyakis dodecasphere*, or DDs. It is the fundamental three-dimensional structure from which many of the worldwide circular religious and cosmologic symbols may be derived. Or, to put it another way, as demonstrated in Section 2 of this book, those two-dimensional symbols appear to have represented the sphere of the earth or other religious and cosmologic entities ever since humans first gazed upon the heavens and sought to understand and express their presence in relationship with the universe.

The disdyakis dodecasphere (DDs) is the curvilinear equivalent of the disdyakis dodecahedron (DD). The symbols illustrated in chapter 2 consist of various geometrical components of a two-dimensional representation of the DDs.

A DDs can be drawn onto a sphere (such as a tennis ball) by marking the six locations where the six vertices (nodes) intersect the sphere. As reference, you can think of those locations being at the north and south poles and along the equator at due north, south, east and west. Then, draw four great circles through each node, with each great circle separated by a 45-degree angle from the adjoining circle at each node. Thus drawn, there is a redundancy in the number of circles made on the sphere, and the actual number of great circles reduces to nine. Viewed from a position over any of the six nodes, the three-dimensional configuration of lines of the DDs projected onto a plane appears to match the description of the disk painted by Wohpe and the configuration of Cloud Peak medicine wheel. Compare Figure 14.5 with Figure 14.1.

 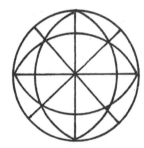

Figure 14.6: a) Disdyakis dodecahedron. b) Disdyakis dodecahedron projected onto a sphere.

Four additional lines become apparent when viewed from a vertex down through the respective node on a DD, but they do not appear on Wohpe's painted disk or Cloud Peak medicine wheel. Those four lines are produced by the four great circles that pass through the four nodes located along the great circle that forms a plane perpendicular to the line of sight (the node extending from the viewer through the center of the sphere). We previously drew representations of those lines as parts of ellipses, but we can now be more precise with their geometry.

Projecting those lines (representing great circles on a sphere) onto a plane, the lines form two ellipses that may be expressed by the equation

$$x2 + 2y2 = 1$$

where the long axis of each ellipse and the diameter of the circle each have lengths of $2x$, and the length of the short axis is 2 times the square root of 5 (written as $2\sqrt{5}$).

Accurate drawing of the ellipse requires solving the equation since it is constructed of a loop with varying curvature. Similarly, accurately depicting curvilinear lines using stones across a medicine wheel would be difficult without understanding the quadrilateral relationship within an orthogonal two-dimensional coordinate system. However, an exception is found where $x = 0$ and $y = \sqrt{5}$ and $-\sqrt{5}$. Those points can be determined by the following procedure (See Figure 14.7).

1. Draw a circle, and divide the circle into four equal quadrants. Label the point on the circle at N270°E (due west) as point A.

2. Draw a line from the center of the circle to a point on the circle located due northeast (azimuth N45°E). Label the point on the circle point B.

3. Draw a line from point B due west to its intersection with the radius extending from the center of the circle to north. Label the intersection point C. The coordinates of the intersection point are (0, v5). The point (0,-√5) can be determined similarly.[2]

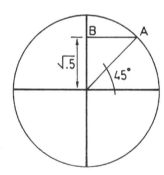

Figure 14.7: Determining a point on one of the ellipses of the DDs.

Drawing one great circle along the equator divides the sphere into two hemispheres. Two additional great circles, extending through the two poles and orthogonal to each other, subdivide the sphere into an octet of equivalent triangular surfaces. In order to retain the symmetry of all surface areas created by the placement of additional great circles on the sphere, a minimum of six additional great circles is needed, for a total of nine great circles. That placement results in the formation of forty-eight equivalent triangular (scalene) surfaces, twenty-four of which mirror the other twenty-four surfaces. The circles are each divided into eight segments by the intersection of at least two great circles at twenty-six points on the surface of the sphere. The nine great circles may be interpreted as representing the earth's equator, meridian lines, galactic equator, and other curvilinear cosmographic images on the earth's surface. Figure 14.8 illustrates these geographic features on the DDs. Those features are incorporated into various circular symbols used by religions around the world since prehistory.

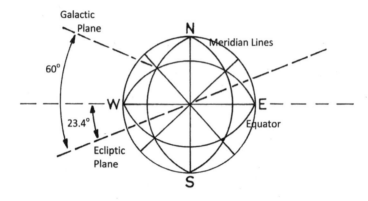

Figure 14.8: Geographic features symbolized by the DDs.

Chapter 13 includes a list of important geometrical facts found in the DD. In similar fashion, here is a summary of important geometrical relationships associated with the DDs:

- It is comprised of a sphere with nine great circles.

- It has forty–eight faces, all equal in size, each consisting of a curvilinear triangle. It is face-uniform, with twenty-four of the faces identical (face transitive) and the mirror image of each of the other twenty-four faces (which are also face transitive). Let's call the first twenty-four faces A and the other twenty-four faces B. As a triangle, each of the faces adjoins three other triangles. Each A face adjoins three B triangles, and each B face adjoins three A triangles.

- It has seventy-two edges: 9 great circles × 8 sides per circle = 72 edges.

- It has twenty-six nodes, where the corners of three triangles meet. It has a face configuration of V4.6.8, indicating that alternating vertices of the triangular faces are associated with four, six, or eight faces each.

- Based on the number of faces, edges, and nodes of the DDs, the Euler characteristic (or Euler–Poincaré characteristic) is x = V - E + F = 26 - 72 + 48 = 2. As noted previously, the Euler characteristic is a topological invariant describing the shape or structure of a topological space. Topology is an area of mathematics addressing the properties of space. The value of 2 is equivalent to the Euler characteristic of the sphere (2), applied identically with spherical polyhedra.

- The dihedral angle (the angle between faces that share a common edge) is 0° 0' 0."

- Each triangular face has corner angles of 30, 45, and 90 degrees.

- The sphere is considered to have perfect symmetry, with infinite planes of symmetry passing through its center, the presence of infinite rotational axes, and yielding the same apparent form under any given amount of rotation.

- The geometrical configuration of nodes and curved faces and edges of the DDs are completely equivalent to those of the DD. Both have full (or achiral) octahedral symmetry (O_h or *432) of order 48. Recall that the O_h group has the same rotation axes as chiral octahedral symmetry, but with the addition of mirror planes (including mirror planes of T_d and T_h). This group is isomorphic to $S_4 \times C_2$ and is the full symmetry group of the cube and octahedron. It includes six orthogonal vertices (perpendicular to the six faces of a cube). Included in the isometric crystal form, the dodecahedron has one of the highest degrees of symmetry. The isometric crystal system includes forms approaching sphericity, or roundness. Remember, however, that the sphere has perfect symmetry. The edges express the hyperoctahedral group for n = 3. The faces have the following conjugacy classes:

  - identity
  - 6 × rotation by 90°
  - 8 × rotation by 120°
  - 3 × rotation by 180° about a fourfold axis
  - 6 × rotation by 180° about a twofold axis

- The faces being curvilinear scalene triangles, if the shortest edge (s₁) of each face has a length of 1, then the lengths of the edges of each face have lengths:

  - S1 = 1.0
  - S2 = 1.5
  - S3 = 2.0
  - This is equivalent to a ratio of 2:3:4

and

  - The inradius (radius of the largest circle inscribed in a face) is 0.5

All spheres have a surface area (S) to volume (V) ratio of 6:1.

Chapter 15

Geopolyhedra

Energy Matters

Energy is mass, Mass is energy
Mass has size, Mass has shape
Mass has volume, Mass fills space
Energy does not fill space
Energy has direction, Energy has angularity
Energy accelerates, Energy Vibrates
Energy is waves of energy, Mass is waves of energy

PAUL BURLEY

Design from Nature

In chapter 1 we saw that R. Buckminster Fuller identified fundamental shapes of space, shapes he called "modules," through his study of polyhedra. He also studied curvilinear spaces resulting from intersections of great circles on sphere-structures he characterized as *geopolyhedra* (from the Greek: *geo*, "earth"; *poly*, "many"; *hedra*, "face"). Those structures are suggestive of curvilinear (spherical) polyhedral forms with many faces. For example, the DDs has forty-eight curvilinear, triangular faces covering the surface of a sphere.

In this chapter we look further at Fuller's understanding of fundamental shapes and three-dimensional modules of space and relate those concepts to the geometry of the Sacred Sphere. Information in this chapter concerning Fuller's studies is drawn from Edmondson's *A Fuller Explanation: The Synergetic Geometry of R. Buckminster Fuller*, which summarizes Fuller's multifaceted career, including his identification of relationships between Euclidean and spherical geometry, and energy in three-dimensional space. Edmondson notes that Fuller was "dismissed as an incomprehensible maverick, but there is a consistent thread running through all the wildly disparate reactions."[1] We will follow that thread as we further our understanding of the Sacred Sphere.

Fuller's study of three-dimensional space was developed using what he called "Design Science," the grammar of a language of images rather than words. Image languages differ from verbal communications in that the former are multidimensional, while the latter use a linear string of symbols, spoken or written. The image language developed by Fuller utilizes natural structure demonstrating aesthetic sensibilities, an intuitive approach to solving scientific problems. This requires good visual interpretive skills, including the ability to perceive and understand information provided in pictorial form. Perception is a complex process. Personal experience and preconceived notions can color our perceptions. For example, we tend to analyze natural structures (such as rocks, trees, rivers, and planets) based on our perception of reality, which rarely envisions a structure consisting of a system of packed spherical atoms. The result is that we fail to perceive the system of organization that actually determines the inherent structural form.

Patterns

If polygons can be classified by what they are, they may also be classified, Kepler argued, in terms of what they do. The geometrical plane has no pattern at all. It has length and breadth but no interior detail. Given an endless supply of squares, however, a geometer may cover the mathematical plane completely by placing one square next to the other, *ad infinitum*. No part of the plane need be left exposed. With the squares in place, the result is a *tessellation* of the plane. By way of contrast, a pentagon cannot tessellate the plane. No matter how its five sides are adjusted, some part of the plane will always show through.

It is this circumstance that suggested to Johannes Kepler a second scheme of classification, and so a second division of regular polygons. Those polygons that could be slotted together to form plane tessellations, which Kepler called sociable; the rest are unsociable. Only three regular polygons are sociable with themselves (self-sociable). They are the equilateral triangle, square, and hexagon. However, different self-sociable polygons may combine with one another to form tessellations of their own—such as squares with triangles, or hexagons with squares. The result is a semi-regular tessellation of the plane.

There is a further level of complexity, because it isn't only the plane that can be tessellated. The Platonic solids fill up a portion of space and so exist in three dimensions. Each Platonic solid is made up of geometrically equivalent faces that are themselves regular polygons. What is a cube, after all, but six squares meeting to form eight equivalent vertices? Solids formed in this way are called uniform polyhedra, and their covering by regular polygons is a tessellation. Similarly, the disdyakis dodecahedron results from a tessellation of forty-eight scalene triangles. However, twenty-four of the triangles are mirror images of the other twenty-four, and so the DD is not classified as a uniform polyhedron. Nonetheless, within it are the geometries associated with each of the Platonic solids, as well as many other natural, fundamental geometrical structures described in chapter 16, each the result of the intersection of nine octagons, or nine great circles, forming the Sacred Sphere.

Fuller developed new ways of looking at the world around us. He discovered how to transform objects of one or two dimensions into three-dimensional forms, and how to transform one three-dimensional form into another. In so doing he discarded conventional metrical systems used by engineers, architects, and scientists, favoring instead natural systems that promoted development of stable yet lightweight structures.[2]

For Fuller, the universe consisted of "a miraculous web of interacting patterns."[3] One of his conclusions was that the world is not what it appears. Clearly, that conclusion has been made by many people throughout the ages, in cultures across the world. Fuller realized that what appears to be solid, lifeless rock is actually a bundle of energetic atomic activity. Similarly, as I noted in chapter 1, rock is not lifeless for the Lakota. It has meaning, it has spirit, and it can speak to us, if only we would listen.

Synergetics

In chapter 1 we saw how synergetics, Fuller's term for the study of spatial complexity, can be applied in engineering design, architecture, and science. But there is a larger significance. Fuller believed that synergetics encourages new ways of approaching and solving problems beyond technical disciplines, since it emphasizes attention to visual and spatial phenomena. We all have the ability to perceive and understand these phenomena. However, Fuller believed that synergetics' holistic approach to problem solving fosters lateral thinking (thinking outside the box) that can lead to creative breakthroughs complementary to or even encouraging further scientific study. Recall that the essence of life is communicating with intent, often with the purpose of problem solving. Synergetics, then, supports the process of communication. As William P. Thurston and Jeffrey R. Weeks state in *The Mathematics of Three-Dimensional Manifolds*, "Experience has shown repeatedly that a mathematical theory with a rich internal structure generally turns out to have significant implications for the understanding of the real world, often in ways no one could have envisioned before the theory was developed."[4] Through necessary and sufficient attention to visual and spatial aspects of natural phenomena, we can improve our understanding of the underlying structure of the universe and how we are related to that structure.

Edmondson calls Fuller "both the pragmatic Yankee mechanic and the enigmatic mystic."[5] She continues:

> Above all, he was driven by curiosity—and found nature a far more compelling teacher than the textbooks. . . . self-directed exploration into pattern and structure became the most powerful influence in his remarkable career . . . [producing] a geometrical system that provides useful background for problem-solving of any kind . . . He concluded that humanity had been on the wrong track all these years.[6]

Fuller believed that derivation of mathematical principles based on experience is an imperative. Edmondson notes that he encouraged starting with the study of real things: "observe, record, and then deduce . . . Working with demonstrable . . . concepts, the resulting generalizations would reflect and apply to the world in which we live . . . [leading] to a comprehensive and entirely rational set of principles that represented actual phenomena. Furthermore, Bucky suspected that such an inventory would relate to metaphysical as well as physical structure."[7]

Fuller proposed that our scientific understanding of reality needs to apply new geometric models, replacing the inappropriate cube and other solids that have deliberately separated science from reality ever since the exposition of geometric principles in ancient Greece. If matter is energy, then there are no "solids." What we perceive as *things* are actually *events*, which Edmondson defines as "transient arrangements of frenetically vibrating atomic motion." She suggests that we should realign our vocabulary and scientific models to be consistent with the energy-event reality of the universe: "an *operational mathematics* must rely on concepts that correspond to reality."[8]

And so the purpose of synergetics is to shed light on what Fuller considered to be invisible events and transformations of Universe—a web of unity, of geometric energy relationships including the physical and metaphysical—applying appropriate models to identify and explain the principles behind the energy events that define Universe and helping to coordinate our senses with a new understanding or paradigm of reality. Fuller's immediate goal was to investigate the geometric system itself, striving to identify structural similarities among physical and metaphysical phenomena.

There being no continuous surfaces in nature, Fuller concluded that there can be no perfect spheres in nature. It is well known that spherical space, although not a physical reality, would allow the greatest volume per unit of surface area and it is therefore the most efficient enclosure for air or any other malleable mass/energy. Mathematicians use spheres as models, assuming all points on a sphere's surface to be exactly equidistant from its center. But this model has no holes: It is impermeable, disconnecting itself from the remaining universe, perpetually energy-conserving, a machine that defies all laws of nature. Fuller believed that this illusion of a spherical system exhibiting a physical continuum results from the limitations of our senses, and that at some level of resolution all physical surfaces and solids will appear, or behave, as discrete particles (Figure 15.1).

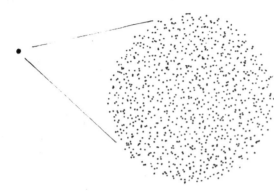

Figure 15.1: At some level of resolution all physical surfaces and "solids" will appear, or behave, as discrete particles.

Surfaces of nature's bubbles are fragmented, constructed from discrete molecules located at an approximately equidistant but non-uniform distance from some arbitrarily defined center. If we could measure the exact distance, the ratio of a bubble's circumference and diameter would be found to be very close to pi, but Fuller would argue that the bubble differs from the Greek ideal sphere. Edmondson clarifies Fuller's argument.

A *real* sphere consists of a large but finite number of inter-connected individual energy events. Moreover, the middle of each of the implied chordal (straight-line) connections between events is slightly closer to the sphere's center than the ends, thereby violating the mathematical definition, and insuring some small departure from the ratio π . . . It is with this reality that Fuller urged widespread recognition of *discrete energy events*, even though they cannot be perceived by human senses.[9]

Fuller found modern language to be anachronistic, tied to dark-ages thinking. For example, he considered *up* and *down* remnants of a belief that the world was flat. Edmondson notes that, "when we say 'look down at the ground' or 'I'm going downstairs' we reinforce an underlying sensory perception of a platform world."[10] Edmondson states that, in Fuller's view, as a result of humanity's perception of an up-down plat-form Earth, humanity became blinded to other rational orders, and that the adoption of the cube inhibited a "sustained and serious attention to the other polyhedra."[11] Up and down are simply not very precise on a spherical planet.

The replacements? *In* and *out*. The radially organized systems of the universe have two basic directions: *in* toward the center and radially *out* in a plurality of directions. Airplanes go out to leave and back in to land on the earth's surface. We go in toward the center of the earth when we walk downstairs. The substitutions seem somewhat trivial at first, but again it is difficult to judge without trying them out. Experimenting with "in" and "out" can be truly reorienting; unexpectedly one does feel more like a part of a finite spherical system—an astronaut on Spaceship Earth. Fuller believed that we need to align our reflexes, our perceptions, with our intellect. Instead of sunset and sunrise, terms that reinforce the sun's apparently active movement, Fuller suggested *sunclipse* and *sunsight*, words more representative of our view of the sun being obscured and the removal of that obstacle, respectively.

Systematic Geometry

Let's forge ahead with additional geometrical considerations related to polyhedra that, under the limiting case, will lead us to greater understanding of the sphere. Envision some unit element that has no substance. The dimensionless point of Greek geometry applies here. Now imagine two points, and there exists otherwise unbounded space between the two elements. The same is true for three non-collinear (not in a straight line) points in whatever arrangement we may envision. Three non-collinear points define the orientation of a plane and the limit of a unique circle (Figure 15.2a).

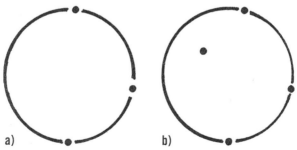

Figure 15.2: a) Three non-collinear points define the orientation of a plane and the limit of a unique circle. b) A fourth point non-coplanar with the other three divides space into two compartments: one within the bounds of the shape defined by the line segments extending between the four points, and the other including the universe beyond. Four non-coplanar points define a minimum system and a unique sphere.

A new situation develops when we introduce a fourth point that is not coplanar with the other three, as space is suddenly and invariably divided into two compartments: one within the bounds of the shape defined by the line segments extending between the four points, and the balance including the universe beyond. With four non-coplanar points we have defined a minimum system and, incidentally, a unique sphere (Figure 15.2b). We find that a minimum of four points (corners, vertices) is required for existence within three-dimensional space. When the points are connected by line segments (edges), the three edges common to a plane form a triangular face. The Greeks called this minimum system created from four points a *tetrahedron*, a name most commonly associated with the polygon consisting of four equilateral triangles. This is the most fundamental shape for enclosing space, and it forms what Fuller discovered to be a minimum module or "quantum."[12] In fact, counting the edges in each of all regular, semiregular, and triangulated geodesic polyhedra (from the cube to the rhombic dodecahedron to the array of geodesics), he found that all of the resulting numbers are multiples of the tetrahedron's six. Thus, take apart any polyhedral system in those categories and we can reassemble its edges into a number of complete tetrahedra!

As Fuller stated, form is "sizeless and timeless."[13] We can envision the size of an object relative to any other object, but size is unlike other parameters such as vertices, edges, and faces. A tetrahedron can have any size and remain a tetrahedron, but the number of vertices, edges, and faces remains fixed. Similarly, a tetrahedron, or any other form we may construct, could have any color, temperature, duration we choose, but it remains a tetrahedron. The number of vertices, edges, and faces define a tetrahedron, while other parameters do not.

An important definition that Fuller used was that of a *system*, any subdivision of Universe. As noted earlier, there are the components of the system, and there is everything else. Consequently, "unity is inherently plural."[14] Fuller explains that "Oneness" is impossible because any identifiable system divides Universe into two parts, and a minimum of six relationships is required to do so. Furthermore, all operations in Universe produce a plurality of experiences. When we pay attention, when we listen to Universe, we discover the existence of otherness, and so, "Unity is plural and at minimum two."[15] Fuller also found that thinking isolates events. Through understanding we interconnect events, and he concludes that "understanding is structure,"[16] meaning that understanding establishes relationships between events. Thus, an example of a system would be a thought interpreted as a "relevant set," or a "considerable set" relating experiences to each other in some manner.[17] Since all other experience is outside the set, thought therefore defines an *insideness* and an *outsideness*, a "conceptual subdivision of Universe"; Fuller discovered "a geometric description of a thought."[18]

Before we delve further into these important geometrical concepts and apply them to specific examples, including the Platonic solids and the DD, take a moment to review the following fundamental ideas resulting from Fuller's studies of geopolyhedra.

- Geometry is the science of systems, which are defined by relationships. Therefore, geometry is the study of relationships.

- A system is necessarily polyhedral, a finite aggregate of interrelated events.

- Relationships can be polyhedrally diagrammed in an effort to understand the behavior of a given whole system.

- Synergetics is the "exploratory strategy of starting with the whole and the known behavior of some of its parts and the progressive discovery of the integral unknowns along with the progressive comprehension of the hierarchy of generalized principles."[19] Edmondson's definition of science is "the systematic attempt to set in order the facts of experience."[20]

- Synergetics describes the extraordinarily important property that the whole is greater than the sum of its parts. In Fuller's words, "Synergy means the behavior of whole systems unpredicted by the behavior of their parts taken separately."[21]

By thinking in terms of systems, isolating systems enables local processes and relationships to be studied and described without reference to size or absolute location within the universe. If we think of the universe as a system in and of itself, one unified system of processes and relationships without regard for size or location of its individual compartments of activity, then we can begin to recognize the sacred relationships necessary for us to function in accord with each other, our world, and the universe. In that light, read again the five fundamental ideas above and see if you can discover PIES levels of meaning within each. The results may seem quite surprising given the overtly physical nature of each statement, just as mythologies have long been held to be nothing more than fanciful stories. However, by opening yourself to the universe, what I call listening to the stones, you will begin to recognize the deeper meanings that each statement relates.

Life and Pattern Integrity

The eighteenth-century mathematician Leonhard Euler discovered that all patterns can be broken down into three elements: crossings, lines, and open areas. He introduced vertices, edges, and faces as the basic elements of structure, elements that underlie all geometrical analysis. Denoting the number of vertices, faces, and edges as V, F, and E, respectively, Euler's formula states that $V + F = E + 2$. Euler's formula states that the number of vertices plus the number of faces in every system (remember Fuller's definition) will always equal the number of edges plus two. This finding is worth emphasizing, for all structures in Universe, from the tetrahedron to galaxies and beyond, obey this simple statement and share this fundamental relationship. The number of vertices of *any* system can be calculated from the number of its faces and edges.

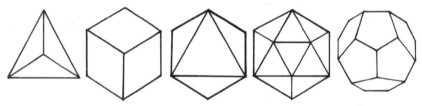

Figure 15.3: The five Platonic solids.

The Platonic polyhedra were defined by Plato in ancient Greece. Their five shapes (tetrahedron, octahedron, cube, icosahedron, and pentagonal dodecahedron [see Figure 15.3] are well known, but what exactly defines this group, and from them what can we deduce about space? There are two requirements for inclusion in this exclusive subset of polyhedra: The faces of a polyhedron must be identical, and the same number of faces must meet at each vertex. For example, the tetrahedron satisfies with its four triangles and four equivalent vertices. As shown above, the minimal polygon is a triangle defined by three points in a plane. Three triangles around one vertex (a fourth, nonplanar point) form a pyramid, the base of which derives from the fourth triangular face. The first regular polyhedron—the tetrahedron—is completed after one step (insertion of the fourth point) since all corners and all faces are identical (Figure 15.4). Euler's formula is satisfied: $4 + 4 = 6 + 2$. The other four Platonic polyhedra can be completed in similar fashion, but then the constraints of space take over, imposing an upper limit based on the two imposed requirements. Edmondson notes, "Invisible, unyielding constraints sound more like mysticism than science."[22]

Figure 15.4: The triangle derived from three points, and a tetrahedron obtained from four points.

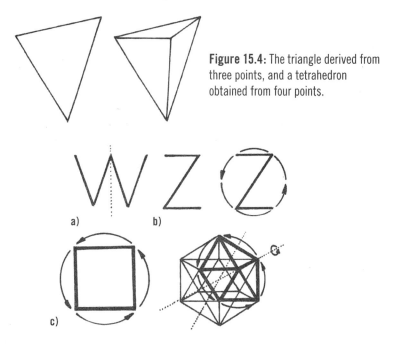

Figure 15.5: a) "W" exhibits mirror symmetry while "Z" does not. b) "Z" has two-fold rotational symmetry. c) Icosahedrons have five-fold rotational symmetry.

Now let's look at another tool for understanding geometrical, that of symmetry, defined as "exact correspondence of form or constituent configuration on opposite sides of a dividing line or plane or about a center or axis."[23] *Mirror symmetry* is the exact reflection of a pattern on either side of a "mirror line" (or plane). The letter W exhibits mirror symmetry while Z does not (Figure 15.5a). *Rotational symmetry* is the ability to rotate a configuration some fraction of 360 degrees without changing the pattern. For example, the letter Z has twofold rotational symmetry, looking the same after a 180-degree turn (Figure 15.5b). Squares exhibit fourfold rotational symmetry about their centers via rotations of 90, 180, or 270 degrees with no apparent change. Icosahedrons have fivefold rotational symmetry about an axis that extends through a pair of opposite vertices (Figure 15.5c). For x-fold rotational symmetry, constituents of a pattern repeat x times about a common center.

Try getting a handle on this quote from Fuller: "Pattern integrities are generalized patterns of conceptuality gleaned sensorially from a plurality of special-case pattern experiences. . . . In a comprehensive view of nature, the physical world is seen as a patterning of patternings . . ."[24] From this idea, says Edmondson, Fuller proceeded to discuss "*the most important and misunderstood of all pattern integrities: life*" [emphasis mine].[25] Fuller believed, "What is really important . . . about you or me is the *thinkable you* or the *thinkable me*, the abstract metaphysical you or me, . . . what communications we have made with one another."[26] This vital conclusion results from the definition of life I proposed in chapter 1, that it is communication or problem-solving with intent. Each of us is a unique pattern integrity, temporarily in the flesh. The flesh is physical, life is metaphysical. Temporarily arranged as cells generating energy, life can be envisioned as pattern integrity transmitted through space and time. All the cells of your body are replaced within every seven years, and yet we exhibit the same physical arrangement, color, and function. We are "self-rebuilding, beautifully designed pattern integrities."[27] Our bodies lose no weight at the moment of death. Though our bodies are physical, our lives are not.

> [A]ll you see is a little of my pink face and hands and my shoes and clothing, and you can't see me, which is entirely the thinking, abstract, metaphysical me. It becomes shocking to think that we recognize one another only as the touchable, nonthinking biological organism and its clothed ensemble.[28]

Fuller believed that the key is consciousness. "Mozart will always be there to any who hears his music," and similarly, "when we say 'atom' or think 'atom' we are . . . with livingly thinkable Democritus who first conceived and named the invisible phenomenon 'atom.'"[29] The qualities of life include awareness and thought, not flesh, not blood. As a unique pattern integrity, each of us evolves with every experience and thought. So, the total pattern of one's life is extremely complex and ultimately eternal. We cannot completely describe this pattern; rather, Fuller leaves the transcription of our lives to the "Greater Intellectual Integrity of Eternally Regenerative Universe."[30]

Angle and Frequency

How can a complexity of energy events produce a stable structural pattern? What provides the coherence? From the perspective of geometry, the answer is simple. Triangles, as we discovered above, are nature's only self-stabilizing pattern. However, recall that shape and size provide case-specific information. Size does not infer shape, and shape does not infer size. In synergetics, shape and size are replaced by *angle* and *frequency*. Fuller's principle of design covariables states that those two factors are responsible for all variation. "Angle and frequency modulation exclusively define all experiences, which events altogether constitute Universe."[31] In short, structure and pattern may be described completely using only two parameters: angle and frequency—another way of saying that the differences between systems are entirely accounted for by changes in angle and length. Again, the goal of such simplification is the demystification of mathematics.

When the words *angle* and *frequency* are used together in a sentence, I am immediately reminded of introductory studies in physics: angular momentum, angular acceleration, spin, frequency, wavelength, amplitude, pendulums, and oscilloscopes. In terms of Fuller's principle of design covariables, however, we will limit our current discussion to the geometry of circles and spheres. Consider a planar closed loop located on the surface of a sphere. The loop forms a perfect circle due to the sphere's uniform curvature. The loop is a great circle if the plane of the loop passes through the center of the sphere. All other circles on a sphere are considered to be lesser circles. And so, the center of any great circle coincides with the sphere's center, and there are an infinite number of great circles for any given sphere. The rate of change in direction (angle) of a point moving along the loop is determined by the sphere's curvature. The time required for the point to complete one full circuit of the loop determines the frequency of that rotation (spin). Earth's equator can be conceived as a great circle if we ignore topographic variations and gravitational forces in the cosmos acting on Earth. An important effect of Earth's spin is the process by which we define and measure units of time.

Great Circles

Here are three interesting and important facts about great circles: (1) a great circle will yield the shortest route between any two points on a sphere—airlines apply this fact every day; (2) any two great circles, defined by two different planes crossing the center of a sphere, intersect twice and only twice, those intersections located 180 degrees apart; and (3) the intersection of two great circles produces two pairs of equal and opposite angles on the sphere's surface; as with two Euclidean straight lines, the two angular values add up to 180 degrees. In fact, only great circles have the geometric characteristics of straight lines on the surface of a sphere. Lesser circles do not.

Spheres symbolize spin, and Edmondson clarifies Fuller's application of spin to understanding polyhedral symmetries.[32] By spinning any system in all directions about its center, the circumscribed form will be a spherical envelope, synergetics' *omnidirectional* form. Placing emphasis on the spinnability of systems, Fuller argued that everything in Universe is in motion; thus different axes of spin inherent in systems are worth investigating. Again, all polyhedra consist of three sets of topological aspects: vertices, edges, and faces. There are three corresponding axes of rotational symmetry, spin, that connect pairs of polar-opposite vertices, mid-edge points, or face centers. An implied great circle is generated at the equator, midway between the two poles, as a polyhedron spins about one of those axes. With an infinite number of great circles, the number of topological symmetries of a given system is also infinite. But the number of symmetries reduces when we look at polyhedra, and each polyhedron has its own great-circle diagram that describes its axes of rotational symmetry—Fuller's "axes of spin."[33] Great-circle patterns generated by related polyhedra may include some of the same circles as a result of common symmetries, but the exact chart of a given polyhedron is shared only by its dual, in which each of the vertices of the polyhedron is replaced by a face, creating the dual form.

Let's look at the relationship between polyhedra and the projection of their faces, edges, and vertices onto the surface of a sphere.[34] By so doing we will discover that great circles help us detect inherent spatial constraints in Universe. The octahedron is a simple, representative system for this exercise. First, interconnect polar opposites, starting with vertices, followed by mid-edge points, and finally face centers. The result is six paired vertices connected by three mutually perpendicular lines, axes meeting at the octahedral center of gravity (Figure 15.6a). Projected onto a sphere, these axes define three orthogonal great circles that divide the sphere's surface into eight convex, triangular areas. Those areas are called octants. Three symmetrically arranged great circles will always form the edges of a spherical octahedron (Figure 15.6b).

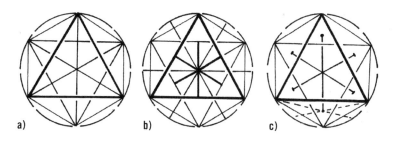

a)   b)   c)

Figure 15.6: a) Six paired vertices connected by three mutually perpendicular axes meet at the octahedral center of gravity. b) Three symmetrically arranged great circles always form the edges of a spherical octahedron. c) The midpoint of each edge of a pair in an octahedron can be connected by six intersecting axes.

The octahedron's twelve edges include six opposing pairs. The midpoint of each edge of a pair can be connected by six intersecting axes (Figure 15.6c). Those axes generate six great circles. But the pattern made by the six great circles does not look like an octahedron. Twenty-four right isosceles triangles outline the edges of a spherical cube, a rhombic dodecahedron, and the edges of two intersecting spherical tetrahedra, this latter form called a "star tetrahedron." This demonstrates a topological relationship between these four systems, and perhaps more importantly, the following aspect of intertransformability[35] between polyhedra and spheres: Great circles generated by a given polyhedron often delineate the spherical edges of its symmetrical cousins.

Finally, when we join the centers of opposite faces of an octahedron (Fig. 15.7), the resulting four pairs of triangles define four intersecting axes, producing four symmetrically arrayed great circles on a sphere (Figure 15.8). As we shall see below, these are the spherical edges of Fuller's cuboctahedron, or vector equilibrium (VE). We can now superimpose the sets of three, six, and four great circles onto one sphere, displaying the "unique topological aspects"[36] of the octahedron: thirteen great circles corresponding to the octahedron's symmetry.

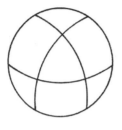

Figure 15.7: Joining the centers of opposite faces of an octahedron results in four pairs of triangles defining four intersecting axes.

Figure 15.8: The four intersecting axes shown in Figure 15.7 allow formation of four symmetrically arrayed great circles on a sphere.

Looking at a cube, we quickly realize that the resulting great-circle pattern is identical to that of the octahedron. The eight vertices generate the same four great circles as the octahedron's faces, the six faces correspond to the three orthogonal great circles, and the twelve edge midpoints are the same as those of the octahedron. This is the effect of duality: the same sets of circles generated by different elements, but with equivalent end results.

In the 1930s Fuller began a detailed study of the vector equilibrium.[37] The VE has two kinds of faces, triangles and squares. However, the above procedure remains applicable. The same four axes as the faces of the octahedron are defined by the VE's eight triangles, and the squares yield three orthogonal (XYZ) axes. The results are seven great circles generated by the axes of the fourteen VE faces. As with the edges of the octahedron and cube, twelve vertices correspond to the same six great circles. Twenty-four edges spin to create a pattern of twelve great circles. We find that the topological parameters of the VE result in a total of twenty-five great circles (Figure 15.9).

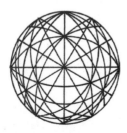

Figure 15.9: Topological parameters of the VE produce twenty-five great circles.

Figure 15.10: The set of great circles resulting from axes defined by the icosahedrons' twelve vertices are "out of phase" with the pattern of six circles produced from the octahedron-VE.

Before we leave our study of the Platonic polyhedra, let's see if we can confirm the applicability of our method for intertransformability between polyhedra and spheres. Six great circles result from axes defined by the icosahedron's twelve vertices. Note, however, that this is an entirely new set of great circles; it is out of phase with the pattern of six circles we produced from the cuboctahedron-VE (Figure 15.10).[38] Rather than the twenty-four isosceles triangles we worked with before, the icosahedrons consist of twelve pentagons and twenty triangles. In the case of the icosa-hedron, the resulting arcs produce the most symmetrical arrangement of six great circles. Fuller demonstrates that, since the icosahedron produces the most symmetrical distribution of twelve vertices on a closed system, the same is true for the corresponding six great circles. Finally, the axes of symmetry connecting the face centers of the icosahedron create ten great circles, and the thirty edges spin to form fifteen more. The total (6 + 10 + 15) is thirty-one great circles in this limit-case pattern (Figure 15.11). In fact, these circles are equivalent to the spherical edges of a pentagonal dodecahedron as well as the octahedron.

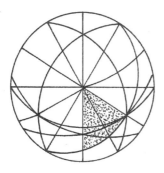

Figure 15.11: Thirty-one great circles result from the icosahedrons' vertices, midpoints of edges, face centers, and mid-edge spins.

Intertransformability is also applicable to the surface characteristics of symmetrical great-circle patterns as they relate to polyhedra and the projection of their faces, edges, and vertices onto the surface of a sphere. Each symmetrical great-circle pattern has a lowest common denominator (LCD) with regard to the shape of the areas delineated by the intersecting circles. For example, the spherical octahedron has eight equilateral faces. Those faces can be split into six asymmetrical triangles, each one 1/6 of 1/8, or 1/48 of the entire spherical surface. Without going into greater detail here, the resulting triangles are LCD units because they cannot be further subdivided to produce equivalent shapes. Interestingly, you might recall that the DD consists of forty-eight isosceles triangles.

To isolate the LCD of twenty-five great circles, we consider the spherical VE, which has both triangular and square faces. The smallest unit that can be reproduced to generate the entire pattern requires 1/6 of a VE triangle joined to 1/8 of an adjacent square. The result is that both aspects of the VE pattern are accounted for in the LCD, and an asymmetrical triangle covering 1/48 of the sphere is the product.

Spiritual Implications

Obviously Edmondson's *A Fuller Explanation* provides a wealth of information in its exploration of Fuller's study of three-dimensional space. Edmondson brought together many facets of Fuller's career, simmered the information, and produced a seasoned and very digestible read. Now we need to boil the pot of information and see if we can find those morsels of creative thought that Fuller expatiated so vigorously and which might shed light on the meaning of the Sacred Sphere. The salts from this boiling process include the following twenty-two statements applicable to our study. Just as before, each statement is worth considering at several levels (PIES) of meaning.

- International image symbols are not specific to any particular verbal language, but generally can be understood by almost anyone.

- Universe consists of "a miraculous web of interacting patterns"[39]; the world is not what it appears, it is a bundle of energetic atomic activity.

- Synergetics emphasizes attention to visual and spatial phenomena.

- Experience is imperative; observe, record, and then deduce; work with demonstrable concepts; results will relate to metaphysical as well as physical structure.

- Be consistent with the energy-event reality of Universe.

- Space has specific properties.

- It is impossible to separate physical from metaphysical; both are experience.

- There are no perfect spheres in nature; a *real* sphere consists of a large but finite number of interconnected individual energy events.

- Form is sizeless and timeless.

- A *system* is any subdivision of Universe, necessarily polyhedral, a finite aggregate of interrelated events; "unity is inherently plural."[40]

- Vertices, edges, and faces are the basic elements of structure.

- Symmetry is "the exact correspondence of form or constituent configuration on opposite sides of a dividing line or plane or about a center of axis."[41]

- "What is really important . . . about you or me is the *thinkable you* or the *thinkable me*, the abstract metaphysical you or me, . . . what communications we have made with one another."[42]

- Life is a pattern integrity; each of us is a unique pattern integrity; each of us evolves with every experience and thought.

- Structures have coherence from triangles.

- Synergetics measures length in terms of frequency.

- "Shape is exclusively angular"; "triangle" describes a concept—three interrelated events without specific length *or* angle.[43]

- A great circle provides the shortest route between any two points on a sphere; any two great circles intersect twice and only twice, those intersections located 180 degrees apart; the intersection of two great circles produces two pairs of equal and opposite angles on the sphere's surface; only great circles have the geometric characteristics of straight lines.

- Spinning any system in all directions will circumscribe a spherical envelope, the *omnidirectional* form.

- Great circles help us detect inherent spatial constraints in Universe.

- There is intertransformability between polyhedra and spheres.

- Each symmetrical great-circle pattern has a least common denominator (LCD) with regard to the shape of the areas delineated by the intersecting circles; that shape is triangular.

The words *reality* and *real* are included in a number of the twenty-two statements listed above. Additional comments concerning Fuller's concept of *reality* are in order. It is quite apparent that Fuller believed that waking experiences are reality, our interconnected individual energy events. What we perceive with our senses, in tandem with what we understand with our minds, is real. Perfect circles and spheres are not real, but a mathematical model that can represent perfect geometries. These ideas conflict with the conclusions of shamans, healers, and prac-titioners of a variety of historical and modern religions that dreams are real, that the visions experienced while in a deep state of meditation, in dreams, or under the influence of natural hallucinogenic chemicals, are reality.

At the same time, Fuller believed that it is impossible to separate physical from metaphysical; both are experience. Of course, human reality is the product of all experience, whether it is the experience of the conscious or unconscious mind. We observe, we record, and then we deduce. In science we work with demonstrable concepts. More often than not, those concepts directly concern the physical structure of Universe, structures that we can use our five senses to experience and enjoy. As to the metaphysical structure of reality, to a great extent this remains in the realm of the individual to identify, experience within one's self, and relate with at all levels of experience—from the microcosmic to the universal. History suggests we may find this reality within the geom-etry of the Sacred Sphere.

Chapter 16
A Round Earth

Moyers: Jung, the famous psychologist, says that one of the most powerful religious symbols is the circle. He says that the circle is one of the great primordial images of mankind and that, in considering the symbol of the circle, we are analyzing the self. What do you make of that?

Campbell: The whole world is a circle. All of these circular images reflect the psyche, so there may be some relationship between these architectural designs and the actual structuring of our spiritual functions.

JOSEPH CAMPBELL WITH BILL MOYERS
THE POWER OF MYTH

Seeing, Being, Becoming

In chapter 3 we saw how structures and symbols of Native science are used as metaphors for native knowledge and creative participation with the natural world. They serve as bridges between inner and outer realities. In practice those realities are expressed in traditional Plains Indian culture by the Sacred Hoop and the medicine wheel structure. Cajete's finding that the science of indigenous people is grounded on an understanding of perspective and orientation is important if we are to

understand the true meaning of the Sacred Sphere. He notes that ceremonial cultural artifacts constructed by indigenous people are "created
with an acutely developed understanding and acknowledgement of the
natural elements from which there were created ... each component ... carefully chosen with regard to its inherent integrity of spirit
and its symbolic meaning within the traditions of a particular tribe."[1]
Obviously those artifacts include medicine wheels built by the Lakota or
other tribes, and I suggest that it would also include such features as the
numerous megalithic stone circles found across Europe and north Africa.
The physiographic context of each medicine wheel and other stone
circles, and the creative process of its maker, is indicative of the unity of
artifact and creator. The wheels became two-dimensional artistic
creations representing the human ability "of seeing ... of being and of
becoming"[2] within the four dimensions of space and time.

The shape of Earth is spherical. Indigenous people know this.
Ancient cultures knew this. The evidence is in Native American medicine wheels and the untold number of megalithic structures found across
Europe, Asia, Africa, Central America, South America, and the Pacific
islands. Tens of thousands of megalithic structures were constructed in
Europe alone, from Scandinavia through France, Spain, and the British
Isles, during the fourth and fifth millennia BC. Using stones weighing as
much as 350 tons, people constructed massive buildings, altars, and
circles, components oriented to represent the form, location, or movement of heavenly bodies.[3]

Measuring the Earth

Researchers Christopher Knight and Alan Butler recently investigated the possibility of a worldwide unit of lineal measurement based on
a theory developed in the 1970s by Alexander Thom, a professor of engineering at the University of Oxford. Based on his study of numerous
megalithic structures, Thom concluded that the basis of lineal measurement for many of the megalithic structures in Western Europe was what
he called a "megalithic yard."[4]

Knight and Butler propose that cultures that designed, constructed,
and used those structures developed a geometry of the circle that included
a total of 366 degrees, rather than the standard 360 in use today. They
contend that the 366-degree circle is based on the natural number of

rotations of Earth in one year, which is, in fact, 366. Knight and Butler reviewed Thom's work and found that the megalithic yard was, indeed, an appropriate standard length of measurement when applied to earth-based measurements and construction of megalithic structures designed to accord with the natural environment. They believed that they had rediscovered the scientific basis for the megalithic yard (equivalent to 2.722 ±0.002 feet or 82.96656±0.061 cm), its relationship to the ancient 30.36-centimeter Minoan foot (circa 2000 BC) based on studies by Princeton University Professor J. Walter Graham, and additional evidence for the basis of unit lengths, areas, volumes, and weights in other areas of Europe as well as in Asia, ancient Sumer, and Africa. Knight and Butler make a strong case for Thom's research and findings, going so far as to conclude that megalithic cultures determined the length of Earth's polar circumference and assigned the length of one second of arc along that distance as equivalent to 366 megalithic yards. Mainstream scholars scoff at that idea, given the lack of evidence that directly supports the claim.

However, from a qualitative standpoint, these findings are completely in accord with what we have seen in the traditions of Sumer, ancient Egypt, Judea, Central America, North America, and other places over the past seven thousand years or more. The fact of Earth's sphericity was observed and measured, and that knowledge was recorded in myth and stone. Ancient people didn't need to walk the polar circumference or measure Earth's surface to know that our planet is a sphere. The evidence was all around them, and it is around us, too. For example, both the light side and the dark side of the moon are readily apparent to us when the moon precedes the sun at dawn. The moon not only appears full, but the line separating the light from the dark is curvilinear, with the lit side obviously facing toward the sun. Immediately you know that the moon is a sphere. Much closer to Earth, we can see clouds in the sky extending beyond the horizon. If we could focus attention long enough we would see those clouds becoming either more visible as they appear to rise above the horizon or less so as they fall below. And in our travels we never encounter a physical boundary barring traveling—no wall or precipice that can't be overcome with either intelligence or brute force. Wherever we are, the clouds continue to rise and fall relative to the horizon. This would not be the case if the face of the world was flat. So, how close were the ancient ones to being correct in believing the earth was round?

- At the equator, Earth's diameter is approximately 7,926.28 miles (12,756.1 km), while from pole to pole the diameter is 7,899.80 miles (12,713.5 km). This is a difference of 26.48 miles, or 0.3352 percent.

- Earth's circumference at the equator is about 24,901.55 miles (40,075.16 km), and the circumference from north pole to south pole and back is 24,859.82 miles (40,008 km). This is a difference of 41.73 miles or 0.1679 percent.

- The circumference of a circle can be calculated as the product of pi times the diameter, or $C = \pi d$. So, if we were assume that Earth's equator was circular, and with an equatorial diameter of 7,926.28 miles, Earth's circumference along the equator would calculate to be $C = \pi \times 7,926.28$ miles, or 24,901.14 miles (40,074.50 km). The difference between the measured equatorial circumference and calculated circumference is only 0.41 miles, or 0.0001647 percent.

- Similarly, the difference between the measured polar circumference and calculated circumference is only 41.87 miles, or 0.1684 percent.

From these facts we can conclude that the earth may be represented as a sphere. It appears that the Ancients were quite right. Admittedly the surface of Earth is bumpy, with hills and dales, mountains and river valleys, and ocean waves and polar ice packs we would have to cross if we were to measure it ourselves. And yet, as we have seen in previous chapters, the Ancients and indigenous peoples around the world knew that they lived on a sphere upon which they could ponder the act of cosmic creation and revel in life itself here on Earth. To believe otherwise is to ignore the ancient graphic symbolism displayed on the surfaces of bedrock, to limit our understanding of mythology as no more than fanciful stories, and to turn away from the few traditions that remain to tell us what it means to live in our universe, on this world, with each other.

It should not be surprising that the Sacred Sphere includes some fundamental geometrical relationships that help us understand its use as a metaphor for all human knowledge and creative participation with the natural world. However, what may be surprising to some scholars is that those geometrical relationships relate the megalithic yard. As Knight and Butler note in their attempt to "demonstrate [the megalithic yard's] mathematical origin and its means of reproduction, using the mass and spin of the Earth, . . . Facts can be tricky things, as the point of view of the observed will always have a bearing on them."[5]

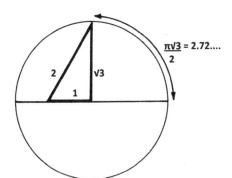

$\frac{\pi\sqrt{3}}{2} = 2.72....$

Figure 16.1: The lineal unit of measurement such as the megalithic yard is a natural consequence of the geometry of circles and triangles (from Franklin, 2004).

Hugh Franklin states that, based on Thom's quantitative analysis of megalithic stone circles, "it should be no surprise to find a unit equivalent to his stated 2.72 ± .003 feet[sic]."[6] Franklin demonstrates that this very unit of lineal measurement is found as a natural consequence of the geometry of circles and triangles (Figure 16.1). Specifically, in a right triangle with acute angles of 30 and 60 degrees, the short leg is half the length of the hypotenuse, and the long leg is the square root of 3 (√3) times longer than the short leg. Let's assume that the short leg has a length of 1 foot. Then the long leg will have a length of √3 feet and the hypotenuse will have a length of 2 feet. Next, we define the center of a circle located at the intersection of the short and long legs (the right angle corner of the triangle). Now, we draw a circle around that point, with the radius of the circle equal to the length of the long leg, or √3 feet. The diameter of this circle will be twice the radius, or 2√3 feet, and the circumference will be equal to pi multiplied by the diameter, 2π√3 or 10.882796 . . . feet. One quarter of the circumference is ½2π√3, or 2.72069905 . . . feet—the length of Thom's megalithic yard! The important ratio is this: be the length of the short side of the 30:60:90 triangle, which defines the circle, is equivalent to 2.72+ units of length of a quarter of the circle. In other words, the number of units of length of the short side, multiplied by four, equals the number of megalithic yards in the circumference! Why is this important?

1. Thom's theory that numerous megalithic structures were constructed using a standard lineal unit of measurement equal

to 2.72 ±0.003 feet is based on measurements he took at a variety of such structures. The length is a statistical average of the lengths he measured and includes an error based on the method of measurement he employed. However, the surprise lies in the significant precision of the unit length common to so many structures built over the course of time and across a large geographic area.

2. As Knight and Butler show, the following list of unit measurements have been used by various cultures through time.[7] The list is certainly not exhaustive, but the data suggest that units of lineal measurement might either a common point of origin, or the method of deriving those units was similar if not the same.

Location	Date	Unit of Measurement	Equivalent Length in Feet
N. Scotland & S. France	3000-1000 BC	megalithic yard	2.722 ± 0.002 feet
India	3300-1700 BC	Indus Inch	2.748 feet
India	2800-1750 BC	Gaz	2.75 feet
Spain	1500 AD	Vara	2.74 feet
Egypt	2600 BC	megalithic yard	2.71 feet
		Average =	2.734 feet

Location	Date	Unit of Measurement	Equivalent Length in Feet
N. Scotland & S. France	3000-1000 BC	megalithic yard × 366	996.252 feet
Crete (ancient Minoa)	2000 BC	Minoan Foot × 1000	996.062 feet
Japan (from China?)	1000 AD	Shaku × 1000	994.094 feet
		Average =	995.469 feet
		=	2.720 feet × 366

The similarity of these measurements might, of course, be coincidental. However, it may also be true that they are related by how they were derived, just as we have seen the tremendous similarity of world mythology. Recall that the length of a quarter of a circle is about 2.72069905 feet when the short side of the 30:60:90 triangle used to construct the circle has a unit length of 1 foot. The calculated length of the circumference will depend on the value of pi used for its calculation. Using pi equal to 3.141592653589793 gives a value of 2.720699046 feet. The following range in the length of a quarter of a circle is based on values of pi used by several cultures.

Location	Date	Value of pi	Length of ¼-circle
Egypt	3000-2000 BC	$4 \times (8/9)2 = 3.1605\ldots$	2.7370....
Egypt	"	$3^{1}/7 = 3.1428\ldots$	2.7128....
Egypt	"	3	2.5981....
Babylonia	2000 BC	3.125	2.7063....
Greece	500 BC	$3 + 8/60 + 30/(60)2 = 3.1417\ldots$	2.7207....
India	600s AD	$\sqrt{10} = 3.1622\ldots$	2.7386....
India	380 AD	$3 \times 177/1250 = 3.1416$	2.7207....
China	130 AD	3.1622	2.7385....
		Average =	2.7105....

From this data can we conclude that each of these cultures based their standard unit length on the length of the small side of a 30:60:90 triangle used to calculate the length of a quarter circle? Of course not. Is it possible nonetheless? Yes, and the error in calculating the equivalent length of a megalithic yard appears to be most attributable to the rather wide range assigned to the value of pi. And even so, the average calculated length is within 99.6 percent of the true value. This is certainly indicative of a potential standard prehistoric unit of lineal measurement.

3. As Franklin points out, it isn't necessary to know the length of the radius of the circle or the value of pi in order to identify and apply the length of a quarter-circle as a standard unit of measurement. In other words, ancient cultures did not need to have any knowledge of irrational numbers to use a standard unit of lineal measurement equivalent to the megalithic yard. This seems hard to understand today, when we are so dependent on irrational numbers in engineering, construction, and the vast array of physical, chemical, and biological sciences. However, it's true. We don't need to know pi in order to construct a

circular structure, whether a modern-day building or a megalithic monument. All we need is a cord of length sufficient to trace the circumference of the circular area we want—in other words, we need the radius.

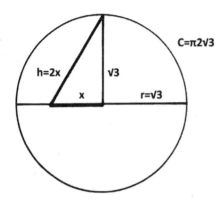

Figure 16.2: Applying the 30:60:90 triangle to create a circle of known radius, using a cord of known length.

Let's call the length of the radius r, and the circumference C (Figure 16.2). Recall that the 30:60:90 triangle used for constructing a circle has a short side; call its length x. The hypotenuse of the triangle is twice the length of x; let's call it h, equivalent to twice the length of the short side, or 2x. We know that the relationship of the lengths of the sides of the triangle is 1:2:√3. So we can define a unit length equivalent to the length of x. The hypotenuse is twice that length. The long side of the triangle has a length of √3 times the length of x, but its value is really of no concern. The important point concerning the long side is that it represents the length of the radius for our circle.

Now, we can take a cord that is three times the length of x. One third of the cord is used to lay out side x of our triangle. Standing at the center of the circle, we extend one third of the cord away from the center of the circle and then extend the other two-thirds of the cord in such a direction as to form a 60-degree angle between the short side and the hypotenuse. We then sight along the line between the two ends of the cord and mark that point on the circle. If we were to set the end of the rope at the due north point of the circle, we would know that side x of our triangle is oriented east–west. The two ends of the rope are then located at the two ends of the line segment that is the radius of the circle.

We can now use a rope of length r, extending from the center of the circle to the point due north, and trace out the circle itself. Lastly, having oriented the short side and long side of the triangle to east–west and north–south, respectively, we can plot the locations of the four points on the circle that are located due north, south, east, and west of the circle's center. In so doing we have set four points, each located at one quarter of the circumference around the circle. The result is that we have constructed a circle using only a length of rope. We have identified a unit length (x), twice that length (h), the radius of the circle (r) having an irrational value $(C/2\pi)$ for its length, the length of the circumference (C), and the length of a quarter of the circumference (M). If $x = 1$, then M equals approximately 2.720699046. If x is 1 foot, M is 1 megalithic yard.

For the purpose of staking out a structure we don't need to know the length of M in feet, only that M is the base length to be used in laying out the structure on the surface. Based on the way we set our circle, we also know that the unit length necessary for the circle's construction is x, the short-side length of the triangle. So, our two important lengths are x and M, and their importance is not in their numerical values, but rather in their relative lengths, the ratio of their physical lengths. This is all we need for our construction. Did ancient cultures use numerical values for design of megalithic structures? Perhaps, but we have little evidence to assert it. Did they need to know those values? No. All they needed to understand were basic, physical, geometric relationships.

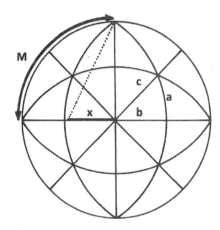

Figure 16.3: There are fundamental geometrical relationships between lengths *x* and *M* and the three different arc lengths found on the DDs.

4. Most significantly, there are fundamental geometrical relationships between lengths x and M and the three different arc lengths found on the DDs (Figure 16.3). Remember that it is length x that is the basic unit of lineal measurement for creating the circle using the 30:60:90 triangle. It is a one, dimensional unit of length upon which lengths h and r are based. Length M, equivalent to a quarter circle (C/4), is based on our choice for length x. If we divide C/4 into three equal parts, then those parts each have a length of C/12, equal to M/3. And as it turns out, M/3 is the exact length of the smallest of the three arc lengths of the DDs. This becomes obvious when we realize that C/12 (or M/3) is one-twelfth of a circle, equivalent to 30 degrees of arc, the same degrees of arc found in the smallest of the three arc lengths of the DDs. Let's call this arc length a. The middle arc length of the DDs (let's call it b) is equivalent to 45 degrees of arc, or one-eighth of a circle. Lastly, the long arc length of the DDs (let's call it c) is equivalent to 60 degrees of arc, or one-sixth of a circle. What we discover is that b is equal to C/8, and c is equal to C/6.

Let's define 1 degree of arc as 1/360 of a circle, so that there are 360 degrees in a complete circle. Knowing the lengths of M, a, b, and c for any given circle, we can then subdivide that circle into any number of arc lengths that are prime factors of 360. This means that we can determine the length of cord equivalent to arc lengths of the following degrees:

1	2	3	4	5	6
8	9	10	12	15	18
20	24	30	36	40	45
60	72	90	120	180	360

We can even determine numerous other arc lengths through combinations of these prime factors, or by halving them, quartering them, and so forth. And there is no need to know the numerical value of pi or $\sqrt{3}$. More to the point, we can subdivide a circle into almost any practical length without the knowledge of numerical values or understanding of irrational numbers. Rather, it is the relationship between the physical characteristic of length and arc that is important. Those relationships become apparent by studying the relative values of a, b, c, x, h, r, d, C, and several fractions of C listed in the following table.

Unit of Length	Value of x					
	$1/\sqrt{3}$	1	3/2	2	3	6
x	$1/\sqrt{3}$	1	3/2	2	3	6
h	$2/\sqrt{3}$	2	3	4	6	12
R	1	$\sqrt{3}$	$3\sqrt{3}/2$	$2\sqrt{3}$	$3\sqrt{3}$	$6\sqrt{3}$
D	2	$2\sqrt{3}$	$3\sqrt{3}$	$4\sqrt{3}$	$6\sqrt{3}$	$12\sqrt{3}$
C	$4M/\sqrt{3}$	4M	6M	8M	12M	24M
C/2	$2M/\sqrt{3}$	2M	3M	4M	6M	12M
C/4	$M/\sqrt{3}$	M	3M/2	2M	3M	6M
C/6	$2M/3\sqrt{32}$	M/3	M/2	4M/3	M	4M
C/8	$M/2\sqrt{3}$	M/2	3M/4	M	3M/2	3M
C/12	$M/3\sqrt{3}$	M/3	M/4	2M/3	M/2	2M
a (=C/12)	$M/3\sqrt{3}$	M/3	M/2	2M/3	M	2M
b (=C/8)	$M/2\sqrt{3}$	M/2	3M/4	M	3M/2	3M
c (=C/6)	$2M/3\sqrt{3}$	2M/3	M	4M/3	2M	4M

The significance of the values listed in the table above is that each arc length of a partial circle can be expressed in terms of length M. This means that the three units of arc length found on the DDs can be expressed in terms of length M. And as we have seen, all fractions and multiples of M are related to the unit length of x. After all, it is x that determines lengths h, r, d, and C. As Hugh Franklin concludes, "Thus any circle or even horizon, can be divided into any number of degrees or equal divisions, by using if you wanted to, any short length of stick or chord as [x, the short side of the 30:60:90 triangle]."[8] The theory developed by Thom and other scholars is that this is not only the means of construction, but also the reason why many circular megalithic structures across Europe are monuments created for observation of the cosmos through application of a fundamental measurement—the megalithic yard—based on the natural functions of time and space. How does the Sacred Sphere apply here? What are the relationships of a, b, and c to x and M? From the data in the table above we find that (Figure 16.4):

$$a + c = 2b = M = \frac{C}{4} \qquad \text{or} \qquad \frac{a + c}{2} = b = \frac{M}{2} = \frac{C}{8}$$

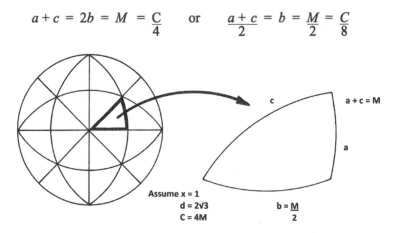

Figure 16.4: Relative lengths of the three arc (edge) lengths, arc length M, and circumference C.

Therefore b, which is also the average of lengths a and c, is equivalent to half of a megalithic yard when x = 1. This also means that 2b and a+c each equal the length of one megalithic yard. The relationships do not end there. There are forty-eight triangular faces on the DDs. Each of those faces is delineated by three edges, namely a, b, and c. Obviously each face has spherical curvature. However, suppose we retain the 90-degree angle between sides a and b and then allow the triangle to lie flat, as though laying a deflated balloon on a table top. What we discover is that the triangle will come to rest in the shape of a 30:60:90

triangle! In fact, each of the forty-eight faces of the DD are close to achieving this form, but the polyhedral geometry does not quite allow the necessary angles of each triangle to be attained.

There are three basic units of length of curvature on the Sacred Sphere—a, b, and c. Length b is equivalent to half of a megalithic yard. Lengths 2b and a + c are equivalent to one megalithic yard, also equivalent to a quarter of the circumference. These relationships are associated with the use of a 30:60:90 triangle to form a circle, with the short side of the triangle having length equal to 1. Also, each triangular face of the Sacred Sphere is representative of a 30:60:90 triangle exhibiting spherical curvature.

In chapter 3 I noted Black Elk's account of the Sioux receiving the Sacred Pipe from White Buffalo Calf Woman. In Black Elk's telling, White Buffalo Calf Woman takes from a sacred bundle the pipe and "a small round stone which she placed upon the ground."[9] She brought the foot of the pipe to the stone and told the Lakota that:

- through the pipe they would be unified with all their relatives (the spirits which dwell in all the universe)

- the red round rock is Earth: Mother (all life upon Earth) and Grandmother (the rocky mass underlying Mother)

- the people were given a red day (the turning Earth brings forth a never-ending supply of new mornings, sacred days, with enlightenment from Wakan Tanka-the Great Mystery, the Creator)

- a red path (a way to live in a sacred manner with all their relatives)

Whether the image is of a red round ball with circles upon it, an altar made of stones, a cross made of stones during a vision quest, a ball made for the game Throwing of the Ball, Wohpe's painted disk, or Cloud Peak medicine wheel, the intent of the symbol is the same: the unification on Earth of people with the universe—all that has been, all that is, and all that will be—and the Creator, in space and through time. The wisdom to walk the red path is gained by experience and through the light provided by the Creator.

Recall Crazy Horse's father saying that Lakota culture and tradition would return when, instead of seven circles, there would be nine circles on the red round stone. During Crazy Horse's vision quest on Bear Butte he laments to the seven sacred directions (including looking inward to his heart) and asks the wind "what is the meaning of the two circles to come?"[10] And then the daybreak star (Venus) appeared, followed by the dawn. The star had nine points representing the nine sacred circles, including two that were yet to come. It was at that moment that he saw a sacred herb begin climbing out of Earth, the symbol of the return of the Lakota nation. The sacred herb also represents a return for all people, a new opportunity to walk the red path.

Plainly the intent of the Sacred Hoop is metaphor for experience within the dimensions of space and time. The spatial dimensions take on spherical form by the six directions (west, north, east, south, up, and down) and the center of the sphere that can be envisioned to be anywhere, yet always found in the heart. The temporal dimension expressed by the four units (day, night, month, and year) of Lakota time is obviously related movements of the movement of Earth, Sun, and Moon, and therefore derived from cyclical, spatial events. Time, therefore, is related to geometry.

There is a geometrical aspect of Cloud Peak medicine wheel that was particularly notable in my exploration of sacred concepts and universal symbolism. The diameter of the ring of Cloud Peak medicine wheel, measured from inside to inside of perimeter stones, is 83 inches along one primary axis and 83 inches along the other primary axis. There are eight approximately equally spaced spokes radiating from the cairn to the stone ring. Orientations of the two primary axes indicate that their orientations likely represent the four cardinal directions. The configuration represents "everything there is in the universe," and the center is "the home of Wakan Tanka."

The importance, the sacredness, of Cloud Peak medicine wheel becomes even more apparent when we look again at its configuration and size (Figure 16.5). The central cairn, eight spokes, and peripheral ring of stones provide the geometrical, two-dimensional basis of the Sacred Sphere. The circle's 83-inch diameter is more than curious. The resulting circumference is $83\pi = 260.75$ inches. Recall that the length of edge b of the Sacred Sphere is one-eighth of the circumference, so that the length of b of the medicine wheel is 260.75/8 = 32.59 inches, or 2.72 feet. *The megalithic yard.*

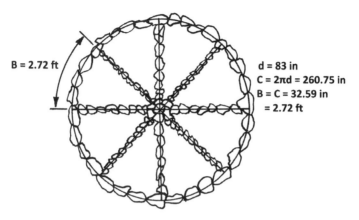

B = 2.72 ft

d = 83 in
C = 2πd = 260.75 in
B = C = 32.59 in
= 2.72 ft

Figure 16.5: The sacred nature of Cloud Peak medicine wheel is apparent by its configuration and size.

We have no evidence of the actual method of sizing Cloud Peak medicine wheel during its design and construction. Nonetheless, the structure is sized to the exact lineal dimension of the megalithic yard. Its diameter reflects the megalithic yard as the fundamental unit of length. To date, archeological and anthropological sciences have no evidence of any knowledge, let alone interaction, between the megalithic cultures of Europe, Africa, Asia, and Australia, and the indigenous people of the Americas. In fact, modern archeology was completely unaware of the possibility, and length of, a megalithic yard until the 1970s. Yet the diameter and circumference of Cloud Peak medicine wheel suggest that the megalithic yard was known to Native Americans at least hundreds of years ago. How can this be explained? I see only four possibilities.

- Coincidence—the megalithic yard has nothing to do with construction of Cloud Peak medicine wheel.

- The technology to determine and apply the megalithic yard was transferred between Native Americans and other cultures long ago. The direction of transfer remains unknown.

- Native Americans and overseas cultures independently derived and applied the megalithic yard, using science and reason to develop the necessary technology to do so.

- There is a cosmic consciousness through which humans recognize the megalithic yard as a fundamental unit of length.

The fact that the medicine wheel is sized to the exact lineal dimension of the megalithic yard strongly suggests that lengths of its diameter and circumference relative to the megalithic yard cannot be coincidence. One and only one length of diameter will reflect this nominal yet fundamental unit of measurement, while there is literally an infinite number of lengths that could be chosen. The probability that the actual diameter would be 83 inches for this particular medicine wheel, therefore, is infinitely small. I conclude that it cannot be coincidence

At the same time, we have no evidence of contact between Native Americans and Neolithic people from the "Old World." Pending discovery of such evidence, we must assume that knowledge of the megalithic yard in America was derived independently. Then, did Native Americans determine the length of the megalithic yard by the same means as Knight and Butler propose for Neolithic Europeans? That scenario requires Native American use of a measured circle hundreds of feet in diameter, divided into subarcs, and watching Venus move from one end of a wooden box of specified length to another while adjusting the length of a pendulum as its swings are counted in such a manner that the megalithic yard can be derived from the length of the pendulum. While such activity could have occurred in America hundreds of years ago, there is no evidence for it.

If Native Americans did in fact derive the megalithic yard as Knight and Butler describe, I will be surprised. So, we are left with either the megalithic yard being derived from another, perhaps more natural and fundamental means, or its length is essentially built into our subconscious, awaiting our call for it when necessary. Frankly, I see little difference between these two possibilities. If we are truly centered, in communication with all our relations, listening with open heart and mind, then it may very well be possible for the fundamental aspects of space and time to be realized. In so doing we are in tune with the cosmic consciousness that we all come to suspect is "out there," and that we feel or recognize "within" us only on the rare occasion.

Native Americans knew that Earth is spherical. The Sacred Hoop represents four dimensions of the universe. All paths lead to the center, Wakan Tanka. However, the difficulty of building a three-dimensional representation of the world from round stones makes such an endeavor impractical. As a result, a subgroup 6 medicine wheel suggests only seven of the nine great circles that form the DDs. What of the remaining two great circles? They would mirror the two great circles that delineate two ellipses formed by the sixth and seventh circles when the DDs is projected onto a plane. They would complete the framework of a sphere. However, the eighth and ninth circles remain hidden and undefined on the Sacred Hoop.

In Black Elk's vision he was taken to the center of the earth, where he "saw that the sacred hoop of my people was one of many hoops that made one circle."[11] Yes, Black Elk understood that the earth was round, a sphere, constructed of many great circles, perfect. He said, "In this manner the alter [sic] was made, and, as I have said before, it is very sacred, for we have here established the center of the earth, and this center, which in reality is everywhere, is the home, the dwelling place of Wakan Tanka."[12] When the last two circles appear on the round stone, the Sacred Sphere will show this to be truth.

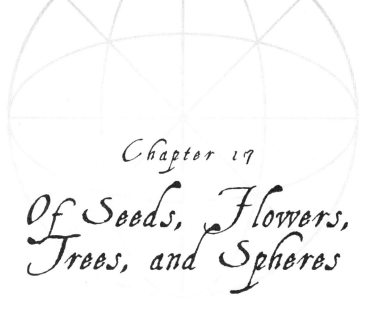

Chapter 17

Of Seeds, Flowers, Trees, and Spheres

*All the basic components of scientific thought and application are
metaphorically represented in most Native stories of creation and origin.*

GREGORY CAJETE
NATIVE SCIENCE

*It was this simple design of the cross within the square . . . that made
this rattlesnake so sacred to the Maya. It was also the basis of their
science, for it taught them geometry.*

ADRIAN GILBERT
2012: MAYAN YEAR OF DESTINY

*If all esoteric traditions have held geometry to be sacred, it is because
their adepts understood it to encode the basic principles on which the
cosmos is founded, and their geometric practice with the compass and
ruler instructed them in its meaning.*

LEONORA LEET
THE SECRET DOCTRINE OF THE KABBALAH

*Mathematicians can tell you what kinds of geometries are possible, but
only measurement can determine the "true" geometry of space.*

LEONARD SUSSKINDA
THE BLACK HOLE WAR

Cycles

Together the four quotes on the previous page outline the process by which cultures separated by space and time have drawn surprisingly parallel conclusions in answer to questions concerning our relationship with Earth, the cosmos, and each other. The process is, in fact, cyclical. World mythology consists of cultural information in the form of a story containing metaphors at physical, intellectual, emotional, and spiritual levels. Geometry found in myth provides keys to understanding this information. Geometry, then, is held sacred as a method of communicating vital information. However, it isn't the symbol that is sacred, but rather the ideas represented by the symbol and containing truth as expressed by tradition. From this truth human beings build a base of knowledge upon which cultural sciences are constructed. These sciences help us understand our origins, our place in the world, and possibilities for the future. Cultures express these understandings in myth. And so the cycle repeats.

It should be expected, therefore, that as data, facts, and truths accumulate, mythology is not static. It evolves as cultural traditions evolve. The Epic of Gilgamesh provides us a snapshot of the cultural traditions of Mesopotamia at the time a scribe imprinted it onto clay tablets more than four thousand years ago. We can be certain that the story evolved over time, perhaps over thousands of years, before taking the form that the scribe prepared. And yet, there is common ground when it comes to geometry—the measure of the earth. Mythologies describe measurement of the world as a metaphor for understanding our place in the universe and our purpose of being here on Earth.

As we continue modifying and managing the world for our own purpose, we forget to consider the impact our actions will have over the next fifty years, or hundreds or even thousands of years. Science will not provide all the answers to our global problems. We have the ability to destroy civilization, but we cannot "save the world." The world will not die, even if we intend to destroy it. Rather, the world will adapt and continue without us, not as it was, but without us. Earth was here before we arrived. It will be here after we are gone. We cannot save the world. We can only save ourselves and promote the well-being of future generations. This is the message and intent of myth. And time and again the message appears inherently tied to an understanding of geometry.

Cultural traditions are filled will allusions to geometry and sacred symbolism. We have seen mythology to be a powerful mode of transmitting a timeless and yet surprisingly uniform understanding of the past, present, and future of humankind. While the modern world has lost the ancient ability to communicate vital information via myth, secular and religious symbolism from ancient and indigenous cultures surround us nonetheless. Many of those symbols are found within the geometry of the Sacred Sphere.

In this chapter we will review the geometry of fundamental sacred symbols, including the Vesica Piscis, the Seed of Life, and the Flower of Life. We will then focus on the configuration of the Tree of Life as expressed in the Otz Chiim, a structure comprised of ten spheres (Sephiroth) and twenty-two channels (Nativoth) or paths between them. The Otz Chiim is most often depicted as a three-dimensional symbol. This third dimension, exhibited by pathways extending between spheres, must have relevance to the meaning of the Tree. The Sacred Sphere is a source for the original configuration of the Otz Chiim. The Tree of Life and the World Tree relate Earth to the cosmos. This relationship provides a clue to a potential Egyptian metaphor for the structure of the mystical Tree of Life. Lastly, there is a strong correlation between the nine great circles of the Sacred Sphere and the physical nature of Earth.

Nine

Ancient Egyptians recognized that the universe was never empty. Matter always existed, but it was in a completely chaotic state until at last a creative force began to put matter into order. Geometry was conceived at that time. The chaotic, watery mass of Nun was separated to create dry land, resulting in two subsets of the universe. Dryness and moisture formed. Sky appeared over Earth, and matter and energy continue to create, destroy, and reform to this very day. The universal geometry has become very complex. However, from the standpoint of Heliopolitan theology, the initial order of things was made manifest in its Ennead. Atum, foundation of all matter and energy in the universe, stood at the head of that pantheon. The rest were Shu, god of air and dryness; Tefnut, goddess of moisture; Geb, god of Earth; Nut, goddess of the sky; Osiris, god of vegetation and the underworld; Isis, goddess of nature, fertility, and magic; Set, god of the desert, darkness, and storms; and Nepthys, goddess of temples and the experience of death.

Other significant deities included Horus, son of Isis and Osiris, the patron of Egypt and representative of war and protection. Thoth was the lunar deity whose attributes later included a role as god of writing and science. Thoth's wife was Ma'at, goddess of truth and morality; she represented the ordering of the universe out of Nun. In chapter 7 I suggested that Ma'at also represented the concept of mercy, an attribute she exhibited as significantly as any other. Another goddess, Seshat, was originally considered a daughter of Thoth but later was perceived as his consort; Seshat was the goddess of astronomy, astrology, architecture, building, surveying, and mathematics.

While many of these deities are discussed in detail in chapter 7, I mention them again here for three reasons. First, the Ennead represents the first nine attributes of the world once the creative force began replacing chaos with order in the universe. It is upon the foundation of these nine gods and goddesses (overtly representing matter, energy, sun, earth, water, air, vegetation, and so forth) that the characters of animals, and humans in particular, are formed. Second, Re, Shu, Tefnut, and Geb represent sun, air, water, and earth, respectively, and are the first four gods created from chaos. Obviously, these deities represent the matter and energy comprising Earth and providing energy to support life in this world. Third, Horus, Thoth, Ma'at, and Seshat represent attributes essential to human civilization, such as protection and the ability to make war when necessary to defend society, the ability to read and write, to understand the world through science and mathematics, to ensure truth and morality within society, to be merciful, to construct the infrastructure needed to support civilization, and to study the cosmos and understand its processes as they impact the world.

We can see that the ancient Egyptians recognized the universal importance of order in the world, facts concerning world geography, and the ways that human civilization best fits into this natural scheme. They unified these understandings in the tradition of mythology. We can see an evolution of the universe from chaos to order in the development of these deities. The initial emphasis was on the gross characteristics of the observable universe followed by institution of order necessary to ensure a thriving culture within civilization. These characteristics are fundamental necessities of any society, ancient or modern.

Remnants of the Sacred

The following is a description of fundamental geometrical figures—pictographic symbols—that have been applied as metaphors for understanding basic relationships between human beings, Earth, and the cosmos. They are ancient constructions, remnants of sacred geometry. No one knows how old they are, dating perhaps to thousands of years before the first known historical records. Each symbol represents sacred aspects of geometry, but this is only one reason why they are addressed here. More importantly, the geometrical evolution of the circle into the Seed of Life and then into the Tree of Life leads to a surprising discovery—the Sacred Hoop, the five Platonic solids, and all other symbols described below are found in the fundamental geometry of the Sacred Sphere. The Sacred Sphere is a metaphor uniting each of these relationships—these principles—into one body of knowledge, a single energy event that may encompass the past, present, and future of humanity.

The Circle

A circle consists of a set of points in a two-dimensional plane, all equidistant from a given point, the point being the center of the circle (Figure 17.1). Each circle expresses perfect symmetry about its center. Shape is independent of size, so the equidistant relationship between all points on a circle and its center give us reason to consider the circle to be a simple symmetrical expansion of the illusory point discussed in chapter 1.

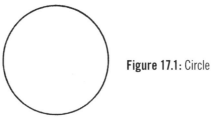

Figure 17.1: Circle

Circles are observed everywhere in nature: in the perimeter of a flowering daisy, the trunk of a tree, the ripples of water extending outward from where a stone enters a pond's surface, the sun and the moon, and Earth itself. The circle's inherently low ratio of surface area to volume gives it high strength. The circle separates interior space from exterior space, and it has been perceived as matter separating the inner and outer spaces of spirit. The equivalent distance between the center of a circle and each point representing a unique location of the circle itself provides the symbolical importance of equality and what Cajete describes as bridges between the inner and outer realities. This is made clear when Black Elk describes the making of a sacred place as simply drawing lines on the ground representing the world or making a round rawhide circle to represent the sun. The focus of the circle is its center, from which all else radiates. "In this manner the altar was made, and, as I have said before, it is very sacred, for we have here established the center of the Earth, and this center, which in reality is everywhere, is the home, the dwelling place of Wakan Tanka."[1] As the center of our solar system, the sun is a metaphor for the power—the creator—of the universe.

It is for these reasons that a circle with a dot at its center is called the *sun circle*. This symbol has ancient origins. It is the Egyptian hieroglyphic symbol for the sun and therefore for Re it is also the ancient Chinese symbol for the sun and day.

How can the center of a circle be everywhere? Isn't the center of a circle located at only one point? The answers are found when we realize that a circle represents all circles, any circle, centered at any location. It may be the center of the Milky Way or the center of the solar system. It may be the center of Earth or centered in ourselves—our hearts. The center of the circle is defined; however, the radius of the circle is undefined. Recall that shape is not a function of size. A circle may have a radius of any size. Using free will we may apply any size radius to our own circle, our self, but that circle must be appropriately centered if we are to benefit in a spiritual manner.

Suppose someone finds he or she is off-center. In other words, he or she is out of balance in terms of mind, heart, body, or relationships with other people or the environment. Where is the center in that case? Imagine a circle drawn with a line of varying thickness. The circle is thin here and thick there: It is out of balance. You might say that the center of mass has shifted away from the center of the circle. This circle symbolizes the off-centeredness of one's heart, mind, body, or relationships. The center of mass remains within the limits of the circle—within the confines of one's world. To regain balance requires a translation of the center of the circle from its original location to a new point, a new center of mass. In so doing we discover our center again, but in a new location. This is experience, growth, and finding new meaning in our lives. The universe is forever changing, and we must adapt to those changes. We find our centers when we meld ourselves with the ebb and flow of the living universe. This center in reality is everywhere. It is the home of the creator and of each of us. The journey to find that home may continue throughout your entire life. Indeed, that journey of discovery defines our very existence.

Two Tangent Circles

Two circles tangent to each other are bound together by a single point (Figure 17.2). Fuller might have referred to that point as a common energy event. Each circle has its own center, yet both have something in common. They express duality and individuality, male and female, old and young. Together they symbolize infinity and the cyclicity of events. The world was created, it exists, and it dies. It is reconfigured, connected, and related to other energies. It lives again in another form. And it will be destroyed again. This is universal truth. This process cannot be changed or altered to fit human wants. We are a part of the process. Mythologies from around the world describe the evolution of human beings in previous worlds, our place in the current world, and expectations for worlds to come.

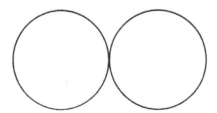

Figure 17.2: Two tangent circles

By representing duality, the two circles bound as one symbolize the separation of unity into parts at the time of creation. Those parts may take the form of matter and energy, light and dark, wetness and dryness, Earth and sky, and so forth. Yet they remain connected to ensure eternal unity. The ancient Egyptian *Shen ring* symbolizes this concept and dates to at least 3100 BC. Shen means "encircle" in ancient Egyptian. Hieroglyphs of the Shen ring show a circular loop of rope with two straight sections of rope, possibly the two end portions, tangent to the loop. The hieroglyph symbolized the concept of infinity, all that is, or all that the sun circles. It bears great similarity to the Greek letter omega (Ω), the last letter of the Greek alphabet and a metaphor for the end, the last in a series, or the limit of a thing. In Revelations 22:13, Jesus describes himself thusly: "I am the Alpha and Omega, the First and the Last."

Vesica Piscis

Two circles with the same radius and placed such that the circumference of each passes through the center of the other form the Vesica Piscis (Latin: bladder or vessel of a fish) at their center (Figure 17.3). A geometrical cousin of the two tangent circles, the Vesica Piscis has been the object of significant interest not only because of its pleasing form and its use as a symbol for the family unit of man, woman, and child, but also because the mathematical relationships between measurements of height and width of the central almond-shaped figure, and the radii of adjoining circles. Since each circle passes through the center of the other, the width of the Vesica Piscis is equivalent to the radius of each circle. So if the radius of each circle is 1 unit of length (unity), then the width of the resulting Vesica Piscis is 1. The ratio of the Vesica Piscis's width to height is 1:√3 (1.7320508 . . . .).

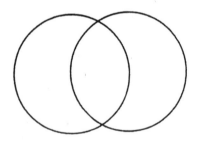

Figure 17.3: Vesica Piscis

From this geometry we find that the width and height of the Vesica Piscis bounds two equilateral triangles stacked one upon the other with a common side between them—one triangle pointing one way, and the other triangle pointing in the opposite direction. Again we see the symbol of duality in this form—up–down, left–right, in–out, above–below. Also, since the diameter of a circle is twice the length of its radius, the ratio of radius to Vesica Piscis height to diameter is 1:√3:2. Those are the dimensions of a right triangle, specifically a triangle with angles of 30 degrees, 60 degrees, and 90 degrees. In this symbol we have, then, the sections of a triangle of great significance to architecture, engineering, and construction, as well as relevance to the study of geometrical relationships in Kabbalah. Furthermore, relative lengths of $\sqrt{2}$ and $\sqrt{5}$ can be easily derived, and so the 45:45:90 triangle can be derived, as well as the right triangle with side ratios of $1:2:\sqrt{5}$. Another significant feature of the Vesica Piscis is that its construction allows the relative length of phi $((\sqrt{5})/2 + 0.5 = 1.61803399 . . . .)$ to be determined (Figure 17.4).

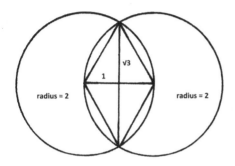

Figure 17.4: Geometry of a Vesica Piscis

With its relevance to accurate measurement of angles in operative masonry, the Vesica Piscis is valued in Freemasonry, too.

In Masonic literature, the Vesica is first stressed by George Oliver. Oliver argues that the Vesica is "a universal exponent of architecture or Masonry, and the original source or fountain from which its signs and symbols are derived—it constituted the great and enduring secret of our ancient brethren." In his Prestonian Lecture for 1931, noted Masonic historian W.W. Covey-Crump calls this statement "quite right," and expresses that "the Vesica Piscis had even from the time of the Primitive Christians possessed a sacred symbolical significance, though the purport of that significance was variously interpreted owing to the secrecy of its transmission."[2]

Of course, it is well known that the Vesica Piscis is associated with femininity and various Christian architectural features including entryways, windows, and flying buttresses seen in medieval cathedrals as well as modern church buildings. Similarly, the ancient Egyptians applied the Vesica Piscis form at passageways between Earth and the underworld. It is a symbol of the Earth Goddess and the resources she provides for sustenance.

Three Circles

When each of three circles of equivalent radii are placed tangent to the other two, they form an equilateral triangle (Figure 17.5). Bisection of the triangle results in two equivalent 30:60:90 triangles. Drawing a bisector from each of the three angles to the center of the opposite side produces six 30:60:90 triangles. The only difference between those triangles is that they are mirror images of each other. Recall that the relative lengths of the sides have ratios of 1:√3:2.

Again, we can move the three circles closer together such that each circle passes through the centers of the other two. In so doing we find that the intersections of the circles produce a tripartite form of the Vesica Piscis, similar to a three-lobed leaf. It is called a triquetara (Latin: *trique-trus*, or "three-cornered") or the Tripod of Life. Lines extended outward from the center of each of the three parts are 120 degrees apart. The form is associated with Northern European runestones and Celtic artwork. It symbolizes the Christian Holy Trinity of Father, Son, and Holy Ghost. The threefold symmetry has also given the triquetara meanings ranging from the three elements of earth, water, and air, to the cyclical nature of birth, life, and death. It also represents fertility, femininity, and the Mother Goddess (Earth). A variation of this trifold symbol is the Valknut, a symbol formed by three interlocking equilateral triangles and most commonly associated with Germanic cultural artifacts from medieval times.

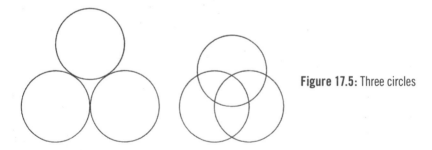

Figure 17.5: Three circles

Four Tangent Circles

Four equivalent circles each tangent to two others form a square (Figure 17.6). If the diameter of a circle has a length of 1 unit, then the distance between centers of two opposing circles is √2. This diagonal is the hypotenuse of two 45:45:90 triangles. An equivalent diagonal can be constructed between the centers of the other two opposing circles, the two diagonals forming an x or cross. Another cross can be made by rotating the x by 45 degrees. The eight endpoints of these two crosses can then be connected to produce an octagon. Eight, of course, is the symbol of infinity and of the cyclical nature of birth, life, death, and rebirth.

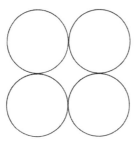

Figure 17.6: Four tangent circles

The four circles have been associated with earth, water, air, and fire. The eight points of the octagon allow the formation of the Templar Cross, also known as the Cross of St. John, Maltese Cross, Iron Cross, Fishtail Cross, or Regeneration Cross. Numerous variations in the form of the cross have been derived from the octagon.

Four Interlocked Circles

We've already seen how the Vesica Piscis and triquetara can be formed by constructing two and three circles, respectively. Similarly, a fourfold rotation of the orthogonal Vesica Piscis is produced from four circles (Figure 17.7). This symmetry allows the 45:45:90 triangle, square, and octagon to be easily drawn. Therefore, these circles provide meanings equivalent to many of those found in the Vesica Piscis and the four tangent circles.

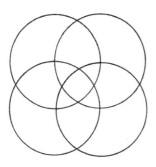

Figure 17.7: Four interlocked circles

The Seed of Life

Six equivalent circles placed tangent to two others form a hexagon (Figure 17.8). This geometry is significant in that a seventh circle fits exactly into the area enclosed by the other six. A true polygonal hexagon can be drawn by connecting the centers of the six exterior circles. By extending six lines from the center of the middle circle to the six centers of the other circles, we divide the center circle into six equivalent areas. Assuming 360 degrees around the circle, we have divided the circle into six curvilinear line segments each extending 60 degrees. Six equilateral triangles extend from the center of the middle circle outward to the line segments located between centers of the other circles. Obviously, bisecting any angle of any of those triangles produces two 30:60:90 triangles. Many of the geometrical aspects of the previous configurations can be derived from this figure. However, there is one additional step we can take to produce the first of three important constructs that have been considered sacred for thousands of years.

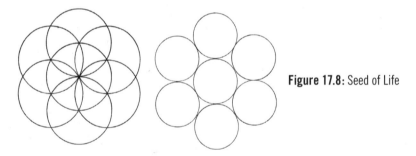

Figure 17.8: Seed of Life

We create a six-sided Vesica by moving each of the six exterior circles toward the interior circle until they each are tangent to the center of the middle circle. The geometrical information contained in this configuration of circles is similar to that found in the previous seven tangent circles, but with an increased number of line intersections and petal-like figures. These seven circles form the Seed of Life. It is the basic two-dimensional construct from which develop larger geometrical forms, each important in the study of sacred geometry. Geometry found in the Seed of Life has been applied in many religions around the world. For example, it has come to symbolize the seven days of creation described in Genesis: The first day is represented by the Vesica Piscis formed from the intersection of two circles, and an additional circle is added to the configuration during each of the following five days. At completion, the seven circles form the Seed of Life. Of course, God rested on the seventh day. The Seed of Life is the fundamental structure from which other sacred symbols are produced, including the Flower of Life, Fruit of Life, and Tree of Life.

The Flower of Life

The Flower of Life is constructed by placing twelve additional circles around the six exterior circles of the Seed of Life (Figure 17.9). The sixfold symmetry remains. All of the previously mentioned geometrical and arithmetic characteristics are contained in the Flower of Life. It is believed to symbolize fundamental aspects of space and time and is therefore valuable in the study of physical and temporal relationships throughout the universe.

Figure 17.9: Flower of Life

The Flower of Life is found across the world, a cross-cultural symbol of significance in artwork, written records, and sacred architecture at Masada, Israel, at Mount Sinai, and in numerous temples in Japan and China, India and Spain. It is associated with sites across southern and southeastern Europe and northern Africa, and in the ancient Assyrian North Palace of Ashurbanipal, which dates to the seventh century BC. The Flower of Life is also drawn on a wall at the ancient Temple of Osiris at Abydos, Egypt. The drawing is undated but appears to be associated with nearby text written in Greek, indicating that the drawing is likely much younger than the temple itself, possibly added while the temple was used as a Coptic nunnery during the early years of Christianity.

This configuration of interlaced circles is found around the world, yet few people would ever consider drawing such an intricate geometrical figure. Why, then, is it found in so many places and associated with many sacred traditions? Esoteric sacred geometry, of course, has much to do with this. According to Flower of Life Research, an organization that disseminates the teachings of Drunvalo Melchizedek concerning the Merkaveh, sacred geometry, and the opening of the heart,

> When the teachings of geometry are used to show the ancient truth that all life emerges from the same blueprint, we can clearly see that life springs from the same source . . . the intelligent, unconditionally loving creative force some call "God." When geometry is used to express and explore this great truth, a broader understanding of the universe unfolds until we can see that all aspects of reality become sacred. The ancients such as the Egyptians, Mayans, and others all knew this truth and incorporated sacred geometry teachings into their mystery schools as a way for anyone to begin to practically understand his or her personal relationship to "God" and the universe.[3]

This statement supports the idea that geometry has been and can be applied to understanding our relationships with each other, the world, and the creator. Geometry is a tool. It relates space to physical forms, and geometrical symbols can express these important relationships. Mathematics provides another tool for similar applications, but rather than express ideas spatially, it uses symbols to communicate numerical relationships. Together geometry and mathematics provide a powerful means to decipher and explain sacred relationships.

> In the ancient world certain numbers had symbolic meaning, aside from their ordinary use for counting or calculating . . . plane figures, the polygons, triangles, squares, hexagons, and so forth, were related to the numbers (three and the triangle, for example), were thought of in a similar way, and in fact, carried even more emotional baggage than the numbers themselves, because they were visual.[4]

This realization—that we could express sacred relationships via numbers and geometry—resulted in the construction of monuments reflecting sacred knowledge. "Sacred geometry is the geometry used in the planning and construction of religious structures such as churches, temples, mosques, religious monuments, altars, tabernacles; as well as for sacred spaces such as *temenoi* [sacred spaces surrounding altars or temples], sacred groves, village greens and holy wells, and the creation of religious art."[5] Cultures apply sacred geometry to produce symbols of their beliefs, incorporating those symbols within traditions. To a significant degree, the specific symbolic and sacred meanings attributed to numbers, proportions, and shapes are culturally specific. Yet the worldwide bond in religious and sacred symbolism expressed by the circle is obvious. From this we can define sacred geometry.

Sacred geometry is the application of mathematics and geometry to express relationships between mankind, the universe, and the creator.

The Fruit of Life

By removing six of the peripheral circles from the Flower of Life we form the structure referred to as the Fruit of Life (Figure 17.10). It is constructed of thirteen circles and retains sixfold symmetry. It has been suggested that the Fruit of Life is a blueprint of the universe, containing the basis for the design of every atom, molecular structure, and life form in existence.[6] Let's call the center of each circle a node, with each node connected to other nodes by a straight-line segment. We will refer to the line segments as links. We find seventy-eight links that form the edges of what can be perceived as a cube in two dimensions. This form is called Metatron's Cube.

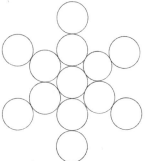

Figure 17.10: Fruit of Life

Geometricians discovered that each of the five Platonic solids can be drawn from the nodes and links of Metatron's Cube. In truth, this is an illusion. Recall that all of the geometric structures we have reviewed so far in this chapter consist of a certain number of circles that are either tangent to each other or overlapping in some particular manner.

Therefore each is a two-dimensional drawing. The cube and the other four Platonic solids cannot be formed from the geometry provided by any of these drawings. True, we can use the concept of artistic perspective and perceive a third dimension within the structure of circles before us. However, the vision is in our mind only and not part of physical space. Even if we replace every circle with a sphere, there remains the issue of proportion. We can envision a cube implied by connecting centers of spheres of the Flower of Life or Fruit of Life, but the relative proportions of the angles and side lengths of the polyhedron are inaccurate. Sides of the cube that we perceive at the forefront should be a little larger than the sides envisioned at the back of the form. Similarly, angles should be adjusted so as to be representative of the viewer's perspective.

As I've noted before, we humans prefer to make things easy. It is easier to create things in two dimensions and envision the third dimension than it is to construct something three-dimensional in the first place. This leads to inaccuracies, false assumptions, and misunderstandings. As symbols the constructions may be sufficient for use as tools for communicating ideas across space and time, but we can improve our communications if our symbols transmit information with greater precision. This includes use of the third dimension. After all, what are ancient megalithic structures such as Stonehenge, the Great Pyramid of Khufu, and the Pyramid of Inscriptions at Palenque but three-dimensional monuments and symbols unto themselves? And what are ancient cuneiform imprinting, Egyptian hieroglyphs, and Mayan temple engravings but representations of three-dimensional symbols in the form of writing, pictures, and sacred geometric symbolism? And what improvements can be seen in the communication of ideas across time—the fourth dimension—when we have the forethought to create symbolism in such a manner!

Plato prepared detailed geometrical analyses of what we now call the five Platonic solids. Kepler considered those ideas and based his model of planetary orbits based on those geometries. It is true that the vertices of each of the five Platonic solids are tangent to a sphere surrounding each polyhedron. Unfortunately, the physics of the universe does not allow the planets to orbit in circles around the sun. The sun is not the center point of each orbit. It is off-center, and each planet continuously adjusts its trajectory with the kinematics of its neighbors, as does the sun itself as it crosses space and pulls the planets along with it. This cosmic dance transcends the third dimension because of the effects of time, a topic I will address in later chapters. For now it is sufficient to limit our discussion of sacred geometry to three dimensions.

The Tree of Life

The Sepher Yetzirah serves as the foundation for the Kabbalah. We've seen that it describes "thirty-two wondrous paths of wisdom engraved by God . . . thirty-two books, with number, and text, and message . . . Ten Sephiroth of nothing, and the twenty-two letters of foundation." The Tree of Life of the Kabbalah is constructed from those ten Sephiroth, spheres of influence linked by twenty-two Nativoth paths of knowledge (Figure 17.11). The structure is intended to help us understand the two-way relationship between ourselves and the creator. Although it is one element of God manifesting into the universe, each Sephira contains infinity within itself. In chapter 10 we saw why Lenora Leet suggests that the origin of the Kabbalistic tradition was with the esoteric knowledge guarded by the ancient Hebraic priesthood: a cosmology linking the processes of cosmic creation to its goal in human transformation, and the sacred science that could demonstrate this purpose.[12] Her particular interest is the consecration of sacred space, and this was our focus in understanding the form of the Tree of Life of the Kabbalah in chapter 10.

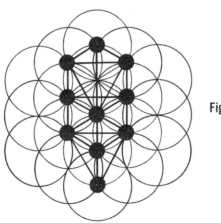

Figure 17.11: Tree of Life

Noting the locations and orientations of the twenty-two Nativoth, which link the ten Sephiroth together, it is apparent that the pathways form a series of triangles. In fact, some of the pathways appear to extend along the forefront of the tree, while others cross behind to connect with other nodes or Sephiroth. This shows that the Tree of Life is a three-dimensional structure. You might think of it as a spiritually engineered truss constructed of gusset plates (Sephiroth) and girders (Nativoth), or ten states of existence connected by relationships extending between our profane world and the universal creator. But why include a third dimension in this construction?

Four Trees

In chapter 7 we saw that Osiris, the *Djed* pillar, represented for the ancient Egyptians the connection between Earth and the cosmos. The *Djed* pillar, then, is equivalent to the World Tree of many other mythologies, as well as Otz Chiim. However, while these and other metaphors for the *axis mundi* are generally depicted as a single item extending between our world and the celestial sphere, there are suggestions that the Otz Chiim may be included in a more complex pattern reflecting human understanding of spiritual astronomy.

In their discussion of the World Tree and astro-alchemy, Jay Weidner and Vincent Bridges describe results of the ancient shaman's journeys of enlightenment.[7] The shaman would return with vital information that was in turn encoded within mythologies, sacred architecture, and graphic symbols. "This spiritual canon, or structures and geometric organization hierarchy . . . almost disappeared with the fall of the ancient world. It survived in fragments and in the quotations of the ancients . . ."[8] Weidner and Bridges find that William Stirling's *The Canon: An Exposition of the Pagan Mystery Perpetuated in the Cabala as the Rule of All Art*, published anonymously in 1897, change this course of ancient mysticism. For example, they write:

> The Canon is based on the objective fact that events and physical changes which are perpetual are never the less completely governed by intrinsic proportions, periodicities and measures.[9]
>
> Stirling's "cabalistic diagram" is the ten-step pattern of unfoldment known to occultists as the Tree of life, and according to his explication, it is the basic pattern of the canon itself."[10]
>
> Stirling tells us that "the ideas which the ancients connected . . . and combined into this figure of ten progressive

steps, appears to form the basis of all their philosophy, religion, art, and in it we have the nearest approach to a direct revelation of the traditional science, or Gnosis, which was never communicated except by myths and symbols . . . this great Word/World tree of geometry."[11]

In chapter 11 I noted Weidner and Bridges' finding that a "geometric pattern emerges from the intersecting lines of the Tree [of Life] like a moiré pattern in a holographic projection."[12] The intersection of four circles stacked upon each other appears to produce locations of the ten Sephira (Figure 17.12), and the three uppermost Sephira (*Keter, Chokmah,* and *Binah*) form the vertices of an upward-pointing triangle. Beneath that triangle is the inverted triangle of *Chesed, Gevurah,* and *Tiferet,* which in turn overlays the triangle of *Netzach, Hod,* and *Yesod.* "The whole pattern is then resolved by, and enfolded into, the last Sefirah, Malkuth. . . . The *Bahir* adds that portions of the celestial sphere can be equated with the spheres of each Sefirah, or globe, on the Tree of Life."[13]

Figure 17.12: Four stacked circles and the ten Sefirah

Based on information provided in the *Bahir,* the Bible, and other sources of esoteric, illuminated astronomy, Weidner and Bridges outline an ancient perspective of the universe perceived in the shape of a sphere within which is a cube of space, referred to as a sphere cube or "Stone of the Wise" (Figure 17.13). The four vertical edges of the cube align with the cusps of Sagittarius/Scorpio, Gemini/Taurus, Pisces/Aquarius, and Virgo/Leo. The constellation Draco appears above the top face of the cube, while the Lesser Magellanic Cluster is situated at the south ecliptic pole. In this scenario the ecliptic pole, Earth's rotational axis, was considered to be the stationary middle pillar of the Tree of Life surrounded by a kinetic world.

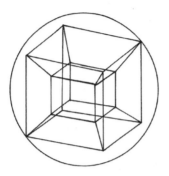

Figure 17.13: The Stone of the Wise (based on Weidner and Bridges, 2003)

Interestingly, Weidner and Bridges note that "Kabbalistic theory suggested, however, that if there was one Tree in the world, there should also be a reflection of all four Trees in the multiverse. The *Bahir* addresses this by implying a projected, jewel-like Tree that covered the surface of the sphere. The center point of the sphere, our sun in the *Tiferet* position, would then be projected outward onto the edges of the cube to form four interlocking Trees" (Figure 17.14).[14] They suggest that the arrangement of four interlocked Trees, each comprising a quadrant of the surface area, provides the basis for the Four World Ages as referenced by Jules Boucher in his article "*La Crois d'Hendaye*," published in *Consolation* in 1936. The transition point between the ages is the sun's precessional alignment with the outward projection of the *Tiferet* centers. The result is that Earth enters a new world age every time either the equinox or solstice crosses one of those points during the 26,000-year precessional cycle.

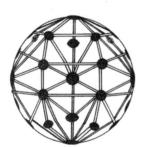

Figure 17.14: Four interlocked Tree of Life diagrams encapsulating a sphere (based on Weidner and Bridges, 2003)

I was quite surprised when I read the ideas Weidner and Bridges laid out in detail with regard to the Otz Chiim. The scheme applies the sphere, cube, and Tree of Life to develop an understanding of cosmology and the concept of world ages. There is additional complexity as the theory relates the Black Stone in the Kaaba ("cube") at Mecca and even to the cubed sphere depicting a tesseract (Fuller's vector equilibrium).

The Spherical Tree of Life

The idea of four Trees of Life covering the surface of a sphere is interesting and no doubt worth serious study for its Kabbalistic implications. However, such a construction is not representative of the worldwide, cross-cultural circular symbolism so pervasive throughout human history. The intricacy of Weidner and Bridges' construct simply does not have the elegance exhibited by traditional symbolism . However, I believe that their ideas have significant merit and are close to the fundamental geometry they and others have searched. I propose an alternative.

In 2007 I prepared two professional papers and presented them at the Plains Anthropological Conference, held at Rapid City, South Dakota.[15,16] The papers presented my findings concerning the symbolic limit of Lakota traditional territory and potential implications of the geometry of the round stone that White Buffalo Calf Woman gave to the Lakota along with the Sacred Pipe, as described by Black Elk. It was during that research that I first recognized the circle as a two-dimensional representation of a sphere and, more specifically, that the Sacred Hoop, the round stone, and numerous sacred symbols from around the world were associated with the geometry of nine great circles enclosing a sphere. I began referring to this symbol as the Sacred Sphere and, indeed, my research continues to affirm the universal and eternal sacred geometry within it. Every mathematical and geometrical aspect of the two- and three-dimensional forms outlined previously in this chapter, as well as the descriptive and graphic symbolism expressed in world mythologies across space and time, are found in the Sacred Sphere. And there is more.

After presenting my papers at Rapid City, I soon recognized that the structure of the Tree of Life was laid out perfectly across the surface of the Sacred Sphere. In fact, the Tree of Life extended around the sphere once, leaving few edges and vertices untouched. Reading Weidner and Bridges' analysis more than two years after this discovery, I concluded that four trees are too many. One tree covers almost the entire spherical surface. The elegance of the Sacred Sphere is explicit in the fact that the geometry of so many sacred symbols fit perfectly within its geometry. It is possible that the Tree of Life of the Kabbalah may be incomplete, that there may be additional Sephira and Nativoth to be discovered. I am certainly not an expert or scholar of Kabbalah (as it is a lifelong endeavor), and I do not claim that a structural revision of the Tree of Life is in fact needed or necessary.

However, all geometric aspects of sacred symbolism throughout the world indicate that our understanding of our selves and the universe remains incomplete. Might our symbols be incomplete as well?

I believe that the Sacred Sphere exists to help us reconnect with the creator and recognize the cyclical relationships we have with each other and the cosmos. Our search may be complete when our symbolism accurately reflects truth. This appears to be one of the messages relayed to us from ancient and indigenous cultures. Indeed, this may be the meaning of the Sacred Sphere itself.

The Sacred World Sphere

The configuration of nine great circles across the disdyakis dodecasphere yields several surprising correspondences with the rotational axis, four lateral directions, equator, meridian lines, ecliptic, and other curvilinear cosmographic images extending on and above Earth's surface (Figure 17.15). Those features are related to various sacred symbols used by cultures around the world since prehistoric times. Are these equivalences direct evidence of the significance of the Sacred Sphere to Earth in particular? Do they help explain the fundamental relationship between ancient traditions and the cosmos? It is difficult to answer these questions based on the available facts. It may be a matter of belief or faith that the Sacred Sphere is symbolic of the physical nature of earth and sky, and therefore of the inner spirit of earthly existence and the human search for the eternal spirit that fills the universe. Recall that the circle separates interior space from exterior space, and so separates the inner and outer spaces of spirit. We can, therefore, envision the sphere serving the same purpose within three-dimensional space.

Figure 17.15: The Tree of Life extending across the DDs. This is a surprising discovery indeed, and further confirmation of the universality of vital, cross-cultural symbolism expressed by the Sacred Sphere throughout time.

Let's look at the geometry of great circles on the Sacred Sphere and identify the corresponding physical aspects of Earth.

- Earth's rotational axis extends from the orb's center outward to the north and south poles. On the Sacred Sphere the two poles may be represented by a pair of vertices directly opposite each other. We can most easily choose two opposed vertices, each located at the juncture of an octet of triangular faces. There are three such pairs, each of which is geometrically equivalent to the other two. The World Tree (Hamlet's mill, cosmic mill, fire stick, world churn) is Earth's rotational axis and may be represented by the referenced vertices on the Sacred Sphere.

- The four vertices remaining after the designation of poles are located at the junctures of unique octets and define the equatorial location of four lateral directions with poles extending outward from the Sacred Sphere's (and Earth's) surface. Normally we would refer to those four directions with respect to the 360 degrees of arc that define a circle, with each of the four directions separated by 90 degrees of arc from the adjoining two directions, and 180 degrees from the opposite direction.

- A unique great circle encompasses the vertices that define north, east, south, and west on the Sacred Sphere and Earth. That circle is the equator. The cosmic equator is an extension of this great circle outward to the celestial sphere.

- Four great circles on the Sacred Sphere pass through the two poles. The circles are equivalent to eight longitudinal meridian lines extending from pole to pole across Earth. If 0 degrees longitude is assigned to one of the meridian lines, then the seven other meridians (each half of a great circle) are located at 45, 90, 135, 180, 225, 270, and 315 degrees, or their equivalent in radians (0, $\pi/4$, $\pi/2$, $3\pi/4$, π, $5\pi/4$, $3\pi/2$, and $7\pi/4$). There are, of course, an infinite number of great circles (meridians) extending between the poles, and the Sacred Sphere may be rotated such that its great circles represent any similar set of eight meridians.

- The ecliptic is the geometric plane defined by the mean orbit of the earth around the sun. Currently it is inclined by about 1.5

degrees with respect to this orbitally defined plane. The ecliptic plane is oriented at an angle of about 60 degrees to the galactic equatorial symmetry plane. It is currently 23.44 degrees from the celestial equator (the Tropic of Cancer and Tropic of Capricorn are each defined as 23.5 degrees north and south, respectively, of Earth's equator); this is the obliquity of the ecliptic, which ranges from about 22.1 degrees to 24.5 degrees over long periods of time (millions of years). Similarly, the Sacred Sphere includes eight vertices where three great circles intersect at an angle of 60 degrees at each vertex, allowing several orientations of the sphere to reflect the geometry of the ecliptic plane relative to the galactic equator. The Sacred Sphere also includes four great circles, which are oriented at an angle of 45 degrees to the axis of rotation. With a current obliquity of the ecliptic of 23.44 degrees, two of those great circles are oriented within 1 degree of the ecliptic, and the other two are oriented within 1 degree of the plane of the Arctic Circle and the Antarctic Circle at this time.

- From Earth's perspective, the two equinoctial points are located where the celestial equator and the ecliptic intersect at opposite directions from the earth. Those two points are situated on the celestial sphere, but can be projected back to Earth's surface at two locations on the equator, each 180 degrees from the other and forming an axis between the celestial sphere and Earth itself. Thus, those two locations can be mapped onto the equatorial great circle of the Sacred Sphere. The equinoctial axis rotates across the celestial sphere at the same rate as precession; this is the meaning of the progression of the equinoxes.

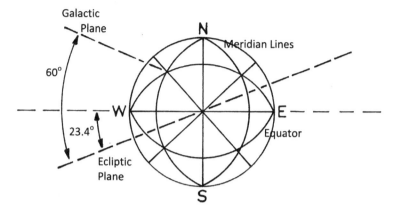

Figure 17.16: The configuration of nine great circles across the surface of the DDs appears to correlate surprisingly well with Earth's rotational axis, four cardinal directions, equator, meridian lines, ecliptic, and other curvilinear cosmographic images extending on and above Earth's surface.

Certainly there is a significant correlation between the physical nature of Earth and the cosmos and the geometry of the Sacred Sphere. Much of this relationship lies buried within world mythologies, each culture providing descriptions of some but not all of the geometry. Are there applications of the entire Sacred Sphere—its twenty-six vertices, forty-eight faces, and seventy-two curvilinear edges—to natural forces found on Earth as well as energy and matter in the universe as a whole? If so, what are the implications for mankind? These are questions I address in the following chapters.

Chapter 18
The Cosmos

The first function [of myth] is the mystical function . . . realizing what a wonder the universe is, and what a wonder you are, and experiencing awe before this mystery. Myth opens the world to the dimension of mystery, to the realization of the mystery that underlies all forms. . . . If mystery is manifest through all things, the universe becomes, as it were, a holy picture. . . . The second [function of myth] is a cosmological dimension, the dimension with which science is concerned—showing you what the shape of the universe is, but showing it in such a way that the mystery again comes through. . . .

JOSEPH CAMPBELL
THE POWER OF MYTH

A Universe of Spheres

For millennia ancient and indigenous peoples not only observed that the cosmos was filled by planets, moons, stars, and other spherical heavenly bodies, they also observed and concluded that the universe as a whole was spherical. Is that conclusion fact or fiction? How do we define the boundary of space, and where is it? Does it change over time? And how do we know? In this chapter we will look at the data gathered from ongoing astronomical studies and see what answers they may provide. Science is a progression of ideas based on testable hypotheses that can

425

lead to theories explaining the physical, chemical, and biological characteristics of the universe. We can debate the question of whether science
can yield truths. However, as we've seen, intellectual and spiritual traditions buried in ancient mythologies were regarded as truths in the minds
of people living in cultures that developed those myths—those overt and
covert storylines explaining the beginning, ongoing, and ending of time
and space.

The first question—whether or not the universe is shaped like a
sphere—can be answered here and now. To date, modern science has not
yet determined the actual shape of the universe. And so the answer to the
question is a resounding "We still don't know." Perhaps I should qualify
that statement by saying scientists don't know, while we've already seen
that ancient cultural traditions described the shape of the universe in
their mythologies. Mythologies express cultural understandings of
creation, operation, and maintenance of the universe and how humanity
fits into universal reality and illusion. As we've seen, the fundamental
shape of the world or universe is often characterized as spherical. The
symbolism of this shape can be interpreted from several perspectives—
physical, emotional, intellectual, or spiritual.

Scientists, on the other hand, are concerned with only the physical
attributes of the universe. Since we cannot physically visit the limits of
the universe, if it is in fact bounded, it is left to scientists to conduct
experiments, gather data, develop, and test mathematical models, and
present results. Those results typically have some measure of reliability,
based on initial conditions and constraints placed on the model used to
represent observable reality. Theories develop from the testable results—
results that can be replicated through further testing. The data is used to
refine the theory, which then undergoes further modeling, testing, and
refining. The ultimate goal of science is to develop a "model of everything," a theory that explains the cause and effect of all observations—from
the microcosmic to the macrocosmic.

But we cannot ask science to provide us with truths. Euclidian geometry is no more filled with universal truths than are Newtonian physics,
Einstein's theory of general relativity, and the theory of quantum
mechanics. Those theories are based on observations, experimentations,
and mathematical derivations. The theories are not necessarily wrong,
but their study remains incomplete. Their universal applicability remains
unproven. Science gives us only what we allow it to provide, and like so

many other aspects of life, it is the journey that is important, not the destination. So, science journeys on a path investigating the shape of the universe, but it has not yet completed that journey.

What have scientists learned so far? To answer that question we first need to understand the constraints that science has placed on itself. Only then can we understand the importance of the conclusions drawn by current science in its quest for a theory of everything, including the size and shape of the universe. First, recall the mythological accounts of the beginning of the world, written down thousands of years ago and expressed in oral traditions for perhaps thousands of years before. Recall detailed descriptions of the formation of the world, philosophies expressed by ancient cultures, understandings of the human perception of reality, and truths held in common across the spectrum of world mythologies. Compare that with the following statement by Stephen Hawking and Leonard Mlodinow, from their book *The Grand Design*, in which they express the current importance of science: "[P]hilosophy is dead. Philosophy has not kept up with modern developments in science, particularly physics. Scientists have become the bearers of the torch of discovery in our quest for knowledge."[1] From those statements we can see scientists ,perception of themselves as champions in the quest to understand the physical nature of the universe. If this self-perceived importance of science is necessary in order to achieve improvements to physical conditions here on Earth, then so be it. The cause is, indeed, a noble one.

What is surprising about the findings of modern science is that they are very much in accord with the teachings of ancient mythology, and yet science misses this correspondence totally, as Hawking and Mlodinow seem to confirm in stating, "The purpose of this book is to give the answers that are suggested by *recent* discoveries and theoretical advances. They lead us to a *new* picture of the universe and our place in it that is *very different from the traditional one*, and different even from the picture we might have painted just a decade or two ago. Still, the first sketches of the new concept can be traced back *almost a century* [italics mine]."[2] As we are about to see, the recent discoveries of science are *not* recent discoveries of humanity, but knowledge well-documented in history. The picture is *not* a new one, and it is certainly *not* very different from the traditional one. Rather, it appears that science is only now recognizing that theoretical, mathematical discoveries reconfirm ancient mythological traditions.

The Shape of the Universe

American mathematician Jeffrey Weeks has written numerous articles addressing the state of scientific study and its understanding of the size and shape of the universe. "Because of the finite speed of light," he writes, "we see the Moon as it was roughly a second ago, the Sun as it was eight minutes ago, other nearby stars as they were a few decades ago, the center of our Milky Way Galaxy as it was 30,000 years ago, nearby galaxies as they were millions of years ago, and distant galaxies as they were billions of years ago."[3] In other words, the universe is very, very large. Incomprehensibly large. And yet, our scientific understanding of the size of the universe is essentially no different from the understanding expressed by our ancestors in the distant past. Recall the mythic descriptions of the beginning of the universe:

- Rig-Veda: Some verses composed four thousand years ago or earlier describe the conception of the egg of the universe: "in the beginning atman alone was this universe; there was nothing else at all to meet the eye. He deliberated: I will create worlds." The atman created Earth, the atmosphere, and the waters above and below, and then brought forth the *purusha* from the waters. The shoots of a tree grew into the earth and created a grove. This ability, a subtle essence to create from oneself, is a universal characteristic "that is the real, that is the soul, that art thou."

- Norse myth: In the beginning there were two worlds: Muspellsheimr and Niflheimr, fire and ice. All else was derived from the melding of heat and cold.

- Heliopolitan theology of Lower Egypt: Teaches the universe began in the utter disorder of Nun, the uniformly dark, watery body of nothing. Out of this unstructured environment came organization in the form of a mound of dry land from which Re is self-created in the form of Atum, the Great He-She, the Complete One.

- Genesis 1:1–3: "In the beginning God created that the heavens and the earth. Now the earth was formless and empty, darkness was over the surface of the deep, and the Spirit of God was hovering over the waters. And God said, 'Let there be light,' and there was light."

- John 1:1 of the New Testament: "In the beginning was the Word, and the Word was with God, and the Word was God."

- Qur'an 7.29: "Say: My Lord has enjoined justice, and set upright your faces at every time of prayer and call on Him, being sincere to Him in obedience; as He brought you forth in the beginning, so shall you also return."

Most astrophysicists hold that time can be traced about 13.7 billion years back to the big bang, when the universe expanded from a singularity containing all matter and energy, and from which all matter and energy observed today is derived. Jeffrey Weeks explains why the energy associated with the big bang is not readily apparent to us:

> If we look still deeper into space, we see all the way back to the final stages of the big bang itself, when the whole universe was filled with a blazing hot plasma similar to the outer layers of the modern Sun. In principle we see this plasma in all directions; it fills the entire background of the sky. So why don't we notice it when we look up at the night sky? The catch is that the light from it—originally visible or infrared—over the course of its 13.7 billion-year voyage from the plasma to us has gotten stretched out as part of the overall expansion of the universe. Specifically, the universe has expanded by a factor of about 1100 from then till now, so what was once a warm reddish glow with a wavelength around 10,000 angstroms is now a bath of microwaves with a wavelength of about a millimeter. So we cannot see the plasma with our eyes, but we can see it with a microwave antenna.[4]

What we know as visible light has wavelengths ranging from about 400 to 700 nanometers, about two thousand times shorter than the smallest wavelength of microwaves. So we can't see microwaves, but if we could, our view of the night sky would be quite different. With the aid of a telescope we would see other galaxies as they were a few billion years ago, as we can now, but visible in the cosmic background would be the post–big bang plasma configured as it was just 380,000 years after the universe began expanding. Obviously, we would be seeing a very early stage in the development of the universe. According to Weeks, information about this cosmic microwave background (CMB) radiation is important because "the plasma holds clues to the universe's birth, evolution, geometry, and topology."[5] The temperature of the CMB is nearly uniform. However, fluctuations in its temperature are measureable and appear to be within a range of about 0.00001 degree Celsius. They are the result of density fluctuations in the plasma (Figure 18.1).[6] In other words, cosmologists can map variations in the density of the early universe by observing the CMB and recording its temperature fluctuations. Density fluctuations throughout space are found at all scales, and the universe appears to consist of an infinite Euclidean space.[7]

As a result, scientists developed a model of the geometry of the universe in which the volume of space resulted from inflation subsequent to the big bang. Think of universal inflation as a balloon being inflated (Figure 18.2). As the volume of air inside the balloon increases, each point within the balloon and on its surface is separated from all other points by a greater distance. Weeks says that "each galaxy sits more-or-less motionless relative to space, and it is space itself that's expanding. So the total distance from galaxy A to galaxy B is increasing, even though galaxy A sits motionless relative to space and galaxy B sits motionless relative to space."[8]

Figure 18.1: CMB depicted as fluctuations of its temperature across the universe (image courtesy of NASA).

Given the readily observable cosmos with its unimaginably large array of stars, galaxies, galaxy clusters, and permeating intergalactic dusts, it isn't surprising that the density of the universe varies with location. But what is meant by "infinite Euclidean space"? "Infinite" indicates the lack of boundary or edge—an unbounded continuity. "Euclidean" suggests a universal geometry understandable in terms of the two-dimensional Euclidean plane and three-dimensional Euclidean space, but with similar applicability in higher dimensions. Working with fewer than four dimensions allows us to envision the geometry of the universe in terms of Cartesian coordinates. We can use real numbers, simple algebra, and calculus to characterize and quantify the size and shape of space, even if it consists of more than three dimensions. However, from our daily experience here on Earth, it is easiest to envision this geometry in terms of no more than three dimensions, such as left–right (i.e., x-axis), up–down (y-axis), and front–back (z-axis). Often the fourth dimension is recognized as time. We are familiar with the relationship of three-dimensional space to time as the change in location of mass or energy over time (velocity) or the change in velocity over time (acceleration). These changes in location are often described in terms of one- and two-dimensional changes in location, such as the distance between points (locations), translations, angles, and rotations. We can also describe the magnitude, or change in magnitude, of some given parameter in space, typically related to mass or energy.

As noted previously, upon identifying the magnitude and direction of any given thing, we have defined a vector. We can therefore describe the physics of the cosmos in terms of vectors corresponding to movement, or change, within Euclidean space. We can measure distance and changes in location (translation) and define rotations based on angles and distances. The area of mathematics addressing the properties of space (from the microcosmic to macrocosmic) is topology, which evolved from the study of geometry and set theory. Cosmologists apply topology to understand the relationships between space, dimensions, and transformations with regard to objects (such as a balloon, or the universe) that can be mathematically conceived as under continuous deformation.

Figure 18.2: Inflation of the universe is similar to an inflating balloon, except that while graphics on a balloon's surface will expand as air fills the balloon, matter does not expand as the volume of the universe increases.

Returning to the realization that there are density fluctuations at all scales in the universe, and that small-scale density fluctuations are indicated by temperature data obtained from the CMB, those observations lead cosmologists to conclude that the universe as we know it is either flat or has a slight curvature (Figure 18.2). Weeks writes,

> More precisely, on a scale where $\Omega < 1$ indicates a hyperbolic universe, $\Omega = 1$ indicates a flat universe, and $\Omega > 1$ indicates a spherical universe, analysis of the WMAP [NASA's Wilkinson Microwave Anisotropy Probe, launched in 2001] data yields $\Omega = 1.02 \pm 0.02$ at the 1σ level. The parameter Ω measures the average mass-energy density of space, which by general relativity completely determines the spatial curvature, with low density yielding a hyperbolic universe and high density yielding a spherical universe. By definition Ω is the ratio of the actual density to the critical density that a flat universe would require.[9]

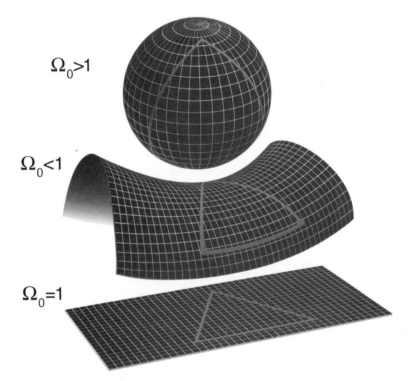

$\Omega_0 > 1$

$\Omega_0 < 1$

$\Omega_0 = 1$

Figure 18.3: a) $\Omega < 1$ indicates a hyperbolic universe, such as a saddle shape. b) $\Omega = 1$ indicates a flat universe, such as a two-dimensional plane. c) $\Omega > 1$ indicates a spherical universe. WMAP data provides a value of about $\Omega = 1.02$, indicating that the observable universe is not flat, but nearly so, with only a slight curvature. This is similar to our view of the local surface of Earth being level, or flat, while we know that on a grand scale Earth's surface approaches the shape of a sphere. (illustrations courtesy of NASA)

What is the implication of Ω, the average mass-energy density of space? As an example of a hyperbolic universe, think of the shape of a saddle. Let's put the saddle out in space, so it is not under the influence of Earth's gravity. Now, roll two marbles in the same direction across the saddle. You'll find that the trajectories of the two marbles deviate away from each other. In a flat universe, as in Euclidean geometry, rolling two marbles in the same direction across a two-dimensional plane results in two parallel trajectories. In a spherical universe, the curvature of space will cause the two marbles to change course and collide at some point in the distance. Since the WMAP data provides a value of about $\Omega = 1.02$, it is inferred that the observable universe is not flat, but nearly so, with a slight curvature tending toward a spherical geometry. This is similar to our view of the local surface of Earth being level, although we know that on a grand scale Earth's surface approaches the shape of a sphere.

There is a complication, however. When astrophysicists take into account Einstein's general theory of relativity, with its implications for cause and effect related to three-dimensional space and the fourth dimension of time, they find evidence suggesting that the observable universe may be better represented by what mathematically is called a 3-manifold. A manifold describes a space that on a small scale appears representative of a Euclidean space within a specific dimension, but can be far more complicated when viewed at larger scales. The geometric line is the simplest manifold, followed by the circle. The triangle, square, and circle each are 1-manifolds within two-dimensional space. The tetrahedron, cube, and sphere are 2-manifolds within three-dimensional space. A 3-manifold, then, is a geometric structure that can be described mathematically within four-dimensional space. Without getting into unnecessary detail, Weeks states that "[o]bservational evidence implies the observable universe is homogeneous and isotropic to a precision of one part in 104, so consider manifolds that locally look like the 3-sphere S3, Euclidean space E3, or hyperbolic space H3."[10]

The WMAP data shows plasma radiation coming at us from all directions. It's like viewing the primordial universe projected on a spherical screen. Cosmologists call the distance from Earth to this screen the horizon radius, and the screen is the horizon sphere. If we assume that we live in a finite universe, then the smallest sphere centered at Earth that encapsulates the universe and intersects itself is referred to as the injectivity radius. Beginning at Earth, the minimum length of the universe's circumference would be two times the injectivity radius. If you're confused, here's the punch line: If we assume that the horizon radius exceeds the injectivity radius, then two different lines of sight will lead us to the same apparent region of space: theoretically we could see a multiple of the same cosmic image (such as distal stars, galaxies, primordial plasma, and so forth) viewed from Earth toward different locations in the sky![11] But, of course, there is no physical or mathematical reason for the universal injectivity radius to be centered on Earth. So the relative length and location of the horizon radius and injectivity radius remains undetermined, if it is real at all. This makes detection of a universally flat topology extremely difficult, even though the estimated average mass-energy density of space ($\Omega = 1.02\pm0.02$) is close to unity. Similarly, Ω is close enough to 1.00 that a hyperbolic universe cannot be ruled out. But the task of detecting its finite hyperbolic topology would not be easy, and the injectivity radius of even small hyperbolic manifolds typically exceeds the horizon radius, a situation that would be contrary to the very way that those concepts are envisioned.

So, from a topological standpoint, it appears that a S3 spherical universe may be more mathematically satisfying that its E3 or H3 challengers. Reviewing Weeks's writing, we can summarize his support for this idea:

- Astrophysicists calculate the horizon radius to be about 46 billion light-years. Recall that the universe has been expanding, inflating, since the big bang. At some point the horizon radius exceeded a distance of 13.7 billion light-years as it expanded. So, while it has taken 13.7 billion years for some of the CMB radiation to reach us, the horizon radius continued to extend outward to 46 billion light-years.

- Geometers prefer working with a horizon radius measured in units of the curvature radius (Rcurv).

- The dimensionless ratio Rhor/Rcurv allows the horizon radius to be represented in radians (306 degrees equals 2π radians).

- Rcurv is strongly dependent on the value of Ω.

- For a spherical universe, Rcurv calculates as 98 billion light-years when Ω is 1.02.

- The current estimate for Rhor is 46 billion light-years, so the geometer's horizon radius Rhor/Rcurv is equal to 0.47 radians.

- Therefore, for $\Omega = 1.02$, the horizon sphere's radius is 0.47 radians, and we can see a rather significant portion of the 3-sphere from our vantage point here on Earth. For cosmologists this is very satisfying, as it means that the topology of a S^3 universe provides a simple and natural configuration for study and an increased potential for detection.

"This potential detectability," Weeks writes, "along with WMAP's observation of $\Omega \approx 1.02$, *has fuelled considerable interest in the possibility of a spherical universe . . .*"[12] [emphasis mine] Of course, Weeks is referring to a four-dimensional (S3) sphere and not our commonly perceived 3-D variety. However, the importance of this finding cannot be overstated.

The topology of a 3-sphere universe has a greater probability of being detected than topologies of other manifolds based on current astrophysical data and mathematical modeling.

While this finding does not prove a spherical universal geometry, it certainly promotes further exploration of cosmic 3-sphere topology. Let's look further at the implications. Weeks states that "any continuously defined field in the physical universe—for example, the density distribution of the primordial plasma—may be expressed as a sum of harmonics of 3-dimensional space."[13] As the name suggests, the CMB consists of waves of energy coming at us from all directions in space. Those fundamental waves of energy have wavelengths of about 1 millimeter. Those waves are expressed by a sinusoidal oscillation moving at the speed of light, with the wavelength (λ) inversely proportional to frequency (f), or $f = 1/\lambda$. A harmonic of a wave is an integer multiple of the frequency, so that the harmonics of a given microwave have frequencies of $2f$, $3f$, $4f$, $5f$, and so forth. Harmonics are periodic at the fundamental frequency, and so the sum of the harmonics is periodic at that same frequency. Also, the sums of harmonics are multiples of the fundamental frequency. Musicians know this to be true with regard to sound waves. The same is true for microwaves. The size and shape of the universe are reflected in the pattern of energy waves we receive as we probe for CMB information coming at us from the horizon sphere. This is the CMB power spectrum, and the modes of spherical S3 space are remarkably regular and predictable.[15]

We can readily see this beginning with the harmonics of a circle, and the mathematics behind oscillations in a circle (S1) parallel numerical analysis of the 3-dimensional sphere (S2), that in turn help characterize oscillations of the 4-dimensional sphere (S3) in terms of homogeneous harmonic polynomials in the x, y, z, and w directions. Is a 4-D universe possible? Mathematicians had classified 3-dimensional spherical spaces by 1932. As it turns out, Weeks states, "The possible groups turn out to bear a close relationship to the symmetry groups of an *ordinary 2-sphere*![16] [emphasis mine]

The result is that cosmologists can transfer symmetries from the 2-sphere to the 3-sphere. Applying a finite group of symmetries of S2 such as the twelve orientation-preserving symmetries of a regular tetrahedron, they could produce the group of fixed-point free symmetries of the 3-sphere. The binary tetrahedral group has order 24. Weeks lists well-known S2 finite symmetry groups, of which we have seen several in the symmetry of the DDs:

- The cyclic groups Z_n of order n, generated by a rotation through an angle $2\pi/n$ about some axis.

- The dihedral groups D_m of order $2m$, generated by a rotation through an angle $2\pi/m$ about some axis as well as a half turn about some perpendicular axis.

- The tetrahedral group T of order 12, consisting of all orientation-preserving symmetries of a regular tetrahedron.

- The octahedral group O of order 24, consisting of all orientation-preserving symmetries of a regular octahedron.

- The icosahedral group I of order 60, consisting of all orientation-preserving symmetries of a regular icosahedron.

We can also transfer those groups from S2 to S3, as noted above, to provide a listing of single action symmetry groups of S3:

- The cyclic groups Z_n of order n.

- The binary dihedral groups D*m of order $4m$, $m \geq 2$.

- The binary tetrahedral group T* of order 24.

- The binary octahedral group O* of order 48.

- The binary icosahedral group I* of order 120.[17]

Armed with an understanding that further analysis of well-proportioned spaces was warranted, and with WMAP's indication of a slightly positive curvature, cosmologists proceeded to evaluate the binary polyhedral spaces of S3/T*, S3/O*, and S3/I*. Each of these polyhedrons—a regular octahedron, a truncated cube, and a regular dodecahedron—has a positive curvature and dimensions with similar magnitudes. Of these, the dodecahedral S3/I* space was considered to be the best candidate because its fundamental domain's inradius of $\pi/10 \approx 0.31$ is within the horizon radius Rhor/Rcurv = 0.47, which corresponds to the estimated universal average mass-energy density (Ω) of 1.02.

The Potential for S^3 Dodecahedral Space

The *S3/I** polyhedral space is known as the Poincaré dodecahedral space, named after Henri Poincaré. Born in 1854 in the French city of Nancy, Poincaré was a mathematician par excellence who attended the École Polytechnique and later the prestigious École de Mines, becoming a mining engineer. After completing his PhD he became an assistant professor at the University of Caen in 1879 and developed an interest in complex variable functions related to the solution of differential equations. Those functions were associated with non-Euclidean geometry. Soon Poincaré entered a rivalry with another gifted mathematician, Felix Klein, concerning the relationship between non-Euclidean geometry and what are called Fuchsian functions. That rivalry ultimately led to a theorem ("the most breathtakingly beautiful in mathematics and, for that matter, in human thought"[18] according to author and professor of mathematics Donal O'Shea) relating topology and two-dimensional geometry. O'Shea summarizes,

> Spheres of different radius are not isometric, nor are spheres with dimples or mountains. However, Poincaré and Klein's work implied that any surface could be given a geometry in which it had a constant curvature (and it followed easily that this geometry had to be unique—if a surface had a flat geometry, we couldn't put a spherical geometry on it).[19]

By 1904 Poincaré, having ambitiously investigated the S3 manifold, wrote: "There remains one question to handle: Is it possible that the

fundamental group of a manifold could be the identity, but that the manifold might not be homeomorphic to the three-dimensional sphere?"
This question became known as the Poincaré conjecture. Poincaré
himself could not solve it. No one could solve it. The question arose after
Poincaré provided an example of a manifold other than the three-dimensional sphere that had the same number of unconnected k-dimensional
surfaces (Betti numbers), with refinements, as the three-dimensional
sphere itself. His example became known as the Poincaré dodecahedral
space, which he described,

> . . . as two solid-holed tori glued together in an appropriate
> manner . . . [with] a pair of paths that do not shrink to a
> point . . . nowadays usually described as the three-manifold that
> we get by gluing together opposite faces of a regular dodecahe-
> dron after a one-tenth counterclockwise turn. Recall that a
> regular dodecahedron is the twelve-sided polyhedron we get by
> assembling twelve regular pentagons of the same size to make a
> closed surface bounding a solid. It is the fifth and last Platonic
> solid. The Pythagoreans would have been delighted.[20]

To be sure, Poincaré was not suggesting that his "two solid-holed
tori glued together in an appropriate manner" was a dodecahedron. Identification of manifolds came several decades later. And, indeed, the
Poincaré conjecture was not solved until 2003, when the Russian mathematician Grigory Perelman provided the proof.

Weeks and his associates numerically computed the modes of binary
polyhedral spaces. Accumulating numerical errors limited the case of the
Poincaré dodecahedral space $S3/I^*$ and a portion of the predicted CMB
power spectrum. However, Weeks provides an interesting summary of
the study's elegant results.[21]

- Observations matched the predicted sequence of expansions
  related to angles on a sphere (including the quadrupole and
  octopole).

- The results best fit in the range of $1.01 < \Omega < 1.02$. This is
  within the WMAP observations of $\Omega = 1.02 \pm 0.02$.

- These were almost no free parameters. The findings suggest
  that the dodecahedral space can only be constructed from a
  regular dodecahedron (zero degrees of freedom), noting that
  the dodecahedral space is globally homogeneous, and appears

the same from all perspectives (again, zero degrees of freedom). Ω was the only free parameter in the initial analysis. While far from a proof, such results were most encouraging.[22]

While Weeks notes that results of the study are not proof of the shape of the universe, the dodecahedral model under study produced three testable predictions: 1) weak large-scale CMB fluctuations that have been documented by the WMAP data; 2) a slight curvature of space supported by the WMAP data and perhaps by forthcoming data from more recent observations by the European Space Agency's Planck satellite; and 3) the appearance of matching circles of cosmic masses in the sky. The third prediction is based on the potential for the fundamental dodecahedron to be smaller than the horizon sphere, in which case the horizon sphere would wrap around the universe, intersecting itself, thereby allowing repeating images of the horizon sphere to appear. These images would appear in opposite directions (antipodal) from Earth and would yield the same temperature patterns if the low-level, but still detectable, CMB temperature fluctuations are purely dependent on the plasma density fluctuations.

Mota, Reboucas, and Tavakol recently searched for such circles in the cosmos, assuming that the universe is flat (Ω=1.00).[23] They calculated the maximum angles of deviation possible for the antipodal pairs of matching circles that would correlate with the shortest closed geodesic for multiply connected, flat, and orientable 3-manifolds. A great circle on an S2 sphere is an example of a closed geodesic on a 2-manifold. They recognized that the upper bound on the deviation from antipodicity could indicate the potential for the antipodicity of matching circles to be significantly greater than zero, the implication being that searches for circles-in-the-sky have been insufficient to exclude detection of a nontrivial flat topology. And so the search continues, while other studies indicate that the Poincaré dodecahedral space may be gaining support.

- In February 2008 a team of cosmologists was reported to have "improved the theoretical pertinence of the Poincaré Dodecahedral Space (PDS) topology," while another team of researchers analyzing the WMAP data discovered that the geometry of the S3 Poincaré dodecahedral space matched a "topological signal characteristic."[24] The report notes that scientists calculated 1.7 billion vibrational modes of the Poincaré dodecahedral space, improving the simulation of the

CMB power spectrum, with a result being an optimal value of $\Omega = 1.018$. The CMB power spectrum predicted by the PDS topology was described as in remarkable agreement with the observed value.

- Boudewijn F. Roukema of the Torun Centre for Astronomy recently suggested that CMB observations may best fit a mathematical model describing a well-proportioned small universe and "the Poincare dodecahedral space is empirically favoured."[25] He notes that the observed effect of residual gravity in the universe may be related to "residual acceleration induced by weak limit gravity from multiple topological images of a massive object on a nearby negligible mass test object. At the present epoch, the residual gravity effect is about a million times weaker in three of the well-proportioned spaces than in ill-proportioned spaces. However, in the Poincare space, the effect is 10,000 times weaker still, i.e., the Poincare space is about 10^{10} times 'better balanced' than ill-proportioned spaces. Both observations and weak limit dynamics select the Poincare space to be special."[26]

An Alternative Model: S3 DD Space

During autumn 2010 I contacted Jeffrey Weeks and inquired about the possibility that the shape of the universe, rather than being represented mathematically by the topology of a Poincaré dodecahedral space, might be better represented by the topology associated with the disdyakis dodecahedron. In other words, rather than modeling the shape of the universe as an S3 dodecahedron, is it possible that the universe exhibits the form of S3 DD space, a four-dimensional DD? Recall that the essential difference between a dodecahedron and a DD is that the dodecahedron is constructed of twelve pentagons, while the DD replaces each of those pentagons with a set of four triangles forming a diamond or lozenge shape; the forty-eight triangles create twelve sets of four triangles each. Weeks responded, saying,

> Poincaré Dodecahedral Space . . . is obtained by starting with a dodecahedron and "gluing" opposite faces to obtain a space that has a finite volume, yet has no boundary. . . . if you

wanted to model the shape of the universe based on a disdyakis dodecahedron, you'd need to devise a way to glue its faces together in pairs, so that its geometry (probably it would need to be hyperbolic) would continue smoothly across the places where its edges meet, and also continue smoothly across the places where its corners come together.

If you could do that (find a consistent way to glue up a pair of faces) then mathematicians would be happy with your result. To make cosmologists happy, though, there'd need to be some observational evidence favoring such a model, and that's the hard part. So far there is only a tentative hint that the real universe may be finite. There's no "smoking gun" evidence that the universe is finite, let alone enough information to deduce its shape.[27]

Would an S3 disdyakis dodecahedral space satisfy the conditions that Weeks lists? If so, it would have to:

1. Be obtained by "gluing" pairs of opposite faces such that the resulting space has finite volume, yet has no boundary; the geometry would continue smoothly across the places where the edges of each face meet, as well as continuing smoothly across locations where the faces' corners come together; and

2. Provide observational evidence favoring such a model.

I suggest that both provisions may be satisfied based on the geometry of the S2 disdyakis dodecahedron and the topology of its 3-manifold, in tandem with the wealth of observations accumulated in recent years by the WMAP and other ongoing cosmological projects. The first provision is satisfied with the appropriate "gluing" of opposite faces (Figures 18.4 and 18.5). The second provision, of course, requires detailed observation and analysis. However, if recent analysis of the Poincaré dodecahedral space as a candidate for the shape of the universe is any indication, then replacing each of the twelve pentagonal faces of the dodecahedron with four appropriately shaped triangles, so as to form the DD, appears to be a minimal modification of shape and may warrant additional study.

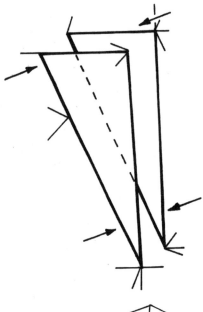

Figure 18.4: "Gluing" together opposite triangular faces of two DDs.

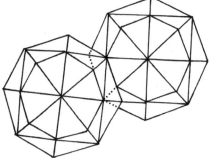

Figure 18.5: Two DDs glued together as a potential model of a 3-sphere universe.

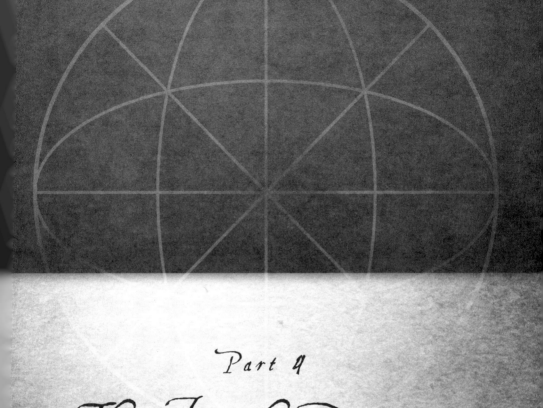

Part 4

The Fourth Dimension

Chapter 19

Time: Temporal yet Eternal, Linear or Circular?

$E = mc^2$

ALBERT EINSTEIN
"DOES THE INERTIA OF A BODY DEPEND UPON ITS ENERGY CONTENT?"

Henceforth, space by itself and time by itself, are doomed to fade away into mere shadows, and only a kind of union of the two will preserve an independent reality.

HERMANN MINKOWSKI
"SPACE AND TIME"

Since we can't step outside our own space-time to view its warpage, the space-time warpage in our universe is harder to imagine. But curvature can be detected even if you cannot step out and view it from the perspective of a larger space. It can be detected from within the space itself . . . The same is true of warpage in our universe—it stretches or compresses the distance between points of space, changing its geometry, or shape, in a way that is measurable from within the universe. Warpage of time stretches or compresses time intervals in an analogous manner.

STEPHEN HAWKING AND LEONARD MLODINOW
The Grand Design

445

Big Time

When we look up at the sky at night we are immediately made aware of an inescapable fact: The universe is very, very large. Great distances separate Earth from Moon, Earth from Sun, sun from sun, galaxy from galaxy, and one side of the universe from the other. As we ponder this observation we realize another fact, that to travel from one point in the cosmos to another requires an incomprehensibly large amount of time. If we could climb into a spaceship and travel from Earth to the center of the Milky Way at the speed of light (186,000 miles per second or almost 300,000 km/s), it would still take us about 26,400 light-years to cross the 1.55×10^{17} miles if we could travel in a straight line to our destination point. In fact, if we aimed our ship directly at the galactic black hole and headed in that direction for 26,400 years, we would miss our target completely because the image we see is of the location of the galactic center as it was 26,400 years ago.

As noted in chapter 8, the Milky Way is traveling through space at about 345 miles per second (552 km/s) relative to the universal cosmic wave background (CMB). So the center of the galaxy has moved about 2.87×10^{14} miles relative to the CMB as we currently perceive it, and it would move another 2.87×10^{14} miles during the 26,400 light-years we expected to travel to the galactic center. Instead, we would need to travel toward the location where the center of the galaxy will be 26,400 light-years from now. And if that distance is farther than 26,400 light-years, due to the distance the black hole has travel in the last 26,400 years, then it will take even longer to arrive at our destination. Obviously the *length of time* needed to travel from point A to point B depends on the distance to be traveled as well as the *rate* of travel. In other words, we should define our course of travel as a function of time and distance, or direction and magnitude of the rate of travel. This would define a vector of our movement through space and time (Figure 19.1).

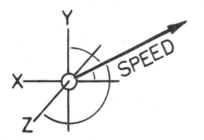

Figure 19.1: Motion defined as a vector with speed and direction through space and time.

What Is Time?

We all experience time. We live our lives in rhythms associated with time. Surprisingly, however, there has yet to be a definition of time satisfying all aspects of the humanities and sciences. Time can be measured by various means. We can use a watch, a pendulum, the number of times the sun appears to revolve around Earth, or a myriad of other means. However, every way that we measure time is associated with a frequency of a particular event—the vibration of a quartz crystal, the period of a swinging pendulum, the number of rotations of Earth, and so forth. Even our hypothetical travel time to the galactic center was measured in light-years, with one light-year equivalent to the distance light travels over the course of one revolution of Earth around our sun.

Time is measured by observation of events that occur sequentially. In order to have an orderly measurement of time, there needs to be a uniform periodicity in the sequential occurrence of those events. We define the length of one unit of time based on that periodicity. For example, one day is equivalent to 24 hours, 1,440 minutes, 86,400 seconds, or 86,400,000 milliseconds. But every one of these packets of time is equivalent to exactly one revolution of Earth about its axis. So a day is defined by the assumed constant and uniform spin of Earth. We can envision this spin as circular motion about Earth's polar axis, a day defined as one rotation. Another way of saying this is that one rotation occurs per unit of time (one day). A pendulum swings in a circular motion about an axis, too. Knight and Butler describe use of a pendulum to define one second of time, the second's relationship to the rate of rotation, and the potential ability of ancient peoples to estimate the size of Earth based on those measurements.[1] A crystal of quartz might not be rotating, but it will vibrate at a constant frequency, or number of vibrations per unit of time. In each case there is a frequency of events on which we rely for the purpose of measuring the rate of other events occurring in the universe—how fast a woman runs ten kilometers, the lifespan of an electron created from "nothing" and then destroyed, or the distance from one end of the universe to the other.

Each of these events is measured in terms of the rate of some event with respect to a unit of time. Perhaps the most common rate of measurement with respect to time is distance. We can think of distance as nothing more than a change in location. We can measure the distance between point *A* and point *B*, or we can measure the distance that point *A* travels from one location to another. We have three parameters in play here. One is the item under study, or what we may call a *source*. The source is usually a thing—either mass or energy—Fuller's energy event. The second parameter is location, or a specified *place*. We are intimately familiar with the concept of place as it relates to the three dimensions we experience every day. Of course the third parameter is *time*, and as we have seen there is a cyclicity to time that is expressed by cultural traditions. Our normal experience is to observe a source at a particular place and to find the source located at another place after some length of time.

Certainly it is far easier to define and measure a unit of time than it is to provide a satisfactory definition of time. Units of time vary depending on tradition. Cultures measuring time in seconds are obviously concerned with sequences of events that occur with relatively high frequency, compared to a culture measuring time in days. Therefore, a universal definition of time ought to exclude mention of measurable units. What is important is that there is a relationship between a minimum of two discrete events in that they do not occur at the same moment in time. Usually the issue is a matter of location—the sequence of change in location or rate of change in location. However, location is not always the concern. In biblical Genesis, God created the world in six days and rested on the seventh. There is no mention of a change in location, only a series of discrete events occurring in a particular order. We also have to recognize that the events either occurred in fact or can be perceived as occurring. We need to relate the occurrence of one event with respect to the occurrence of another. If two events occur at the same time, then time has no meaning with their regard. They are simply two events occurring simultaneously. Perhaps, then, for our purpose we can define time and its measurement as follows:

Time is recognition of two discrete events. It is measured by comparing the occurrence of those events with respect to the periodicity of a third event.

The Process of Creating

Western science is particularly adept at measuring and recording events related to temporal relationships between observable mass and place, or energy and place. This may be the cause for Hawking and Mlodinow's belief that "Scientists have become the bearers of the torch of discovery in our [human] quest for knowledge."[2] However, Western science has difficulty with, or even an aversion to, unobservable or immeasurable relationships between source, place, and time. Rather, those relationships are associated with religious studies in general and spirituality in particular. From the perspective of indigenous cultures, there is not only sacredness in creation, but sacredness as well in the *act* of creating. After all, we can view creation as the reordering of mass and energy from one place to another through time to form new relationships. Is this not what God is described as doing during the first six days of biblical creation? And so it is for the indigenous artist reaching into the recesses of conscious being to discover dimensions of reality that Western science considers illusory, but which the artist expresses as sacred truth. The result is transformation of energy and mass at the disposal of the artist into expressions of the artist him- or herself transformed. The sacred journey taken by the artist to create—this is what is important, not the artifact itself. Cajete describes the journey as a process that must begin with purification.

> This is primarily a conscious effort to simplify, to become aware, to sharpen the senses, to concentrate, to revitalize the whole being . . . There is a guiding spirit, or "consistent adherence to original intent," the notion of applying one's will to concentrate one's whole being into a task, a creation, a song, a dance, a painting, an event, a ceremony, a ritual.[3]

Cajete finds that after preparing for the journey the artist proceeds in a general pattern that attends to sources and involves an adherence to patterns of form, time, and place.

> **Sources.** Attention to the nature of the sources of raw materials to be used in the creation of an artifact, especially one for ceremonial purposes, is essential. Not only is the quality of materials important, but also how and where they are obtained.

> **Adherence to patterns.** Forms and designs are adhered to while also being transcended. Cajete finds, "Generally, ceremonial artistry acknowledged the inherent mystery, the intrinsic integrity of both medium and material, but within the parameters and adherence to a sometimes strict cultural convention."[4]

Time. Time itself becomes an artistic/creative ingredient as a culturally defined dimension combined with the intuitive and spiritually conditioned sense of timing applied by the master artist in the production of ceremonial art. Artists concerned themselves with the timeliness of creating throughout the process. Through the entire creation process there were a series of right moments or phases that might be suggested by a certain smell, a quality of the raw material, a feeling, emotion or dream, which might indicate whether or not to proceed to the next phase of creation.[5] In short, the artist sought an alignment of knowing.[6]

Right place. Time and place are integrally related. Therefore, the place of creation often becomes a consideration in the creation of ceremonial artifacts. Certain places were considered by indigenous peoples to be conducive to certain endeavors. Such places might be characterized by invisible qualities, the availability of appropriate materials, environmental areas conducive to heightened awareness and creativity, dreaming places, healing places, dancing places, living/settling places, singing places, and creative places. Part of the reason for preparation before the creation of ceremonial artifacts is the need to locate an appropriate place of creation through sensitivity to feedback at several levels of sensation. Location includes sensitivity to the metaphor of spatial orientation and alignment of self to the environment with the material used. Space and location-orientation are important considerations in all indigenous people's activities.[7]

We can see in the artistic process *a priori* emphasis on place as it relates to the artist's centering of self—body, mind, and energy—with respect to the universe. This is purification. This is followed by relating the artist to a creative environment—artistic materials, their derivation and location, and the freedoms and constraints placed by cultural traditions. These represent sources—resources—for the artisan. Of course, artistic endeavors take time. However, they must also be created at the *right* time, when circumstances are favorable for creation as an expression of the sacred. Thus it isn't only a matter of how long it takes to create, but also of *when* to create. There is a relationship—a unity—formed in time and place, and the artist must know this. There is intangibility to the artistic process that is indicated by the temporal and observable characteristics of location. We can see the place of creation, we can measure the time over which creation occurs, but we can neither experience nor measure the cause of creation as it is experienced and applied by the creator. Science can only observe and measure the effects.

The Geometry of Time

Much has been written about the geometry of time and, more specifically, whether time is linear or cyclical (Figure 19.2). There is little need to argue the perspective of Western culture. It is second nature for us to count second by second, minute by minute, and so forth as the years march on. We number years with regard to the estimated date of the birth of Christ, and we count the years forward or backward along a timeline that assumes no end in either direction. Time then may be viewed as one-dimensional. It has no width and no breadth, but is eternal and provides a benchmark for the measurement of progress, however progress may be defined.

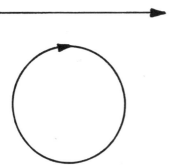

Figure 19.2: Two simple models for the geometry of time—linear and circular.

For many indigenous cultures, however, time is cyclical. It is measured in terms of the regular appearance of the sun (day and night), the moon (months), the growth and decline of nature here on Earth (seasons), and changes in the cosmos (years). The sun appears to rise and set each day, the moon waxes and wanes each month, animals and plants experience changes in size and form throughout the year, and the planets and stars dance across the sky in a predictable fashion year after year. These are natural cycles, and so time is measured with respect to events occurring within those cycles. Even our births, lives, and deaths are cycles. For indigenous cultures, an individual's life is less important than that individual's relationship with the environment (people, the world, and the creator). Value is placed on good relationships and experiencing the world in ways that promote the welfare of not only the individual but all people, and acquisition of knowledge leading to wisdom and understanding and appreciation of our place in the universe.

If we believe that time is linear, moving from distant past to present and then on to some as-yet-unknown finality, then it makes sense to support an ideology—a tradition—emphasizing the benefits of progress. We assume that it is natural and preferable for hunter-gatherers to become farmers and civilians, to organize themselves into ordered societies, moving from the Paleolithic to the Neolithic, through the Copper, Iron, and Bronze Ages, the Dark Ages and the Middle Ages, the Renaissance to the Space Age, and ultimately to our chosen destiny-going where no one has gone before. The emphasis is on the progression of technology, from stone tools to increasingly hardened metal tools, and now with an ever-increasing use of synthetic materials and electronics with an acceleration toward nano size and a biomechanical synthesis uniting humans with environment. Why is this the preferred path? Is this real progress or illusion? What are we progressing toward, and what are leaving behind?

An alternative to linear time is cyclical time. This is experiencing life in harmony with the natural rhythms of the world. There is less emphasis on technological progress and perhaps less development of ideological progress. Western culture may consider this to be stagnation, a self-imposed limit on all that humanity can be. However, this does not appear to have been the conclusion of our ancient predecessors and indigenous peoples around the world. When we understand time to be cyclical, then we foresee departure and return. What goes around comes around. If we make mistakes, we can right them at the proper time, in the right place, under the appropriate conditions. We know that these conditions will occur because they have been before. If we make the right decisions under the proper circumstances then both humanity and environment benefit. The only measure of progress is the individual's ability to consistently center him-or herself with the world at all times, in all places, under any conditions. If we are unable to do so, if there is a break in the cyclicity of time, then purification is at hand and progress will be measured at a universal level. Cleansing and renewal creates opportunities for the world to come into accord with the universal order.

Cangleska

Our reality, of course, lies between these two extremes of linear and cyclical time. The Lakota say we have a choice. We can walk in harmony with the world, or we can choose the road of imbalance. Since everything in nature appears to have a curvature of form, and the four traditional periods of time—day, night, moon, and year—cycle without end, it is a simple matter to conclude that time, like everything else in the world, is cyclical. Obviously, then, it is important to live in a manner that promotes a natural progression throughout our life yet is in balance with this understanding of universal cyclicity.

The Native American medicine wheel reflects the pervasive "roundness" observed throughout the world. As with many other indigenous people, the Lakota recognize that to be in balance with nature means to live *with* nature—as *part of* nature—that humans are *related* to all life, and that the source of life is a mystery. This is why the medicine wheel is one of the most popular and well-known designs found in Lakota art and cultural artifacts, as it reflects this understanding. In his book *The Lakota Way*, Joseph M. Marshall III explains the importance of the Native American medicine wheel as it relates to balance and centering one's life.

> The Lakota word for it is cangleska, "spotted wood." This literal description is from the four colors painted on the wheel, or hoop, which is made of wood. The shape and the colors used represent the power of life, hence the translated term medicine wheel; having pejuta or medicine can mean possessing a certain power or ability.[8]

The medicine wheel, with its circle and four quadrants resulting from two lines intersecting and passing through its center, is simple in form yet powerful in symbolism, not the least of which is related to the number four. Four cardinal directions. Four colors (black, white, red, yellow). Four seasons. Four elements (Earth, Water, Air, Fire). And many other quadripartite attributes of the world, all with sacred meaning. For Marshall, however, "[t]he greatest principle the circle symbolizes . . . is the equality that applies to all forms of life. In other words, no one form of life is greater or lesser than any other form. We are different from one another certainly, but different is not defined as 'greater than' or 'lesser than.' And we share a common journey, the *maka wiconi*, or 'life on

Earth'—in English, the Circle of Life . . ."⁹   Science can identify differences in the chemical, physical, and mental aspects of the multitude of life forms on Earth, but those differences do not change the fact that each life begins, experiences a temporal existence, and dies. However, what sets humans apart from all other life on Earth is that we understand that we can make choices, we express freewill by the intent by which we live and how we relate to others. Our historic intent for continual progression has impacted other lives to the extent that we now recognize that our decisions have affected virtually the entire planet.

By viewing time as a linear progression of events, we have challenged ourselves as if life is a race against time to improve conditions on Earth for our own benefit—ensuring our own health and safety, providing more comforts, and living as long as we possibly can, risking all other life in the process. Is our self-induced progression that which nature and the creator also intended? If so, then how do we explain the fact that as we increase our management of the world for our own benefit, Earth loses its naturalness, unable to cope with the ever-increasing rate of change we force upon it? Does that change remain an unintended consequence of our actions, or do we understand and accept the change as a natural consequence of our actions? Are we living *with* Universe, or *against* it? Perhaps, as Marshall explains, we can discover some of the wisdom symbolized by *cangleska*:

> The medicine wheel is circular with a balanced cross of two intersecting lines in its center; the ends of the lines connect the wheel at four points. The circle, of course, represents life, and the two intersecting lines represent the two roads in life: the good road, usually painted in red, and the bad road, usually painted black. The good road is also referred to as the Red Road; it is the most difficult to travel. The bad road, the Black, is a wide, easy way to go. These are the two basic choices in life, and we choose one in every situation: the good or the bad.¹⁰

Time, then, is precious for each of us. Each life begins and ends. Life is constructed of a chain of experiences related to various frequencies of cyclical events by which we measure time. When life ends, time ends too. It might be, however, that it isn't the journey that is important after all. Perhaps our lives are not intended for the purpose of receiving information, gaining knowledge, and solving problems. After all, all other life forms do these very things. But humans have the capacity to learn from solving problems, develop wisdom from the events of our lives, and pass that wisdom on to later generations. In each of our lives the universe is what we make of it. Each journey requires time. Will our time end eventually, or will it return to a beginning again?

Dimensions Are Everything

If we live in a universe of three spatial dimensions, why is there only one dimension of time? Calling time the fourth dimension may be a misnomer, as it implies a timeline like that preferred by Western culture, a linear series of events with a beginning, a past, a presence, and a future. Time envisioned using the medicine wheel suggests two dimensions—like a circle—with a beginning, a presence that is constantly changing, and an ending where it began and begins again. We can think of time, linear or circular, as a vector (Figure 19.3). Linear time can be thought as having a universal constant unit of magnitude, regardless of the perspective of individual experience, and moving in a constant direction—from past through present and into the future. Circular time, in contrast, may consist of the same unit of magnitude as that of linear time (for example, one day) but its direction is constantly changing until it arrives at its point of origin and begins again.

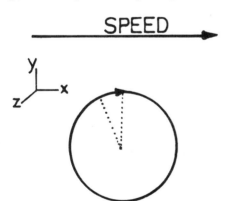

Figure 19.3: Time can be viewed as a vector having speed and direction—linear or circular.

We have seen that the circle is a two-dimensional symbol for the sphere. The Circle of Life, our life on Earth, is in reality a dance in the four dimensions of time and space. This can be envisioned as time having some cyclical frequency based on whatever observations we chose—seasonal changes, moons, winter solstices, and so on—and at the same time moving forward from a beginning to an end. Whether or not this ending is another beginning, we don't know. However, this model of time is like a coil having uniform curvature and extending eternally (Figure 19.4). We can modify the model if we wrap each cycle of time around a sphere, each cycle represented by a great circle. The Sacred Sphere includes nine great circles, and as we have seen, in many traditions nine is believed to represent completeness—an endpoint and yet the point of beginning. The nine great circles of the DDs, therefore, may represent the infinite number of great circles of time that together form the sphere itself.

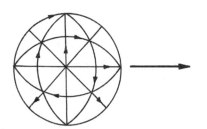

Figure 19.4: A model of time in which time is similar to a coil having uniform curvature and extending eternally. The nine great circles of the DDs may represent the infinite number of great circles of time that together form the sphere itself.

At the Speed of Time

As a symbol representing three-dimensional time, the Sacred Sphere may be seen in the relationship between energy and mass. Recall that Albert Einstein's theory of special relativity postulates that laws of physics are the same for all observers in uniform motion relative to one another, and the speed of light in a vacuum is the same for all observers, regardless of their relative motion or the motion of the source of light. One result of these ideas is that there is equivalence between mass (m) and energy (E):

$E = mc^2$ where c is the speed of light   **Equation 19.1**

We can rewrite this relationship as $E/(mc^2) = 1$. Also, recall that the equation of a circle can be written as:

$x^2 + y^2 = 1$ where the radius of the circle is unity (1)

From these two equations we find that:

$$x^2 + y^2 = \frac{E}{mc^2} = 1 \text{ or } (x^2 + y^2)(mc^2) = E$$ **Equation 19.2**

Equation 17.2 states that Equation17.1 is satisfied if and only if $x^2 + y^2 = 1$. In other words, $x^2 + y^2$ must lie on the surface of the circle. Next, we can rewrite Equation 17.2 as:

$$(x^2 + y^2)(c^2) = \frac{E}{m}$$

Equation 19.3

We assume that energy and matter cannot be created or destroyed, and further, the ratio of energy to matter in the universe is constant. If $x^2 + y^2 = 1$ then Equation 17.1 is satisfied. However, if $x^2 + y^2 < 1$ then the value of c^2 must increase in order to satisfy Equation 17.1. This means that the speed of light (c) must increase inside the circle (the radius is less than 1). Conversely, if $x^2 + y^2 > 1$ then the value of c^2 must decrease to satisfy Equation 17.1. In other words, the speed of light must decrease outside of the circle (the radius is greater than 1).

We can perform a similar analysis of the equation $E = mc^2$ with respect to a sphere with a radius of unity. The equation of a sphere using Cartesian coordinates can be written as:

$x^2 + y^2 + z^2 = 1$ where the radius of the sphere is unity (1)

From these two equations we find that:

$$x^2 + y^2 = \frac{E}{mc^2} = 1 \text{ or } (x^2 + y^2)(mc^2) = E$$

Equation 19.4

This equation states that Equation 17.1 is satisfied only if $x^2 + y^2 + z^2 = 1$. In other words, $x^2 + y^2 + z^2$ must lay on the surface of the sphere. It follows that:

$$(x^2 + y^2 + z^2)(c^2) = \frac{E}{m}$$

Equation 19.5

Again we assume that energy and matter cannot be created or destroyed, and the ratio of energy to matter in the universe is constant. If $x^2 + y^2 + z^2 = 1$ then Equation 17.1 is satisfied. However, if $x^2 + y^2 + z^2 < 1$ then the value of c^2 must increase, and the speed of light (c) must increase inside the sphere. Also, if $x^2 + y^2 + z^2 > 1$ then the value of c^2 must decrease and the speed of light must decrease outside the sphere (the radius is greater than 1).

The conclusion of this exercise is this: In a spherical universe where the ratio of energy to mass is constant (which scientists say must be the case in our universe), the value of the speed of light is a constant at a radial distance from a given location, such as Earth. The speed of light is greater than that value at a smaller distance from the center and lesser at a distance greater than that radius. This means that the speed of light can have the observed value it has in our universe only on the surface of the sphere upon which we are located. If there is a center of the universe, the speed of light increases if we move in that direction and decreases in the opposite direction. However, the speed of light as observed by scientists appears to be uniform in every direction away from Earth. This indicates that there is no *center* of the universe, or that, indeed, the center is everywhere, just as we saw in terms of the universe expanding in the way of an inflating balloon—an S3 balloon at that!

The velocity of energy and matter can be modeled as a vector with a speed and a direction. Therefore the movement of a photon can be considered to be a vector. Since the speed of light appears to be uniform everywhere, the direction of movement is immaterial and we can focus our attention on the speed. Speed is simply a ratio of distance versus time. Speed increases if the distance increases for a given amount of time and decreases if the distance decreases in that same amount of time. We know that the speed of light is a constant in our universe. Therefore, the distance traveled by a photon is a constant for a given length of time. It follows, then, that a defined unit of time must be constant everywhere. (Of course, as Einstein showed in the theory of special relativity, the passing of time may be different depending on the relative motion between two masses or energies. The passing of time as observed by each energy event will appear to be equivalent. However, this complication is of little concern in the realm of mundane reality.)

Ultimately we find that time is uniform in any direction we look. It does not pass slower or faster at any location we place ourselves with respect to any other location we might go. We can conclude that the passing of time at a given location is constant and that some energy events in the universe express a cyclicity that allows us to place sequences of events in a logical order. The combination of cyclicity and sequencing produces a pattern, the coiling of time. Just as indigenous artisans acknowledge the mystery of the medium and material they use within the sphere of cultural convention, so we recognize the adherence of time to form and design, which can be transcended within the sphere of our reality. This is the experience of the shaman.

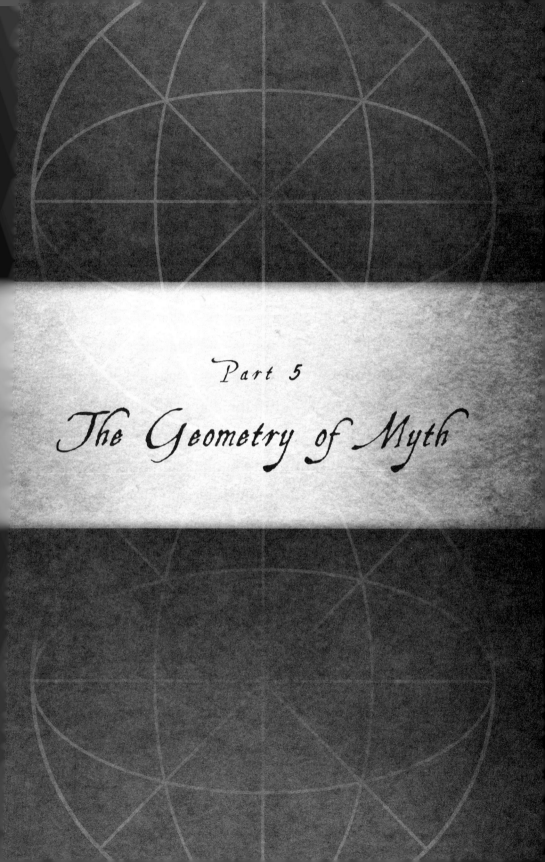

Part 5

The Geometry of Myth

Chapter 20

Pineal Games

In terms of such "structuralist" analysis, of which the most distinguished modern exponent is Claude Levi-Strauss, the symbols of any religious system conform to a grand logical design, of which those who use and believe in the symbols are unaware, and which cannot be perceived by ordinary observational methods of anthropology alone.

RAYMOND FIRTH
Symbols: Public and Private

The mushroom religion is actually the generic religion of human beings, and all later adumbrations of religion stem from the cult of ritual ingestion of mushrooms to induce ecstasy.

TERENCE MCKENNA
"HALLUCINOGENIC MUSHROOMS AND EVOLUTION"

I fell into a dream state for what seemed like a very long time, and as with most dreams I now find it hard to remember the details. All I can confirm is my absolute certain conviction that something happened—something of lasting importance to me.

GRAHAM HANCOCK
SUPERNATURAL

Mythogenesis

Recall the Epic of Gilgamesh, the oldest myth we have in written form. The story is set at the location of one of the earliest known civilizations and focuses on the beginnings of civilization, religion, and cosmography. Early in the myth a harlot-priestess named Shamhat (Akkadian: to be magnificent) meets a wild man named Enkidu ("Enki's creation") at a watering hole where wild animals gather. Shamhat offers herself to Enkidu to end his wild ways and brings him to Uruk "to live like a man." By accepting her offer he immediately loses his strength and wildness. Shamhat replaces the wild beasts as a more suitable companion for him, and Enkidu's animal companions reject him after he mates with Shamhat for six days and seven nights. Enkidu, no longer the wild man, gains knowledge, understanding, and wisdom. He has "become wise like one of the heavenly gods."[1] Having trapped the wild man Shamhat builds up Enkidu's ego, complimenting his good looks and equating him to the gods. She induces him to question his life on the edin, and prompts him to follow her to the city—to leave desolation and discover his place with other god-like humans.

Remember also the Akkadian epic of Atra-Hasis in which man is first formed from clay. The birth goddess Belet-ili receives clay from Enki so that she can create man. The Igigi spit on the clay and Enki treads upon the clay in Belet-ili's presence. A lesser god brought from heaven is sacrificed, and his flesh and blood are combined with clay to create man, who is destined to continue the physical labor previously provided by the gods.

Parts of these myths are strangely familiar. Robert Graves and Raphael Patai note that "Some elements of the Fall of Man myth in Genesis are of great antiquity . . . The Gilgamesh Epic . . . describes how the Sumerian love-goddess created from clay a noble savage named Enkidu."[2] Shamhat then completes Enkidu's transformation into a man, wise like the gods. The similarities in this story and the first thirty-eight verses of biblical Genesis are numerous (all emphasis mine).

Genesis 1
The Beginning

[1] In the beginning *God created the heavens and the earth.* [2] Now the earth was formless and empty, darkness was over the surface of the deep, and the Spirit of God was hovering over the waters.

[3] And God said, "Let there be light," and there was light. [4] God saw that the light was good, and *he separated the light from the darkness.* [5] God called the light "day," and the darkness he called "night." And there was evening, and there was morning—*the first day.*

[6] And God said, "Let there be a vault between the waters to separate water from water." [7] So *God made the vault and separated the water under the vault from the water above it.* And it was so. [8] God called the vault "sky." And there was evening, and there was morning—*the second day.*

[9] And God said, "Let the water under the sky be gathered to one place, and let dry ground appear." And it was so. [10] *God called the dry ground "land," and the gathered waters he called "seas."* And God saw that it was good.

[11] Then *God said, "Let the land produce vegetation: seed-bearing plants and trees on the land that bear fruit with seed in it, according to their various kinds."* And it was so. [12] The land produced vegetation: plants bearing seed according to their kinds and trees bearing fruit with seed in it according to their kinds. And God saw that it was good. [13] And there was evening, and there was morning—*the third day.*

[14] And God said, "Let there be lights in the vault of the sky to separate the day from the night, and let them serve as signs to mark sacred times, and days and years, [15] and let them be lights in the vault of the sky to give light on the earth." And it was so. [16] *God made two great lights—the greater light to govern the day and the lesser light to govern the night.* He also made the stars. [17] God set them in the vault of the sky to give light on the earth, [18] to

govern the day and the night, and *to separate light from darkness*. And God saw that it was good. [19] And there was evening, and there was morning—*the fourth day*.

[20] And God said, "Let the water teem with living creatures, and let birds fly above the earth across the vault of the sky." [21] So *God created the great creatures of the sea and every living thing with which the water teems and that moves about in it*, according to their kinds, *and every winged bird according to its kind*. And God saw that it was good. [22] God blessed them and said, "Be fruitful and increase in number and fill the water in the seas, and let the birds increase on the earth." [23] And there was evening, and there was morning—*the fifth day*.

[24] And God said, "Let the land produce living creatures according to their kinds: the livestock, the creatures that move along the ground, and the wild animals, each according to its kind." And it was so. [25] *God made the wild animals according to their kinds, the livestock according to their kinds, and all the creatures that move along the ground according to their kinds*. And God saw that it was good.

[26] Then *God said, "Let us make mankind in our image, in our likeness*, so that they may rule over the fish in the sea and the birds in the sky, over the livestock and all the wild animals, and over all the creatures that move along the ground."

[27] So *God created mankind* in his own image, in the image of God he created them; *male and female he created them*.

[28] God blessed them and said to them, "Be fruitful and increase in number; fill the earth and subdue it. Rule over the fish in the sea and the birds in the sky and over every living creature that moves on the ground."

[29] Then God said, "I give you every seed-bearing plant on the face of the whole earth and every tree that has fruit with seed in it. They will be yours for food. [30] And to all the beasts of the earth and all the birds in the sky and all the creatures that move along the

ground—everything that has the breath of life in it—I give every green plant for food." And it was so.

³¹ God saw all that he had made, and it was very good. And there was evening, and there was morning— *the sixth day.*

Genesis 2

¹ Thus the heavens and the earth were completed in all their vast array.

² *By the seventh day God had finished the work he had been doing; so on the seventh day he rested from all his work.* ³ Then God blessed the seventh day and made it holy, because on it he rested from all the work of creating that he had done.

Adam and Eve

⁴ This is the account of the heavens and the earth when they were created, when the LORD God made the earth and the heavens.

⁵ Now no shrub had yet appeared on the earth and no plant had yet sprung up, for the LORD God had not sent rain on the earth and there was no one to work the ground, ⁶ but streams came up from the earth and watered the whole surface of the ground. ⁷ *Then the LORD God formed a man from the dust of the ground and breathed into his nostrils the breath of life, and the man became a living being.*³ [emphasis mine]

Enkidu mated with Shamhat for six days and seven nights, just as God made the earth and the heavens over the course of six days. On the seventh day Enkidu lost his strength and was unable to keep up with the wild animals. By the seventh day God finished the work of creation and rested. The Sumerian gods created man from clay; God created man from dust of the ground. The Sumerian gods made man to dig and maintain irrigation canals. There was no one to work the ground before God made man, although streams watered the whole surface of the ground. It was then that God breathed the breath of life into man's nostrils, and man became alive.

These mythologies are related in another way. God created the heavens and earth in six days. During each of those days He created by sundering the universe and bringing life to the world.

- On the first day He separated the light from the darkness.

- On the second day He separated the water under the vault from the water above it.

- On the third day He let dry ground appear and gathered the waters under the sky in one place. Then God created seed-bearing plants, and trees that bear fruit with seed in it.

- On the fourth day He made two great lights to separate light from darkness.

- On the fifth day He created every living thing with which the water teems, and every winged bird.

- On the sixth day He also made mankind—male and female.

God then separated Adam and Eve from the Garden of Eden after the couple had tasted of the Tree of Knowledge of Good and Evil. The Sumerian gods made man to work the ground too, but it was left to Shamhat to coax man to Uruk, where he would work for Gilgamesh and maintain the canals. Thus Shamhat is the one who *separates* man from the *edin* after he gains knowledge from her. Why would the final task of forming man into a civilized, knowledgeable, understanding, and wise being be left to a harlot-priestess? Just as God removed Adam and Eve from the Garden of Eden, the gods and people of Uruk were compelled to remove Enkidu from the *edin*. Why? To answer these riddles we begin with understanding the reason for Adam and Eve's banishment from Eden, the Fall of Man. As we read in Genesis 2:

The Fall

[1] Now the serpent was more crafty than any of the wild animals the LORD God had made. He said to the woman, "Did God really say, 'You must not eat from any tree in the garden'?"

² The woman said to the serpent, "We may eat fruit from the trees in the garden, ³ but God did say, 'You must not eat fruit from the tree that is in the middle of the garden, and you must not touch it, or you will die.'"

⁴ "You will not certainly die," the serpent said to the woman. ⁵ "For God knows that when you eat from it your eyes will be opened, and you will be like God, knowing good and evil."

⁶ When the woman saw that the fruit of the tree was good for food and pleasing to the eye, and also desirable for gaining wisdom, she took some and ate it. She also gave some to her husband, who was with her, and he ate it. ⁷ Then the eyes of both of them were opened, and they realized they were naked; so they sewed fig leaves together and made coverings for themselves.

⁸ Then the man and his wife heard the sound of the LORD God as he was walking in the garden in the cool of the day, and they hid from the LORD God among the trees of the garden. ⁹ But the LORD God called to the man, "Where are you?"

¹⁰ He answered, "I heard you in the garden, and I was afraid because I was naked; so I hid."

¹¹ And he said, "Who told you that you were naked? Have you eaten from the tree that I commanded you not to eat from?"

¹² The man said, "The woman you put here with me—she gave me some fruit from the tree, and I ate it."

¹³ Then the LORD God said to the woman, "What is this you have done?"

The woman said, "The serpent deceived me, and I ate."

¹⁴ So the LORD God said to the serpent, "Because you have done this, "cursed are you above all livestock and all wild animals! You will crawl on your belly and you will eat dust all the days of your life. ¹⁵ And I will put enmity between you and the woman, and between your offspring and hers; he will crush your head, and you will strike his heel."

¹⁶ To the woman he said, "I will make your pains in childbearing very severe; with painful labor you will give birth to children. Your desire will be for your husband, and he will rule over you."

¹⁷ To Adam he said, "Because you listened to your wife and ate fruit from the tree about which I commanded you, 'You must not eat from it,' cursed is the ground because of you; through painful toil you will eat food from it all the days of your life. ¹⁸ It will produce thorns and thistles for you, and you will eat the plants of the field. ¹⁹ By the sweat of your brow you will eat your food until you return to the ground, since from it you were taken; for dust you are and to dust you will return."

²⁰ Adam named his wife Eve, because she would become the mother of all the living.

²¹ The LORD God made garments of skin for Adam and his wife and clothed them. ²² *And the LORD God said, "The man has now become like one of us, knowing good and evil. He must not be allowed to reach out his hand and take also from the tree of life and eat, and live forever."* ²³ So the LORD God banished him from the Garden of Eden to work the ground from which he had been taken. ²⁴ After he drove the man out, he placed on the east side of the Garden of Eden cherubim and a flaming sword flashing back and forth to guard the way to the tree of life.⁴ [emphasis mine]

Having the free will to do so, Adam and Eve eat from the Tree of Knowledge of Good and Evil. Man and woman become like the gods. God, concerned, concludes that humans cannot be allowed to use their free will, for if they do so they will live forever, like the gods. So God banishes Adam and Eve from the garden. Essentially they are imprisoned beyond Eden, sentenced to hard labor for their survival.

God could not allow humans to live forever, which given the choice we may decide to do. Why was the lifespan of human beings such a great concern for God? We can only assume that the gods (God refers to himself and others as "us") benefit by limiting the human lifespan, or humans benefit from a limited lifespan and God knows this to be true. Adam and Eve (and all of us) are sent out east of Eden and assigned laborious work in order to keep us from scheming to re-enter Eden and live forever.

The imagery of the cherubim and the flaming sword warrants consideration. First let's understand the nature of cherubim.

> The word *cherub* (*cherubim* is the Hebrew masculine plural) is borrowed from the Assyrian *kirubu*, from *karâbu*, "to be near"; hence it means near ones, familiars, personal servants, body-guards, courtiers. It was commonly used of those heavenly spirits who closely surrounded the majesty of God and paid Him inti-mate service. It eventually came to mean as much as "angelic spirit." (The change from *K* of *Karâbu* to *K* of *Kirub* is nothing unusual in Assyrian. The word has been connected with the Egyptian *Xefer* by metathesis from *Xeref*=*K-r-bh*.) A similar metathesis and play upon sound undoubtedly exists between *Kerub* and *Rakab*, "to ride," and *Merkaba*, "chariot."[5]

For the Babylonians, such servant-spirits to the gods took form as the *shedu*: human-headed, winged bulls. *Shedu* statues were placed at the doorways of important buildings to symbolically tie important places into the power of the universe. A related creature, the *lammasu*, took the form of a human-headed, winged lion. From an astrological standpoint these spirits represent the four corners or fixed signs of the zodiac: Taurus the bull, Leo the lion, Scorpio the scorpion (or eagle), and Aquarius the water-bearer. So we have spirits represented by celestial figures guarding entry to the cosmological Eden, abode of the gods, source of everlasting life. These ideas are carried forth by the cherubim in Genesis. In that case, however, the spirit guardians are placed at the east entrance. As we have seen earlier, the east-facing temple doorway allows the rising sun to shine into the sacred abode constructed for God. The east is the source of enlightenment. Humans gained knowledge by eating the fruit of the tree, but that knowledge is insufficient to regain entry into the garden. In fact, partaking of the fruit was the cause of the Fall. The east entrance is barred to re-entry. How can we get back into Eden? Is there another way?

God also placed a flaming sword between Adam and Eve and the entrance to the Garden of Eden. What is this gleaming sword? I cannot see it as anything other than a body of water, its surface ruffled by the wind, the light of the sun reflected back at us in a dance of fire. To re-enter Eden we must first cross this fiery sword. From a purely earthly standpoint, is this simply something physical like the Red Sea? Or might it mean something at a deeper level, an intellectual level, such as the watery cosmos? Recall that during the second day of creation God "separated the water under the vault from the water above it" and "called the vault 'sky.'" The flaming sword, then, is the Milky Way extending across our night sky, stars shining like billions of flaming jewels. Certainly the imagery of Eden lying so far across space from Earth gives us reason to pause, a sense of deep emotion and a need to return to our proper place in the universe. To recognize this need approaches the spiritual. This is indeed sacred knowledge.

A Taste for Knowledge

Returning to the Epic of Gilgamesh, we recall that Enkidu's name is derived from the name of the god Enki, lord of the earth, god of knowledge and magic, and god of the seas, of fresh water upon the earth and groundwater beneath. Enkidu, then, is Enki's creation. Enkidu is of Earth. He is the receiver of knowledge and magic, associated with the waters of the world. He is taken from edin by Shamhat, who gives him not only of herself but clothes to wear, beer to drink, and bread to eat. And he goes with her to Uruk, toiling on Gilgamesh's behalf and sacrificing his life to protect a king symbolizing all humanity, which through free will seeks everlasting life. The imagery in the epic is more overtly physical than that of Genesis and so perhaps more easily digested. However, whether or not Adam and Eve discovered good and evil by eating fruit obtained directly from the tree of knowledge, or Enkidu gained knowledge by eating processed vegetation in the form of bread and beer, it was nonetheless the ingestion of vegetable matter that brought knowledge, just as Innana ate cedar, cypress, herbs, and any other plants she could find in order to receive knowledge of sexual intercourse, kissing, and other "womanly duties."

Obviously the Sumerians knew that eating certain plants led to knowledge. And so it is that the oldest recipe for making beer, written in cuneiform script on a four-thousand-year-old seal, is from the Sumerians. Titled "Hymn to Ninkasi," the script outlines the brewing process. Ninkasi is the goddess of alcohol, she is the last of the eight children borne by Ninhursig after she absorbed the pain Enki felt at eight locations in his body, as outlined in chapter 4. Ninkasi is also known as Ninti (Lady Rib or Lady Life) and is, of course, Ninhursig herself.

This association between Sumerian brewing, Shamhat the harlot-priestess, and eating vegetable matter comes full circle when we understand the purpose of the harlot-priestess in ancient Sumerian culture. Prostitution in the ancient Middle East was a vital component of society and culture. The mysteries of sex and sensuality were celebrated by priestesses of love and sensuality at temples of Inanna across Sumer. Anders Sandberg outlines the goddess Innana's attributes as follows:

> Inanna was the Goddess of the city Uruk, having brought the sacred laws (the me) to the people there by stealing them from her grandfather Enki, the god of water and wisdom. She was the goddess of love, fertility and war, revered for her power and feared for her temper. She was said to have a rapacious appetite for men and didn't take "no" for an answer. Many myths tell about her revenges against lovers who refused her or people who treated her badly.[6]

We saw previously Inanna's determination to understand womanly duties prior to her marriage to the shepherd Dumuzi, subsequently the king of Uruk. Shamhat represents the priestesses, perhaps the high priestess, of the primary temple of Inanna (the House of Heaven) at Uruk. The temple priestesses might have been required to go through an initiation in which they played the part of Inanna, including her courtship with Dumuzi and the possible re-enactment of her search for sexual knowledge prior to marriage. Dumuzi was the god of fertility and of the annual growth and decay of vegetation. His cycle of growth, maturity, decline, and death was played out in the Sumerian ritual of sexuality, wherein Dumuzi ultimately regained his fertility. The following year's growth was then assured, and the king's power was renewed. Of course, the priestesses might have been included in a ritualized *ius primae noctis*, the right of the first night exercised by the king and the initiates. Sandburg notes, "This ceremony developed, and the priestesses of Inanna became sacred prostitutes, ensuring the fertility of the land by giving themselves to the worshippers. There were also male prostitutes, representing Dumuzi for the female followers."[7] The ritual also likely included consumption of beer and other beverages to raise the attendee's spirits, which might also have benefited "the holy taverns which surrounded the temples."[8]

Beyond sex and drink, it is well known that priests and priestesses (whom I will from here on refer to collectively as shamans) have used hallucinogenic compounds throughout history for the purpose of experiencing a oneness with the universe and gaining knowledge of the mysteries of life. This oneness—direct communication between an earthbound human and metaphysical, cosmological gods—was derived from consumption of a variety of plants, as well as a few choice fungi or secretions from animals such as toads. The effect is much more intense than that of imbibing alcohol. Inhalation or consumption of certain natural organic compounds yields hallucinogenic effects once the chemicals make contact with the human brain. This method of inducing shamanic ecstasy has been in use for tens of thousands of years.[9] Those effects have likely contributed to cosmic consciousness experienced by shamans and the worldwide recognition of circular and spherical symbolism.

John Allegro suggests that mushrooms were used to induce shamanic experiences by the early period of Sumerian civilization.[10] Of course, use of narcotics in the name of religion has been a worldwide phenomenon. Knight and Lomas add,

> Narcotic drugs have been used in religious ceremonies in almost every ancient human culture and it would be surprising if such an advanced culture as that of the early Egyptians did not possess very sophisticated knowledge concerning their use. The expected method for a man to reach the heavens in death was to traverse the bridge of life, usually with the aid of narcotics.[11]

They also quote Mircea Eliade's statement that,

> The funerary bridge, a link between the Earth and Heaven which human beings use to communicate with the gods, is a common symbol of ancient religious practices. At some point in the distant past such bridges had been in common use, but following the decline of man it has become more difficult to use such bridges. People can only cross the bridge in spirit either as a dead soul or in a state of ecstasy. Such a crossing would be fraught with difficulty; not all souls would succeed, as demons and monsters could best those who were not properly prepared. Only the "good" and the skilled adepts who already knew the road from a ritual death and resurrection could cross the bridge easily.[12]

Knight and Lomas find that these ideas concerning Shamanism equate to known Egyptian beliefs on every level. As the Osiris, the dead pharaoh's passage to the Underworld would be completed in complete safety because in life the pharaoh lived in accord with Ma'at, and the funerary bridge the Osiris travelled had been made known to pharaoh through the process of being made Horus. "The new king could then follow the dead king across the heavens," continuing a cyclical process that Knight and Lomas suggest ensures that the new king has learned, "the way so that he could in turn lead the next king at his own death . . ."[13]

This line of thought led Knight and Lomas to conclude "that the Egyptians adopted much of their theology and technology from the secrets of the city builders of Sumer and that the Sumerians were extremely well versed in the use of drugs for religious purposes."[14] Hallucinogens are attractive to shamanic works because they appear to increase consciousness and improve the ability to communicate with and be receptive to the powers of the universe. McKenna adds that narcotics "catalyze consciousness, that peculiar, self-reflecting ability that has reached its greatest apparent expression in human beings. Consciousness, like the ability to resist disease, confers an immense adaptive advantage on any individual who possesses it."[15]

Terence McKenna goes so far as to suggest that the evolution of human behavior included a feedback loop with plant hallucinogens, something he referred to as a "hidden factor."[16] Beyond human companionship, the three basic necessities of life are food, shelter, and clothing. We have two sources of food: animal and vegetable. We have three sources of shelter materials: mineral (earth), animal, and vegetable. And we have two sources of clothing materials: animal and vegetable. Obviously we are highly dependent on other life forms for our sustenance, necessities, and amenities. However, plants have always served as the main source of support of human life.[17] Our primary necessity has always been food. In search of nutrition, particularly during times of hunger, humans must have experimented with every available plant material. Some plants were found to be poisonous and were quickly removed from further consideration. However, ingestion of some plants, while not necessarily pleasant to the palate, "induced physical and mental states not at all unpleasant and oftentimes of startling unreality. Man had then become familiar with narcotic plants."[18] With those discoveries made by people around the world, ingestion of specific plants was seen as a method of encouraging relationships with certain spirits. Plants were considered "to be efficacious as an intermediary between man's world of humdrum reality and the supernatural or spirit realm."[19]

Leaves, flowers, bark, roots, fungi too—all were tried for potential effects. Richard Schultes found that all plants exhibiting a narcotic effect have at some time in their history been linked to religion or magic. Obvious examples include tobacco, peyote, coca, and opium. It has even been proposed that hallucinogenic mushrooms might have been the primary driver of spirituality and the genesis of religion.[20] Of course, some plants and fungi were found to be poisonous and unsuitable for the intended purpose unless, perhaps, sacrifice was involved. Even animals were consumed for potential toxicological effects and undoubtedly there were people who suffered sickness or death by ingestion. Nonetheless, experimentation continued and long-term benefits emerged.

The Shaman's Journey

From studies in iconography and ethnobotany, we know the ancient Maya radically altered their perceptions in order to have visions of the underlying nature of reality. They achieved this heightened awareness through the use of hallucinogens collected from sacred plants or the *Bufo marines* toad. Some powerful hallucinogens, particularly psilocybin mushrooms and Dimethyltriptamine (DMT and its derivative 5-MEODMT), can enhance one's visual acuity, giving one brief access to subtle energy fields not normally visible to the eye. If the Maya cosmonauts gazed into the night sky while in such a state . . . the Galactic center might have become overwhelmingly obvious, a blazing energy-knot in the night sky.[21]

This quote from John Major Jenkins is all the more pertinent to us as a result of the discovery in November 2010 of twin, potentially spherically shaped "bubbles" of gamma ray–emitting gas extending above and below the galactic plane of the Milky Way (Figure 20.1).[22] The lobes of gas are linked to the black hole located at the galaxy's center, each having a diameter of 25,000 light-years. NASA found both low-energy X-rays and microwaves that might indicate the outline of each bubble. I'm not suggesting that Maya priests indulged in hallucinogenic visions of galactic gamma waves, X-rays, and microwaves. Perception and recognition of the Milky Way's center by indigenous and ancient shamans is amazing in and of itself. Yet, can it be totally dismissed without testable evidence to the contrary? And exactly how would we go about testing a shamanic vision from a thousand years ago, anyway?

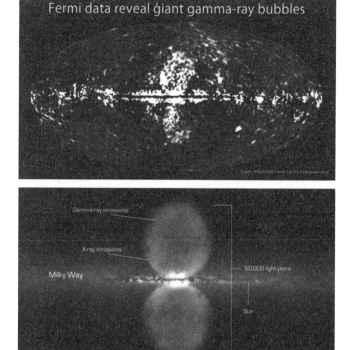

Figure 20.1: Twin "bubbles" of gamma ray–emitting gas extending above and below the galactic plane of the Milky Way. The bubbles, moving at the speed of light, were discovered by NASA in 2010 (illustration from NASA, 2010).

Franco Fabbro, professor of physiology at the University of Udine, has studied the cerebral representation of language in individuals, relationships of the brain and states of consciousness, the application of neuropsychological methods to the study of poetry, and literary studies concerning the neuropsychological basis of religion. Results of his well-documented studies suggest there is a long history of the use of hallucinogens for religious and spiritual purposes across the Middle East and Southeast Asia. He finds that "ritual use of hallucinogenic substances has been widely documented for various shamanic cults of Asian, American and African culture. In the so-called major religions hallucinogens have also been found to play a relevant role in ritualistic practice."[23] Examples include:

- Hymns of the Rig Veda composed under the influence of "soma," identified as the mushroom Amanita muscaria (fly-agari)

- Spreading use of soma from early Indian religion to Zoroastrianism in which an intoxicating drink called haoma was used in rites

- The probable practice of ingesting hallucinogenic substances to attain ecstasy and visions during religious ceremonies that may have influenced Jewish sacerdotal environments, including exiles of Jewish people to Babylonia during the 6th century BCE; Fabbro suggests that some Jewish sacerdotal groups may have become acquainted with hallucinogens during captivity, influencing prophetism as in the biblical books of Ezekiel and Zechariah, as well as apocalypticism found in the book of Daniel, and Enoch and Ezra of the Old Testament Pseudepigrapha. Fabbro's points to "typical features of ecstatic experiences induced by hallucinogens" in Ezekiel 1–3 and Ezra 9:23–28.[24]

Does use of narcotic plants induce escape *from* reality or a connection *with* reality? From illusion or to illusion? Richard Schultes argues that they create a temporal escape from reality, and this certainly is the opinion of the secular scientific community. Yet, the sheer variety and availability of narcotics to pre-Columbian shamans of Mesoamerica and South America is astounding. Schultes knew this, as his personal experience shows in his essay "Hallucinogenic Plants of the New World," published in *Harvard Review* in 1963. He admits to taking many of those hallucinogens while studying native users *in situ*, in Oklahoma, across Mexico, and into the Amazon basin and the Andes mountains in northwestern South America.[25]

Based on his extensive study of the hallucinogens used by indigenous people located throughout the Americas, Schultes's findings include a startling array of botanical psychomimetics, or what we may commonly refer to as psychedelic drugs; he notes that German toxicologist Louis Lewin categorized these narcotic plants as *Phantastica*.

Differing from the psychotropic drugs, which normally act only to calm or to stimulate, the phantastica or psychedelic agents act on the central nervous system to bring about a dream-like state marked by extreme alteration in consciousness of self, in the understanding of reality, in the sphere of experience, and usually by serious changes in perception of time and space; they almost invariably induce a series of visual hallucinations, often in kaleidoscopic movement, usually in rather indescribably brilliant and rich and unearthly colors, frequently accompanied by auditory and other hallucinations and a variety of synesthesias.[26]

The following summarizes Schultes's listing of New World hallucinogenic plants and their effects.

Schultes's conclusions after studying the effects of these *Phantastica* are not what we might expect from a scientist. However, from the perspective of the sacred intent by which these and other intoxicants have been used by shamans around the world, Schultes is emphatic: "We can no longer afford to ignore reports of any aboriginal use of a plant merely because they seem to fall beyond the limit of our credence. To do so would be tantamount to the closing of a door, forever to entomb a peculiar kind of native knowledge which might lead us along paths of immeasurable progress."[28]

From Journey to Geometry

A fascinating aspect of indigenous and ancient use of hallucinogens within the shamanic sphere is that many of the persons so connected to a greater consciousness were also the ones observing the cosmos and recognizing the inherent relationships between humans and space-time. One of the first large pyramids constructed in Mesoamerica is a cone-shaped structure at the early Olmec city of La Venta near the Gulf of Mexico. Archeaoastronomical investigation of the pyramid has shown that the structure had to be periodically reoriented toward specific stars of Ursa Major (the Big Dipper) as a result of precession. Jenkins suggests that such evidence of detailed attention paid to the cosmos by the Olmec and their apparent need to recalibrate their infrastructure to reflect the observed changes "represents a watershed event in the development of Mesoamerican cosmology."[29] These concerns are also reflected in Olmec symbology and its interrelation with their mythology and understanding of the cosmos.

Figure 20.2: Depiction of paintings made by Chumash Indians in a cave located in the hills above Santa Barbara, California. The paintings are estimated to be no older than 1,000 years.

That shamans made significant contributions to the culture and traditions of religions around the world is without question. On a very personal level our prehistoric ancestors recorded their visions as cave art, drawn as early as tens of thousands of years ago and as recently as within the past hundred years. We can observe their contributions today. They drew not only bison, horses, deer, and other animals, but also the sun, stars, and other observed cosmological phenomena. They also drew supernatural beings and abstract configurations such as grids, wavy lines, nets, ladders, zig-zag lines, and spirals (Figure 20.2). These forms are surprisingly similar to entoptic phenomena, abstract geometrical patterns recorded during controlled neuropsychological experiments conducted under laboratory conditions, suggesting that much prehistoric cave art may be associated with the recording of hallucinogenic experiences.[30]

Contemporary Indian shamans record their visions, too. Pablo Amaringo has received international acclaim for his artwork, which is inspired by his use of ayahuasca, a psychoactive drug prepared from a vine (*Banisteriopsis spp.*) and often mixed with leaves of shrubs containing dimethyltryptamine, still used by shamans in South America. Hancock provides detailed descriptions of Amaringo's works, and there is no need to reiterate his commentary here.[31] However, it is significant that Amaringo's art consists of vibrant geometric forms, often on a dark background. Serpents, crocodiles, jaguars, monkeys, birds, people, supernatural life forms, urban and jungle scenes, and abstract patterns of lines, curves, circles, circles within circles, and dots fill his canvases. The paintings suggest vibrations, wavelengths, frequencies, angularities, motion, and above all, the play of light energy on the physical and metaphysical universe. One painting shows what appears to be a crystal refracting light in all directions, multicolored, with numerous triangular and rhombohedral faces, twenty-four visible faces on half of the polyhedron, exemplary of the disdyakis dodecahedron (Figure 20.3). Geometry, including symmetries found in the Sacred Sphere, obviously serve an important role in visions, shamanistic or otherwise.

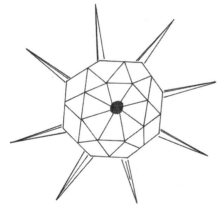

Figure 20.3:.An image of a crystal with twenty-four triangular and rhombohedral faces, and resembling the form of a disdyakis dodecahedron (based on a painting by Pablo Amaringo). The form may be inspired by the artist's use of ayahuasca, a psychoactive drug still used by shamans in South America.

In his quest to pass through the portals of the shaman, to glimpse the spiritual life of the ancients, author Graham Hancock partook of a variety of psychoactive drugs and recorded results of his research and experiences in his book *Supernatural: Meetings with the Ancient Teachers of Mankind*. His comments concerning the geometries he experienced during ayahuasca-induced visions are insightful.

- I close my eyes, and without fanfare a parade of visions suddenly begins, . . . geometrical and alive, . . . a pulsing, swirling field . . . of solar systems revolving, of spiral galaxies . . . nets and strange ladder-like structures . . . multiple square screens . . . immense patterns of windows . . . their real function is to announce the arrival of something else.

- . . . complex interlaced patterns of geometry . . . the skin of a snake . . . a circle in the center of each rectangle . . . spinning . . . swirling and turning . . .

- . . . geometry and ladders . . . a vast dome, . . . patterns of nested curves . . .

- . . . a very large inverted bowl rising up . . . glowing with light.

- The visions begin with 20 minutes of geometry . . . .

- . . . planets . . . immense and surrounded by rings or discs . . . a transparent earth sphere . . . the fragility, texture, and glittering iridescent colors of a soap bubble . . . The sphere is rotating and seems to float in space between two cupped hands.

- . . . multiple rows of green pyramids . . . a sphere, a cube, and a triangle, . . .

- . . . pyramid shapes built around a lattice or framework of some kind.[32]

Hancock argues that although there is an unspoken assumption that the brain essentially manufactures hallucinations, and while our modern, scientific presumption is that the brain is simply a receiver of data from the outside world that we process, categorize, and act on, there is another possibility—that the brain can operate as a receiver with hallucinations, too. This suggests that inducing altered states of consciousness, such as use of hallucinogens or other means, might result from the brain's ability to temporarily tune in to a realty of "frequencies, dimensions and entities" that are not normally manifest in the world we normally perceive to be "reality."[33]

Evolution of Mind

Hancock's conception of shamanism is a search for, and entry into, other realms of existence where benefits can be obtained for the good of society. If, in fact, benefits to society are derived from the shaman's journeys into altered states of consciousness, and if those benefits are attributed to information gained from the spirit world, then the reality of such worlds and the beings that occupy them becomes reality in metaphysical terms at the very least. Otherwise, we must assume that the information gained through the experience, the knowledge that may be applied, and the wisdom to know how best to apply that knowledge for the greatest benefit must have been derived not from spirit worlds, a third-party intelligence, but unlocked from captivity within the shaman's own brain, his DNA. And since humans all share the same DNA, Hancock concludes, it would be understandable that tapping natural resources via trance states would help us access that common information base and help to explain commonalities such as universal motifs, imagery, and phenomena that hallucinating individuals, shaman or otherwise, experience. But beyond this, Hancock suggests that through visits to spirit worlds, or the harnessing of information that has been hidden within our own being, shamanism might have been a driving force in the accelerated pace of human evolution:

> Gifted and experienced shamans the world over really do know more—much more—than [scientists] do . . . the possibility that the spirit world and its inhabitants are real, that supernatural powers and non-physical beings do exist, and that human consciousness may, under certain circumstances, be liberated from the body and enabled to interact with and perhaps even learn from these "spirits." In short, did our ancestors experience their great evolutionary leap forward of the last 50,000 years not just because of the beneficial social and organizational by-products of shamanism but because they were literally helped, taught, prompted, and inspired by supernatural agents?[34]

With this understanding of shamanism, including Olmec and Mayan use of natural hallucinogens to assist the priesthood in their understanding of human relationships with the cosmos, let's return to the subjects of Mayan cosmography and calendrics to see how modern Maya apply geometry to illustrate that relationship. Recall that the Maya conceived the earth as having four cardinal directions. The ground surface was perceived to be square or circular, and the solar equinox and solstice events were important for aligning buildings and roadways and for tuning in to the forces of nature. The Maya could imagine the earth represented by a field of maize or, alternatively, as a turtle floating upon water. This latter analogy is found throughout Native American cultures. Each of the four cardinal directions was represented by a color, bird, tree, and god with a particular aspect. Cosmographically, some Mayan mythologies divide the sky into thirteen concentric layers, or hemispheres, inhabited by thirteen gods. Earth lay below, and the underworld includes nine layers inhabited by nine deities. A Tree of Life (*yaxche*, or "ceiba") is located at the center of the world as a means of communication between the various spheres. During the Mayan Classic period the earth and sky were perceived to be serpents and dragons extending outward. These creatures were commonly two-headed and at times were covered by feathers—the plumed serpent. Gods used similar snakes and dragons as a means of travel between spheres. Dragons with features of the snake, crocodile, and deer included "star" insignias identifying them as the night sky or the Milky Way.

Fungi, Anyone?

This vision of the world is indeed an amalgamation of cosmological interpretation and mythological truth, expressions of physical observations, intellectual understanding, emotive release, and spiritual uplift. It is vision filled with vital communication, as if a shaman has made a voyage into the unknown, to the far side, and returned to tell us what we need to know—information supportive of our existence and encouraging us to move forward. By this I do not mean technological or scientific progress, but encouraging us to continue progressing in our lives, discovering who we really are. The cultural impact of this knowledge, including the work of past artisans, architects, and engineers, is apparent in the ruins of Palenque. McKenna explains,

> In southern Mexico, coincident with the Mayan cultural area, natives use a number of psilocybin-containing mushrooms: *Psilocybe mexicana*, *P. aztecorum*, *P. maztecorum*, and others. These mushrooms constitute the Mexican mushroom complex

discovered by Valentina and Cordon Wasson in the early fifties. *Psilocybe cubensis* also occurs in these areas, being especially prolific at Palenque. Palenque is the site of the ruins of one of the most exquisite cities of the Mayan climax. Many people have taken the mushrooms at Palenque and have had the impression that they were ingesting the sacred sacrament of the people who built this fabulous abandoned 7th-century Mayan city, but this notion is disputed by modern botanists.[35]

So, can the use of such a sacred sacrament be totally dismissed without testable evidence to the contrary? In light of the apparently continuous, worldwide space-time conformity indicated by the symbol of the Sacred Sphere—associated with geometry and worldwide cultural and religious traditions—can we afford to be so sure that it can't?

As we've seen, Franco Fabbro finds a long history of the use of hallucinogens for religious and spiritual purposes in the Middle East and Southeast Asia. Should it be surprising to discover use of sacred sacraments in the early Christian church? This is Fabbro's hypothesis, and he finds corroboration in the worship hall of the ancient Basilica of Aquileia dating prior to 330 AD. The basilica exhibits a floor mosaic that attests to an early Christian habit of ingesting mushrooms during religious ceremonies.

> Religious habits in early Christianity had so many things in common with early Judaism, and in particular with apocalypticism, that a transmission of the techniques to reach ecstasy and visions from early Judaism to early Christianity is most likely to have occurred. Philological studies of the past suggested that some early Christian groups also made use of Amanita muscaria as a hallucinogenic substance during specific religious rites. . . .[36]

The oldest part of the basilica is the oratory of the northern hall. Fabbro found the floor of the oratory to be a mosaic depicting various animals and symbols, as well as something quite curious: a basket filled with red mushrooms and another basket containing nine snails. The mosaic includes an epigraph stating that religious ceremonies were conducted in this part of a basilica. At least eight mushrooms held in the first basket have dark red caps and other characteristics of the hallucinogenic mushroom Amanita muscaria, common to the region around Aquileia. Fabbro notes that their collection in a basket suggests to some scholars that they may have been included as edible substances during the ceremonies. Also, he states that characteristics of the snails contained in the other basket are indicative of the type Helix (Helix) cincta, likely eaten together with the mushrooms. As to the effects of such ritual cuisine, Fabbro says,

> It has been suggested that the two baskets containing edible plants and animals hint at ritual meals and agapae enjoyed by early Christians in places of worship.
>
> The ingestion of 1 to 4 fly-agarics may induce an intense feeling of joy and excitation with a reduction of the sense of fatigue and an enhancement of verbal production. By taking in more pieces (5 to 9), subjects first become very agitated and have vivid hallucinations, then they fall in a narcotic-cataleptic state, characterized by a deep sleep, from which they cannot be roused, and a very intense dreaming activity.[37]

Fabbro concludes that mushrooms symbolized in the floor mosaic in the Basilica of Aquileia yielded psychotropic properties beneficial to religious rites during early Christianity. The cults conducting those rites likely required secrecy because their activities were associated with mystic ecstasy through ingestion of hallucinogens.

How common was this practice? Fabbro notes that Roman authorities often accused early Christians of sorcery and use of hallucinogenic substances. In contrast, Irenaeus, Bishop of Lyon in the second century AD, considered the practice limited to heretical cults, which would have included Gnostic churches. Fabbro suggests that the hallucinogenic mushrooms depicted in mosaic tile on the floor of an ancient Christian church could be a key to understanding mysterious rites referred to as the "discipline of the arcanum," characterizing the oldest of Christian liturgy and defining the nature of what needed to be kept secret, as it was only handed down orally to disciples who had been properly initiated.[38]

A Return to the Garden, or *Edin*

Earlier we saw how God separated Adam and Eve from the Garden of Eden after the couple tasted of the Tree of Knowledge of Good and Evil. That discussion suggests that the type of fruit provided by this "tree" may be worthy of reconsideration. However, we know that the task of removing Enkidu from edin was left to a woman, Shamhat, the shamanic harlot-priestess from the temple of Inanna at Uruk. In ancient Sumer the process of creating a knowledgeable, civilized human being from a wild man was left to a beer-drinking, narcotic-using, quasi-theocratic prostitute. Why?

From our review of the similarities between biblical Genesis and the Epic of Gilgamesh, and the documented worldwide shamanic use of drugs to induce direct communications with the spirit world, the answer is now readily apparent. Adam and Eve expressed their free will by eating from the Tree of Knowledge of Good and Evil and in so doing became knowledgeable and recognized that they had become different from everything else in creation. They covered themselves. Similarly, Enkidu expressed his free will by acting on Shamhat's offer of sex, as well as by eating bread and drinking beer. Narcotics likely were a part of Shamhat's toolbox. So man put clothes on himself and became knowledgeable, realizing full well what he had done. With no other recourse available to him, he became civilized, or at least became part of civilization.

Shamhat is a metaphor for the attractions that have always drawn humans from the country to the city. However, the Sumerians understood that the proper place for humans is not in the city. Humans belong in the edin. Gilgamesh prides himself in his city walls of fired brick, but are the walls designed to keep the edin out or the civilians in? Genesis helps provide the answer, for just as Adam and Eve are imprisoned in the world beyond the garden, Enkidu is taken from the edin and imprisoned within the walls of civilization; there is no turning back. The Epic of Gilgamesh is both comedy and tragedy. It is a story about the best of times and the worst of times. There is no source of eternal, youthful life for humans. Our lot was cast when we decided to enter Uruk, coaxed by Shamhat's promise of a better life. In nature we are in unity with the world. The ancient Sumerians knew that they had separated themselves from the rest of the world, and the only way to retain some connection with nature was through their cultural traditions, mythology, and religious cosmology. This information was deemed vital, and they recorded it as their legacy for the future.

However, this is not the end of the story. God cannot allow humans to live forever, which given the choice we may decide to do. All world mythologies detail belief in world ages, the birth, life, destruction, and rebirth of the world. The benefit to the gods (Universe) is discovered in the process of purification and renewal. A limited lifespan ensures that this process will occur. The problems of the previous world are destroyed and an opportunity is created for a better world. For some religious traditions this is reincarnation. For others it is resurrection. Our free will does not determine the fate of the world. Purification and renewal will continue. With a higher consciousness, a progressive relationship with the universal spirit, and a realization that we are part of unity, we find knowledge and understanding that this is reality.

Recall that Adam and Eve were sent out eastward from Eden. God assigns cherubim to keep us from re-entering Eden, and we are tasked with the laborious production of children and food in order to distract us from scheming to live forever. Similarly, Enkidu is enticed to Uruk, enters the city, and becomes Gilgamesh's alter ego, destined to sacrifice his self to save his self. In either case, entrance back into Eden (*edin*) is guarded on the east side, the side of knowledge and enlightenment. Can we get back in? How?

God (the gods, Universe), of course, left a way open to our return, pending purification and renewal. There are four rivers flowing from Eden, and there are four great rivers of the ancient world. The cherubim guard the gateway, the river to the east. But there are other rivers that we can follow back to Eden (*edin*). The flaming sword points the way. Shamans have seen the way. They *know* the way. Whoever wrote the Epic of Gilgamesh knew the way. It is hidden in cuneiform script. It is represented by the medicine wheel, the mandala, and the Sacred Sphere.

Chapter 21
Conclusions and Beginnings

Mind is the . . . uniquely human faculty that surveys the ever larger inventory of special-case experiences stored in the brain bank . . . from time to time [it] discovers one of the rare scientifically generalizable principles running consistently through all the relevant experience set.

R. BUCKMINSTER FULLER
Synergetics

Symbolization is a universal human process . . . Pervasive in communication, grounded in the very use of language, symbolization is part of the living stuff of social relationships.

RAYMOND WILLIAM FIRTH
Symbols: *Public and Private*

Links and Nodes

During the course of this book I have demonstrated that ancient and indigenous cultures around the world, during all ages, communicated a common understanding of relationships between human beings, the world, and the universe. These relationships involve the cyclical nature of creation: birth, life, death, and rebirth. Each of us experiences this cycle. The world itself experiences a constant cycle of change that we experience every day in own our lives. The sun was born, it burns brightly, and one day will end its term as the center of our solar system. The center of our Milky Way continues to create, destroy, and recreate suns, planets, and other components of our galaxy. We are part of this recycling of creation, destined to play out our role in Universe.

We communicate our understanding of these vital processes and relationships using symbolism. Communication can be verbal and virtually instantaneous, or it can be graphic (drawn, painted, engraved, carved, or engineered and constructed from cobbles or megaliths) and transmitted across thousands of years. Every culture uses geometric constructs linking fundamental relationships between the individual, society, the world, and the creator. The symbols are found painted on cave walls and etched on rock outcrops, constructed of stones placed on the ground, described in cuneiform text on baked clay tablets, preserved in oral traditions, and built into the architecture of temples and cathedrals around the world.

Traditions include various means and media to create sacred symbolism in one, two, or three dimensions. Cave paintings and medicine wheels are prime examples of the sphere illustrated in two dimensions as a circle. Another example is use of the word *round* to describe shape. Does round mean circular or spherical? Of course it can mean either, or both. A third example is the use of lineal dimensions to describe size and shape. If I say that an object has a length of 1 cubit, width of 1 cubit, and height of 1 cubit, is that object a cube, or am I providing the length of each of three orthogonal diameters of a sphere or some other form? What would be the best perspective when depicting a three-dimensional object on paper or a cave wall? Unfortunately the ambiguity can result in a miscommunication of the symbolism. Or is it possible that the ambiguity is intentional?

Aspects of sacred symbols exemplify both spatial and temporal continuity. The symbols vary between cultures and exhibit flux through time. Nonetheless, as we've seen, each symbol is representative of a particular facet of the Sacred Sphere. Together these facets form a framework of sacred and secular understandings, and the geometry then unifies all cultures across space and time. Certainly no culture—ancient, indigenous, or otherwise—intentionally identified and applied a facet of achiral octahedral symmetry of order 48 for the purpose of expressing a universal principle. Such a conclusion would be ludicrous and lacks any evidence whatsoever. However, as we've seen, our relationships with each other, with all animate and inanimate features of this world, and with the cosmos, are universally expressed by a specific component of that symmetry.

After briefly reviewing analogous historical relationships between Hebraic sacred science and the Egyptians, Indians, and other ancient cultures, Lenora Leet considers a spiritual concept—a cosmological principle—found in the sacred science of all ancient and indigenous cultures, that which she refers to as the Secret Doctrine of the Son:

> [T]his most essential core of the Jewish religion has consistently been both covertly proclaimed for those who could understand it and as openly denied. But from the Bible, through Merkabah mysticism and the Kabbalah, to Hasidism, the central mystery of the son, as of the cosmic process, has ever had but one meaning, that ultimate unification of the human and divine that could affect both the personalization of the divine and complementary divinization of perfected humans.[1]

Unification between humankind and the creator, and the transformation of both when unity is achieved, is what I believe is at the heart of the intended, yet hidden and previously unrecognized meaning of the four-dimensional structure of the Sacred Sphere. Two-dimensional geometric structures—those ancient symbols, communications, energy events—are found within the various symmetries of the disdyakis dodecasphere. In all cases there are four key components of those sacred communications. By now we are quite familiar with each of them. They are life, geometry, symbol, and religion.

- Life is communicating/problem-solving with intent.

- Geometry is applied mathematics addressing shape, size, position, and properties of space.

- Symbol means putting together that which was divided.

- Religion links us back to what has been sundered, our relationship with the creator; symbolism, then, is an important aspect of religion.

These four components of communication, whether as sender, receiver, or means for transmitting a message, come together in the sacred tradition of every ancient and indigenous culture. The message is simply this: *A centered life is symbolized by sacred relationships.* The message is uncomplicated, elegant, and universally applied. Our task is to decipher the symbolism communicated to us via geometrical relationships at levels of the physical, intellectual, emotional, and spiritual. Each culture perceives a particular pattern and instills a specific, multilevel geometric relationship in its sacred traditions, and each pattern is a component of the fabric that clothes the world community. *In toto* the patterns create a unified symbol of what it means to be human and, perhaps, the intended purpose of the universe.

We all have the same vision when we communicate vital, sacred information. The surprise is that the circle is the perfect symbolic expression of *each* relationship, whether subatomic or universal in scale. And so the sphere is the perfect symbol of *all* relationships throughout time and across three-dimensional space. The Sacred Sphere communicates this unity of mind, body, and spirit. It holds the key to our past, present, and future.

How do we relate the Sacred Sphere to our own lives? That question I cannot answer for you. Each person must choose his or her road and then embark on a personal and unique journey to find the answer. For myself, I believe that there has always been human *intent* to identify, communicate, and rekindle our relationship with the world and our creator. Buried deep within us is a longing to return to nature, to be part of and not master over the world. The Ancients recognized that there can be no *unity* until we link back to a balanced relationship with each other and the world around us. The traditional Lakota lifeway, like many other indigenous cultures, constantly encourages balanced relationships, the recognition of the sacredness of life, and an awareness of the Great Mystery that surrounds us. We can forge that link or allow it to be forced upon us as a matter of universal principle. Either way, Shiva *will* dance.

Unfortunately, two-dimensional symbols have not effectively communicated the vital importance of universal unity to succeeding generations. And, as Lenora Leet notes, even use of three-dimensional constructions such as the Platonic dodecahedron is not sufficient to accommodate the intent of the communication.

[I]t is only in this three-dimensional world that the dodeca-hedron and icosahedron can come into a true dualing relationship with each other . . . [T]he two-dimensional world of Yetzirah . . . [can accommodate] accurate orthogonal projections of the tetra-hedron, octahedron, cube, and icosahedrons, but [not] an accurate orthogonal projection of the dodecahedron . . . [A]s a model of the creation . . . this form cannot be perfected in the two-dimensional supernal model, but, for its true expression, must await and contain the three-dimensional cosmos and, in particular, the earth.[2]

The Sacred Sphere, with its perfect three-dimensional symmetry and geometrical unity resulting from a unique combination of links and nodes, and universal application of specific facets of its geometry to symbolize sacred relationships, represents the balance we desire and the path we must follow if we are to achieve this goal. The evidence for this is all around us. Our ancestors, brothers, and sisters, remind us in many ways, time and again:

- The universe—from micro- to macrocosmic—operates in accordance with relationships between space and time, matter and energy, and physical law.

- The cycle of creation, existence, destruction, and recreation occurs at all scales in the universe.

- Universal relationships are expressed in geometry.

- Unity is expressed in perfect symmetry.

- The spiral of creation is exhibited in the eternal linearity and cyclicity of time.

- Experience is the effect of interrelated geometrical energy events in space and time.

- Our perception of duality in the universe is illusion.

- Human separation from the world was the Fall of Man.

- Shamanic experience connects to a greater consciousness in which inherent relationships between humans and space-time are observed and recognized.

- Sacred architecture (such as the Tree of Life, the Preserver of Life, the pyramids, and many others) reflects universal relationships, linking humankind with the creator.

- The Sacred Hoop is a metaphor for all that is natural and where power can be maintained.

- The Preserver of Life is metaphor for purification, recreation, and unification of all life with the world.

- Sacred symbolism is applicable in religious and secular life.

- The shape of the universe, although unknown, is theorized to be dodecahedral.

- All mathematical relationships in traditional sacred geometry are found in the geometry of the Sacred Sphere.

- We will return to a balanced relationship with the world when the last two circles appear on the round stone.

Together these findings and the synergy they create in balance and unity parallel the Sacred Sphere as metaphor for universal sacred relationships. At the same time these ideas are unified in another way. Taken individually or as a whole, like the Sacred Sphere itself, these truths are completely independent of materialism or technological progress. Is it possible that Universe defines the human condition by our spiritual relationships, rather than our self-assumed mental and physical superiority? Shouldn't our progress be measured in terms of our relationship with Universe? Can we dig ourselves out of a hole, or is it time to fill the hole back up, plant some seeds, and water the garden?

Listen

Cultural traditions develop from an evolving matrix of human experience. The Sacred Sphere is the implicit, fundamental geometric structure encompassing many cultural traditions ranging from entire world mythologies to pictographic symbolism. After reviewing specific attributes of symbols used by cultures from across the world over the course of tens of thousands of years, the disdyakis dodecasphere surfaces as the archetype from which many of those sacred symbols are derived. Its geometry symbolizes mankind's common understanding of the relationship between the three dimensions of space and the fourth dimension of time. The structure's symmetry reflects spatial attributes common at all scales in the universe.

The Sacred Sphere expresses the cosmic consciousness within each of us. Its inherent geometry of nine great circles intersecting to form seventy-two curvilinear line segments, forty-eight triangles, and twenty-six vertices, contains all the information people have used to symbolize their relationship with each other, the world, and the creator. It is a framework from which different cultures across time and space have taken geometrical subsets to create traditional symbols associated with the sacredness of life. Its geometry contains numerical values of 1 through 9 and beyond; the value of π, which defines the relationship between a straight line and circular curvature; the value of ϕ binding together the chemical, physical, and biological nature of the universe; two-dimensional forms including the circle, Vesica Piscis, triangle, square, pentagon, hexagon, and octagon; three-dimensional forms including the sphere, tetrahedron, cube, dodecahedron, and Fuller's cuboctahedron or vector equilibrium; the sacred geometries of the Platonic solids and the Great Pyramid at Giza; the shape of the ark and the structure of the Tree of Life; the sun circle and medicine wheel; the spin of a photon and the axis of Earth; the fundamental shape of space and the limits of the universe; the beginning and the end; and a new beginning. The symbolism awaits our efforts to discover the boundless and timeless relationships between each other, the universe, and the creator.

Individual experience and cultural perspectives mask recognition of the true meaning of our purpose for being in this world. The result is our incomplete understanding of what *reality* is and what *illusion* is. Shamans continue their search for the answer. Myth suggests they may never know. Scientists study the realms of the subatomic and extragalactic for the answer. The Heisenberg uncertainty principle infers they may never know. However, pieces of the puzzle are evident everywhere, and together they make it apparent that the Sacred Sphere represents the complete picture. It contains the answers that we humans have uncovered to date, and I must conclude that it contains the key to additional understanding and ultimate wisdom.

The search for answers is left to each individual. Knowledge can be found in a book, but experience cannot. The Sacred Sphere provides a framework or context for our journey. The Ancients and indigenous cultures provide clues for us. The clues are recorded in the Epic of

Gilgamesh and the spherical ark called the *Preserver of Life*. The medicine wheel is *maka wiconi*, the Circle of Life. The nine worlds of Norse mythology teem with an array of life and relationships. The universe creates, lives, dies, and is reborn. This, indeed, is the circle of life, demonstrated by the gluons from which we are made and the galaxies dancing their imperceptibly slow dance before our eyes.

More than anything else, the Sacred Sphere represents unity. It is the source of the mystic's contemplative esotericism, the artisan's inspiration, the geometrician's metrics of the world, the mathematician's model of the universe, and the shaman's conception of truth. Each experiences life, identifies patterns, and interprets the data. But it's all the same data-energy events—seen from different perspectives. Free will colors our perspective. It can lengthen or shorten our journey through life. It can enlighten us and it can leave us in the dark. We can choose the narrow and difficult red road or the wide and easy black road. At the end of the road, however, we discover that we are all the same. And the journey begins again.

Why have humans applied a specific spherical geometry time and again to express relationships between the animate and inanimate; the mundane and spiritual; the past, present, and future; creation, existence, purification, and re-creation; birth, life, death, and rebirth? The answer, of course, is visible in the geometry of Earth, its spin, and its relationship with the physics of the cosmos. It is also associated with the Fall of Man and with past worlds, the present world, and expectations of the world to come. An understanding of astronomy requires prolonged observation, detailed recording of data, an understanding of mathematics and principles of linear and spherical geometry, and the ability to condense all of this information down to physical principles that explain the observations. Evidence of our understanding comes from the accumulation of additional data that confirms predictions made on the basis of the historical record. The basic geometrical relationship between Earth, moon, sun, and planets was understood long ago. Ancient and indigenous oral traditions and mythologies document human understanding of cosmic harmony over thousands of years.

This understanding of cosmology is reflected in the dimensions and forms of megalithic structures such as the pyramids at Giza, Stonehenge, and the works of early civilizations across the Americas. It is also evident in the elegant structure of the Sacred Hoop. Just as the fundamental shape of space is not dependent on size, so the symbols that represent vital sacred concepts are independent of size, shape, and material. They can be created anywhere at any time, so long as the creator finds justification though relationships—positive energies—between environmental, spiritual, and cognitive conditions. In its two-dimensional representation as a circle, the Sacred Sphere continues to have far-reaching implications in Western sciences, including astronomy, archeology, and anthropology, and in Native science as well. It provides insight into the evolution of religious thought and our place in the universe.

During your personal journey of discovery—whether you find yourself in the walled city of Uruk or the realm of the *edin*—take time to look about. Pay attention. Listen. The Sacred Sphere is all around us.

The metaphor is the mask of God through which all eternity is to be experienced.

JOSEPH CAMPBELL
The Power of Myth

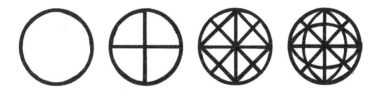

Figure 21.1: Progression of the circle to the Sacred Hoop, to the Arahuaco geometric form of Father, Mother, emergence of humans, and our responsibility to maintain equilibrium on Earth, and ending with the Sacred Sphere.

Notes and References

Chapter 1

1. De Loof A. 2008 Nov 17. Definition of Life: At last "What is Life?" can be answered, simply and logically. *SciTopics.* < http://www.scitopics.com/Definition_of_Life_At_last_What_is_Life_can_be_anwered_simply_and_logically.html> Accessed 2011 Mar 2.

2. Schejter A, Agassi J. 1994. On the Definition of Life. Journal for General Philosophy of Science 25: 97–106.

3. De Loof.

4. Ibid.

5. Ibid.

6. Ibid.

7. Ibid.

8. Firth RW. 1973. Symbols: Public and Private. Ithaca (NY): Cornell University Press. p. 49.

9. McKechnie Jean L. ed. 1977. Webster's New Twentieth Century Dictionary. Unabridged. 2nd ed. New York: Collins World.

10. Firth, p. 15.

11. Campbell J, with B Moyers. 1988. The Power of Myth. Betty Sue Flowers, ed. New York: MJF Books. p. 268.

12. Firth, p. 49.

13. Edmondson AC. 1986. A Fuller Explanation: The Synergetic Geometry of R. Buckminster Fuller. Boston: Birkhäuser. <http://www.angelfire.com/mt/marksomers/40.html>

14. Kobychev VV, Popov SB. 2005. Constraints on the photon charge from observations of extragalactic sources. Astronomy Letters 31: 147–151. DOI 10.1134/1.1883345.

15. Raman CV, Bhagavantam S. 1931. Experimental proof of the Spin of the Photon. Indian Journal of Physics 6: 353. <http://dspace.rri.res.in/bitstream/2289/2123/1/1931%20IJP%20V6%20p353.pdf>. Accessed 2010 Sept 4.

16. For example, see Section 1.3.3.2 in Burgess, C and G Moore. 2007. The Standard Model: A Primer. Cambridge (MA): Cambridge University Press.

17. Fuller RB. 1982. Synergetics. New York: Macmillan.

18. Edmondson, p. 6–9.

19. Ibid., p. 9–12.

20. Ibid.

21. Ibid.

22. Ibid., p. 189–193.

23. Ibid.

24. Ibid., p. 127–130.

25. Ibid.

26. Ibid.

27. Ibid., p. 193–197.

28. Ibid.

29. Ibid., p. 9–13.

30. Ibid., p. 200–205.

31. Ibid., p. 6–9.

32. Ibid., p. 9–12.

33. Ibid.

34. Ibid.

35. Wolf L. 1999. Practical Kabbalah: A Guide to Jewish Wisdom for Everyday Life. New York: Three Rivers Press. p. 32.

36. Wolf, p. 33.

37. Firth, p. 15.

38. Ibid., p. 60.

39. Ibid., p. 81.

40. Edmondson, p. 9–12.

41. Campbell provides numerous examples throughout in which geometry plays an important role in world mythology. Refer to examples on p. 38–39, 100, 107, 110–112, 268–274.

42. Campbell, p. 14.

43. Campbell, p. 44–45.

Chapter 2

1. Coomaraswamy A. 2004. Guardians of the Sundoor: Late Iconographic Essays. Louisville (KY): Fons Vitae. P. viii–ix.

2. Tellinger M, Heine J. 2009. Temples of the African Gods. Waterval Boven, South Africa: Zulu Planet. p. 31.

3. Ibid., p. 53.

4. Ibid., p. 31.

5. Ibid., p. 53.

6. Ibid., p. 6–16.

7. Wilford JN. 2010 Feb 15. On Crete, New Evidence of Very Ancient Mariners. New York Times. NY edition. p. D1.

8. Hancock G. 2002. Underworld: The Mysterious Origins of Civilizations. New York: Three Rivers. p. 322–323.

9. Hancock (2002), p. 360.

10. Ibid., p. 360–361.

11. Ibid., p. 362.

12. Ibid.

13. Harrod JB. 2010. OriginsNet: Researching the Origins of Art, Religion, & Mind. <www.originsnet.org>. Accessed 2011 Jan 4.

14. Harrod JB. 2010. Hominid Cognitive, Artistic and Spiritual Diasporas Out-of-Africa. <http://www.originsnet.org/abouteras.html>. Accessed 2011 Jan 4.

15. Ibid.

16. Ibid.

17. Ibid.

18. Harrod JB. Oldowan Era. <http://www.originsnet.org/eraold.html>. Accessed 2011 Jul 1.

19. Harrod JB. Early Paleolithic Art, Religion, Symbols, Mind (circa 1.4 million to 100,000 years ago). http://www.originsnet.org/mindep.html. 2010.

20. Harrod James B. Upper Paleolithic Art, Religion, Symbols, Mind (circa 90,000 to 10,000 years ago). http://www.originsnet.org/mindup.html. 2010.

21. Ibid.

22. Ibid.

Chapter 3

1. Cajete G. 2000. Native Science: Natural Laws of Interdependence. Santa Fe (NM): Clear Light; p. 14.

2. Ibid., p. 2.

3. Ibid.

4. Ibid., p. 98.

5. Ibid., p. 2–3, 13.

6. Hough, Louis, ed. 1971 [1909]. The Spanish Regime in Missouri: A Collection of Papers and Documents Relating to Upper Louisiana Principally within the Present Limits of Missouri during the Dominion of Spain, from the Archives of the Indies at Seville, etc. 2 vols. Chicago: R.R. Donnelly and Sons. Reprinted in 1 vol., New York: Arno Press.

7. Mails TE, Chief Eagle D. 1990. Fools Crow. Garden City (NY): Bison Books.

8. Hyde GE. 1975. Red Cloud's Folk: A History of the Oglala Sioux Indians. Norman (OK): University of Oklahoma Press; p. 3–7.

9. Hennepin L. 1903. A New Discovery of a Vast Country in America. American Journeys. Madison: Wisconsin Historical Society. <http://www.americanjourneys.org/aj-124a/index.asp>. Accessed 2011 Apr 15.

10. See chapter 1 in Hyde for detailed discussion of the Teton crossing of the Missouri River at this location.

11. Marshall III, JM. 2006. The Day the World Ended at Little Bighorn. New York: Viking Penguin; p. 35.

12. Marshall, p. 39–40.

13. Doerner J. 2007. Road to the Little Bighorn. Garry Owen (MT): Friends of the Little Bighorn Battlefield & Bob Reece.

14. Mails, Chief Eagle.

15. Mallory G. 1893. Picture-writing of the American Indians. 10ᵗʰ Annual Report of the Bureau of [American] Ethnology [for] 1882–'83. Washington: Smithsonian Institution, U.S. Government Printing Office; p. 3–882.

16. Utley RM. 1993. The Lance and the Shield: The Life and Times of Sitting Bull. New York: Ballantine Books.

17. Black Elk N, Neihardt JG. 1979. Black Elk Speaks: Being the Life Story of a Holy Man of the Oglala Sioux. Lincoln: University of Nebraska Press.

18. Ibid.; Brown, 1970; Mails & Chief Eagle, 1990; Steinmetz, 1990; Walker, 1991; Walker, 1992; Walker 2006.

19. Walker JR. 2006. Lakota Myth. Elaine Jahner, ed. Lincoln: University of Nebraska Press; p. 58–89.

20. Ibid., p. 251–369.

21. Brumley JH. 1988. Medicine Wheels on the Northern Plains: A Summary and Appraisal. Archeological Survey of Alberta, Manuscript Series No. 12. Alberta Culture and Multiculturalism, Historical Resources Division, Edmonton, Alberta.

22. Cajete, p. 73.

23. Brumley, p. 98.

24. Brumley, p. 67.

25. Burley PD. 2006. Initial Site Investigation: Cloud Peak Medicine Wheel, Cloud Peak Wilderness Area, Big Horn County, Wyoming. Report on file at Wyoming State Archeologist's office, Laramie, WY.

26. Kuehn DD. 1988. The Swenson Site, 32DU627: A Medicine Wheel in eastern Dunn County, North Dakota. Reprint. Journal of the North Dakota Archeological Association 3. Grand Forks (ND): University of North Dakota.

27. Ibid.

28. Abbott J, Ranney W, Whitten R. 1982. Report of the 1982 East River Petroform Survey. Report on File, South Dakota Archeological Research Center, Fort Meade (SD).

29. Rood RJ, Overholser Rood, V. 1983. Report on the Class I and Class II Cultural Resource Investigations of a Portion of the Cendak Water Project Area, eastern South Dakota. Volume 2. Report on File, South Dakota Archeological Research Center, Pierre (SD).

30. Bray KM. 2006. Crazy Horse: A Lakota Life. Norman (OK): University of Oklahoma Press; p. 407.

31. Hassrick RB. 1964. The Sioux: Life and Customs of a Warrior Society. Norman (OK): University of Oklahoma Press; p. 30, 61.

32. Hassrick, p. 53–54.

33. Cajete, p. 47.

34. Frank F. 1981. Art as a Way: A Return to the Spiritual Roots. New York: Crossroad.

35. Powers WK. 1982. Oglala Religion. Lincoln: Bison Books; p. xv.

36. Ibid., p. 206.

37. Cajete, p. 306.

Chapter 4

A. Coomaraswamy A. 1997. The Door in the Sky. Princeton (NJ): Princeton University Press.

1. Neihardt JG. 1985. The Sixth Grandfather: Black Elk's Teachings Given to John G. Neihardt. Lincoln: University of Nebraska Press.

2. Walker JR. 1991. Lakota Belief and Ritual. Raymond J. DeMallie and Elaine A. Jahner, eds. Lincoln: University of Nebraska Press.

3. Utley, 1993.

4. Brown JE. 1989 [1970]. The Sacred Pipe: Black Elk's Account of the Seven Rites of the Oglala Sioux. Norman (OK):University of Oklahoma Press.

5. Mails & Chief Eagle, 1990.

6. Steinmetz PB. 1990. Pipe, Bible, and Peyote Among the Oglala Lakota: A Study in Religious Identity. Knoxville (TN): University of Tennessee Press.

7. Brown JE, p. xvii.

8. Ibid. The symbolism of the sacred pipe is described throughout Black Elk's account of the gift of the pipe to the Lakota.

9. Ibid., p. xvii.

10. Ibid., p. 5–6.

11. Tyon in Walker, 1991, p. 120.

12. Brown, 1989, p. 5.

13. Ibid., p. 5-6.

14. Ibid., p. 7.

15. Ibid.

16. Ibid., p. 7. Refer to footnote 11 for a drawing of the stone with seven circles.

17. Hassrick, p. 259.

18. Brown, 1989, p. 32.

19. Ibid., p. 39, footnote 1.

20. Ibid., p. 49.

21. Ibid., p. 59.

22. Ibid., p. 67.

23. Ibid., p. 72.

24. Ibid., p. 92.

25. Ibid., p. 108.

26. Ibid., p. 115.

27. Ibid., p. 123.

28. Ibid., p. 135.

29. Neihardt, p. 8.

30. Brown Vinson. 2004. Crazy Horse, Hoka Hey! The Story of Crazy Horse: Legendary Mystic and Warrior. Happy Camp (CA): Naturegraph; p. 65.

31. Ibid., p. 60.

32. Ibid., p. 107.

33. Kaltreider K. 2004. Amercian Indian Prophecies, Conversations with Chasing Deer. Carlsbad (CA): Hay House; p. 86–87.

34. Brown, 2004, p. 112–113.

35. DeMallie, RJ. in Neihardt, p. 80.

36. Seven Rabbits in Walker, JR. 1982. Lakota Society. Lincoln: University of Nebraska Press. First Bison Books printing: 1992; p. 107.

37. See Walker, 1992; and Brown, 1989.

38. Lame Deer, J (Fire) and Erdoes, R. 1994. Lame Deer, Seeker of Visions. New York: Simon and Schuster; p. 110–111.

39. Brown, 1989, p. 69.

40. Walker, 2006, p. 324.

41. Brown JE. 2007. The Spiritual Legacy of the American Indian; With Letters While Living with Black Elk. Bloomington (IN): World Wisdom; p. 26.

42. Brown, 1989, p. 33.

43. Neihardt, 1985, p. 55.

44. McKechnie Jean L. ed. Webster's New Twentieth Century Dictionary.

45. Buechel E, Manhart P. 2002. Lakota Dictionary. Lincoln: University of Nebraska Press.

46. Powers, 1984, p. 65.

47. Lame Deer and Erdoes, p. 111.

48. Fools Crow in Mails & Chief Eagle, 1990.

49. Walker, 2006, p. 369.

50. Brown, 1989, p. 33.

51. Ibid., p. 9.

52. Ibid., p. xvii.

53. Neihardt, 1985, p. 55.

54. Brown, 1989, p. 138.

55. Brumley, 1988.

56. Wyoming State Historic Preservation Office. 2011. Medicine Wheel National Historic Landmark, Big Horn County (WY). <http://wyoshpo.state.wy.us/NationalRegister/Site.aspx?ID=60>. Accessed 2011 Apr 18.

57. Hunter C, and Fries A. 1986. The Bighorn Medicine Wheel, A Personal Encounter. Meeteetse (WY): Big Horn Medicine Wheel Research.

58. Wilson M. 1981. In the Lap of the Gods: Archeology in the Big Horn Mountains, Wyoming. Archeology in Montana 17(1–2): 33–34. See also Wilson, M. 1981. Sun Dances, Thirst Dances, and Medicine Wheels: A Search for Alternative Hypotheses in Megaliths to Medicine Wheels: Boulder Structures in Archeology. Wilson, Road, and Hardy, eds. Proceeding of the Eleventh Annual Chacmool Conference. University of Alberta, Calgary, Alberta.

59. Eddy JA. 1974 Jun 7. Astronomical Alignment of the Big Horn Medicine Wheel. Science 184(4141): 1035–1043.

60. Stanford SOLAR Center. 2006 Bighorn Medicine Wheel. <http://solar-center.stanford.edu/AO/bighorn.html>. Accessed 2006 May 6.

61. Royal Alberta Museum. 2005. What is a Medicine Wheel? <http://www.royalalbertamuseum.ca/human/archaeo/faq/medwhls.htm>. Accessed 2010 Nov 10.

62. Brumley, 1988. p. 80.

63. Burley, 2006.

64. Royal Alberta Museum, 2005.

65. Ibid.

66. Brown, 1989, p. 36.

67. Ibid., p. 89.

68. Walker, 2006, p. 87.

69. Burley, 2007a.

70. Neihardt, 1985, p. 33.

71. Black Elk and Neihardt, 1979, p. 108.

72. Cajete, p. 13.

Chapter 5

1. Wyse E, ed. 1999 [1989]. Mesopotamia: Towards Civilization. In Past Worlds, Atlas of Archaeology. Ann Arbor (MI): Borders Press in association with HarperCollins.

2. Campbell J. 1991 [1964]. Gods and Heroes of the Levant in The Masks of God: Occidental Mythology. Arkana, New York: Viking Penguin. p. 103.

3. Lambert WG. and A. R. Millard 1969. Atra-hasis: The Babylonian Story of the Flood. Oxford: Clarendon Press. P. 42–105.

4. Ibid.

5. George, AR., trans. 1999. The Epic of Gilgamesh, The Babylonian Epic Poem and Other Texts in Akkadian and Sumerian. London: Penguin Books.

6. Sanders, NK, trans. 1972. The Epic of Gilgamesh. London: Penguin Books. p. 61–69.

7. Graves, R, and Patai R. 1983[1963]. Hebrew Myths: The Book of Genesis. New York: Doubleday; p. 78–79.

8. Rosenberg D. 1994. World Mythology: Anthology of the Great Myths and Epics. New York: Mcgraw-Hill; p. 194.

9. Speiser EA 1958. The Epic of Gilgamesh. In The Ancient Near East, An Anthology of Texts and Pictures. Pritchard, JB, ed. Princeton (NJ): Princeton University Press; p. 44.

10. Rosenberg. p. 177.

11. Speiser. p. 44.

12. Sanders. p. 30.

13. Ibid., p. 194.

14. Ibid., p. 196

15. Ibid.

16. Ibid.

17. Ibid.

18. Ibid., p. 197.

19. Ibid.

20. Ibid., p. 199.

21. Ibid.

22. Ibid., p. 200.

23. Ibid., p. 201–202.

24. Ibid., p. 202.

25. Ibid., p. 203.

26. Ibid., p. 199.

27. Enki and Ninhursag: How Enki surrendered to the Earth Mother and Queen. <http://www.gatewaystobabylon.com/myths/texts/retellings/enkininhur.htm>. Accessed 2010 Dec 12.

28. Ibid.

29. Ibid.

30. Kramer SN. 1985. BM 23631: Bread for Enlil, Sex for Inanna. Orientalia 54: 117–132.

31. Dalley S. 1991. Myths from Mesopotamia: Creation, The Flood, Gilgamesh and Others. New York: Oxford University Press; p. 126–127, note 20 to the Epic of Gilgamesh.

32. Leeming D. 2004. Jealous Gods and Chosen People, the Mythology of the Middle East: A new perspective on the ancient myths of modern-day Iraq, Turkey, Egypt, Syria, Lebanon, Israel, Palestine, Jordan, Yemen, the Gulf States, and Saudi Arabia. New York: Oxford University Press. p. 46–47.

33. Leick G. 1991. Inanna and Utu. A Dictionary of Ancient Near Eastern Mythology. London: Routledge; p. 91.

34. MDidea. Pine Nuts or Pinus edulis, Nutritional Supplement Source and Good Nut for Art of Love! <www.mdidea.com/products/proper/proper02306.html>. Accessed 2011 Jul 1.

35. Kennedy M. 2010 Jan 1. Relic Reveals Noah's Ark was Circular. London: Guardian New and Media Limited. <http://www.guardian.co.uk/uk/2010/jan/01/noahs-ark-was-circular>. Accessed 2011 Jul 1.

36. Ibid.

37. Ibid.

38. Ibid.

39. Ibid.

40. Beckman P. 1993. A History of Pi. New York: Fall River; p. 22.

41. Van Duzer C. 2006. The Mythic Geography of the Northern Polar Regions: Inventio fortunata and Buddhist Cosmology. Culturas Populares. Revista Electrónica 2 (Mayo-Agosto 2006). <http://www.culturaspopulares.org/textos2/articulos/duzer.pdf>. Accessed 2010 Jun 23.

42. Ibid., p. 3.

43. Ibid., p. 4.

44. Ibid., p. 8.

45. Ibid., p. 9.

46. Ibid.

47. Ibid., p. 10.

Chapter 6

1. Hinduism. 2011. In Encyclopaedia Britannica. Retrieved from <http://www.britannica.com/EBchecked/topic/266312/Hinduism> Accessed May 26, 2011.

2. Deussen P. 1966. The Philosophy of the Upanishads. New York: Dover. p. vii.

3. Ibid., p. 40.

4. Kant, quoted in Deussen, p. 41.

5. Ibid., p. 46.

6. Ibid., p. 237.

7. Ibid., p. 51.

8. Ibid., p. 52.

9. Ibid., p. 107–108.

10. Ibid., p. 114.

11. Ibid., p. 216.

12. Ibid., p. 227

13. Ibid., p. 228–229.

14. Ibid., p. 231–232.

15. Ibid., p. 398–399.

16. Ibid., p. 355-356.

17. Ibid., p. 315.

18. Ibid., p. 402.

19. Ibid., p. 216.

20. Baker IA. Mandala. In Glossary of Terms, Shrestha, Romio. 2009. Celestial Gallery. New York: Fall River Press.

21. Ibid., Compassion.

22. Ibid., Wisdom.

23. Ibid., in Foreward by Deepak Chopra.

24. Nirukta 10.7.

25. Deussen, p. 107.

26. Kramrisch S. 1994. The Presence of Siva. Princeton, New Jersey: Princeton University Press. p. 18.

27. Kumaraguruparar (ca 1628—1688). Chidambra Mummani Kovai. Kumaraguruparar was a poet-saint from Tamil Nadu; he founded monastery at Varanasi and was a strong proponent of Saiva Siddhanta philosophy.

Chapter 7

a. Gimbutas M. 1999. The Living Goddess. Miriam Robbins Dexter, Ed. Berkeley: University of California Press. p. 1.

1. Rosenberg, P. 260–263.

2. Ibid., p. 257.

3. Ibid., p. 263.

4. Mackenzie DA. 1928. Buddhism in Pre-Christian England. London: Blackie.

5. Ellis PB. 1999. The Ancient World of the Celts. New York: Barnes & Noble Publishing. p. 170.

6. Duir A. 2005. Who is Cernunnos? Association of Polytheist Traditions. <http://www.manygods.org.uk/articles/essays/Cernunnos.shtml>. Accessed July 19, 2010.

7. Ibid.

8. Ibid.

9. Ibid.

10. Ibid.

11. Ibid.

12. Ibid.

13. Ibid.

14. Ibid.

15. Murray M. 1931. The God of the Witches. London: Sampson, Low, Marston & Co; p. 17.

16. Duir. 2005.

17. Ibid.

18. Ibid.

19. Jones M. Dagda. Jones Celtic Encyclopedia. <http://www.maryjones.us/jce/dagda.html>. Accessed July 16, 2010.

20. Ibid.

21. Ibid.

22. Green M. 2001[1992]. Symbol & Image in Celtic Religious Art. London: Routledge. p. 88.

23. Van Duzer, p. 9.

24. West DC. 1996. Inventio fortunata and Polar Cartography 1360–1700. Presented at the conference De-Centering the Renaissance: Canada and Europe in Multi-Disciplinary Perspective 1350–1700, Victoria University in the University of Toronto, March 7–10. In Van Duzer, p. 9.

25. Van Duzer, p. 10.

26. Ibid., p. 9.

27. Ibid., p. 3. Quoted from Fridtjof Nansen. In Northern Mists, Vol. 2.

28. Ibid., p. 13–15.

29. Ibid., p. 6.

30. Firth, p. 15.

31. Van Duzer. p. 15–16.

32. Ibid, p. 16.

33. Prose Edda.

34. Rosenberg, p. 208–213.

35. Brodeur AG. trans. 1916, 1923. Snorri Sturluson's Prose Edda: Skáldskaparmál. Northvegr.

36. Ibid.

37. Ibid.

38. Ibid.

39. Ibid.

40. Rosenberg, p. 208.

41. Grimm J. Teutonic Mythology. 1883, 1888 [2004]. James Steven, ed. and trans. 4 vols. New York: Dover.

42. Lindow J. 2002. Norse Mythology: A Guide to the Gods, Heroes, Rituals, and Beliefs. New York: Oxford University Press.

43. Faulkes A, transl. and ed. 1987. Prose Edda (Snorri Sturluson). London: J.M. Dent & Sons.

44. De Santillana G, and von Dechend H. 1992. Hamlet's Mill. Boston: David R. Godine.

45. Ibid., p. 87.

46. Ibid., P. 361–362.

47. Ibid., p. 90.

48. Baird B. 2010. Parallelism in Sioux and Sami Spiritual Traditions. <http://www. utexas.edu/courses/sami/diehtu/siida/ religion/paralellism.htm>. Accessed August 3, 2010.

49. Ibid.

50. Ibid.

51. Ibid.

52. Ibid., Text in quotes from Valkeapää, Nils-Aslak. 1997. The Sun, My Father. Seattle: University of Washington Press.

Chapter 8

1. Knight C, and Lomas R. 2001. The Hiram Key: Pharaohs, Freemasons and the Discovery of the Secret Scrolls of Jesus. Beverly, Massachusetts: Fair Winds Press.

2. Ibid., p. 115.

3. Hancock G. 1995. Fingerprints of the Gods. New York: Three Rivers Press. p. 381.

4. Knight and Lomas, p. 102.

5. Ibid., p. 101.

6. Schmidt E. 1899. Solomon's Temple. In The Biblical World, Vol. 14. William Rainey Harper, ed. Chicago: The University of Chicago Press. p. 164.

7. Ibid.

8. Miskin M. 2010. Dig supports King Solomon's Temple. The Jerusalem Connection Report. <http://www.thejerusalemconnection.us/columns/2010/02/24/dig-supports-king-solomons-temple.html>. Accessed: Feb 24, 2010.

9. Ibid.

10. Milstein M. 2010. King Solomon's Wall Found—Proof of Bible Tale? National Geographic Daily News. <http://news.nationalgeographic.com/news/2010/02/100226-king-solomon-wall-jerusalem-bible/>. Accessed: Feb. 26, 2010.

11. Jewish Encyclopedia. 2011. Temple of Solomon. <http://www.jewishencyclopedia.com/view.jsp?artid=129&letter=T>. Accessed: March 4, 2011.

12. De Vaux R. 1961. Ancient Israel: Its Life and Institutions. John McHugh, trans. New York: McGraw-Hill.

13. Finkelstein I, and Silberman NA. 2001. The Bible Unearthed: Archaeology's

New Vision of Ancient Israel and the
Origin of Its Sacred Texts. New York:
Free Press.

14. Schmidt, 1899.

15. Clegg S. 1850. Lectures on Architecture:
Lecture IV—Pelasgic Remains in
Greece, Italy, Asia Minor. Architecture
of the Jews. In William Laxton, ed., The
Civil Engineer and Architect's Journal
13: 113.

16. Schmidt, 1899.

17. Ibid.

18. Schick C. 1896. Die Stiftshutte,
der Tempel in Jerusalem und der
Tempelplatz der Jetztzeit. Berlin.

19. Schmidt, 1899.

20. Ibid.

21. Knight and Lomas, p. 207.

22. Knight and Lomas, p. 217.

23. Budge E.A.W. 1969. The Gods of the
Egyptians. Volume 1: New York: Dover
Publications. Originally printed 1904.
p. 407-415.

24. Book of the Two Ways. Coffin text 1130.

25. Allen JP. 2000. Middle Egyptian: An
Introduction to the Language and
Culture of Hieroglyphs. Cambridge
University Press, Cambridge,
Massachusetts. p. 116.

26. De Santillana and von Dechend,
Appendix 39—Excursus on Gilgamesh.
p. 440–451.

27. Rosenberg, p. 194.

28. Ibid., p. 196.

29. De Santillana and von Dechend,
Appendix 32—Excursus on Gilgamesh.
p.412–417.

30. Ibid., p. 417.

31. Ibid., p. 140–141.

32. Budge E.A. Wallis. 1969. The Gods
of the Egyptians. New York: Dover,
Publications; p. 503.

33. De Santillana and von Dechend,
p. 128–129.

34. Ibid., p. 222–223.

35. Ibid., p. 135

36. Hancock G, and Bauval R. 1996. The
Message of the Sphinx. New York: Three
Rivers Press.

37. Jenkins JM. 1998. Maya Cosmogenesis
2012. Rochester, Vermont: Bear &
Company.

38. De Santillana and von Dechend, p. 135.

39. Ibid.

40. Ibid., p. 390.

41. Berlinski D. The Secrets of the Vaulted
Sky. New York: Harcourt, Inc. p. 43.

42. Ibid.

43. Ibid., p. 175.

44. Kepler Johannes. 1999. Mysterium
Cosmographicum-the Secret of the
Universe. Norwalk, Connecticut: Abaris
Books.

45. Berlinski, p. 189–190.

46. Ibid., p. 190.

47. Ibid.

48. Budge EAW, p. 92-94.

49. Hancock, 1995. p. 264.

Chapter 9

1. Hancock, 1995. p. 432.

2. Ibid., p. 433–434.

3. Thompkins P. 1995. Secrets of the
Pyramid. New York: Harper & Row.
p. 437.

4. Hancock, 1995, p. 440–441.

5. Bauval R, and Gilbert A. 1995. The
Orion Mystery. New York: Three Rivers
Press. p. 77.

6. Ibid., p. 122.

7. Ibid., p. 135.

8. Ibid., p. 195.

9. Leiden Museum Papyrus No. 344,
p. 6-12. In Bauval and Gilbert, p. 215.

10. Ibid.

11. Ibid., p. 122.

12. Brown Vincent. 2002. Pyramid of
Man—The House of Going Forth by
Day. http://www.pyramidofman.com.
Accessed: October 19, 2010.

13. Ibid. Introduction.

14. Maspero Sir G. Quoted in Brown, 2002.
Introduction.

15. Schwaller de Lubicz RA. 1998. Temple
of Man. Rochester, Vermont: Inner
Traditions.

16. Robins G. 1994. Proportion and Style in
Ancient Egyptian Art. Austin, Texas:
University of Texas Press.

17. Ibid., p. 3.

18. Ibid., p. 37.

19. Schwaller de Lubicz. 1998.

20. Robins, p. 70.

21. Ibid.

22. Brown V. 2002. Introduction.

23. Dilke OAW. 1992. Mathematics and Measurement. Berkeley, California: University of California Press.

24. Levy J. 2005. The Great Pyramid of Giza: Measuring Length, Area, Volume, and Angles. New York: Rosen.

25. Edwards IES. 1986. The Pyramids of Egypt. New York & London: Viking Penguin. p. 285.

26. Cole JH. 1925. Determination of the Exact Size and Orientation of the Great Pyramid of Giza. Cairo: Government Press. Survey of Egypt, Paper No. 39.

27. Petrie Sir WMF. 1883. The Pyramids and Temples of Gizeh. London: Field & Tuer.

28. Ibid., p. 39.

29. Ibid.

30. Verner M. 2003. The Pyramids: Their Archaeology and History. London: Atlantic Books. p. 70.

31. Brown V. 2002. Introduction.

32. Ibid.

33. Pinch G. 2004. Handbook of Egyptian Mythology. Oxford & New York: Oxford University Press. p. 114–115.

34. Ibid.

35. Applegate ML. 2001. The Egyptian Book of Life: Symbolism of Ancient Egyptian Temple and Tomb Art. HCI. p. 173.

36. Brown V. 2002. Introduction.

37. Ibid.

38. Ibid.

39. Allen JP. 1993. Reading a Pyramid. Hommages Ã Jean Leclant. BdE 106/1. In Vincent Brown, Introduction.

40. Lehner M. 1997. The Complete Pyramid. London: Thames & Hudson Ltd. p. 9. In Brown, 2002. The Concept of the Pyramid.

41. Brown V. 2002. Introduction.

42. Brown V. 2002. Osiris-Djed.

43. Ibid.

44. Brown V. 2002. The Measures.

45. Petrie (1883) in Brown V. 2002. The Measures.

46. Robins, p. 259.

47. Brown V. 2002. The Measures.

Chapter 10

a. Jenkins JM. 2009. The 2012 Story. New York: Tarcher Penguin.

b. Men H. 2010. The 8 Calendars of the Maya. Rochester, Vermont: Bear & Company.

1. Gilbert, AG., and Cotterell, MM. 1996. The Mayan Prophecies. New York: Harper Element. p. 1–2.

2. Ibid., p. 5.

3. Ibid.

4. Gilbert A. 2008. 2012: Mayan Year of Destiny. Virginia Beach, Virginia: ARE Press. p. 262–268.

5. Ibid., p. 266.

6. Landa D de. Gates WE, trans. 1937. Yucatán Before and After the Conquest.

7. Gilbert and Cotterell, p. 111.

8. Ibid., p. 113.

9. Ibid., p. 118.

10. Rogers K. 2008 October 2. Crossing Over. <http://www.kathleenrogers.co.uk/work/>. Accessed: July 1, 2011.

11. VMG. 1998. Calendar origin. <http://geocities.ws/alma_mia/calendar/loose.html>. Accessed: July 1, 2011.

12. Gilbert and Cotterell, p. 119–120.

13. Ibid., p. 120–121.

14. Rosenberg, p. 202.

15. Gilbert and Cotterell, p. 123.

16. Men H. 2010. The 8 Calendars of the Maya. Rochester, Vermont: Bear & Company. p. 50.

17. Ibid.

18. Jenkins, 1998. p. XXXVII.

19. Ibid., p. XXXVIII.

20. Ibid., p. XXXVIIII.

21. Ibid.

22. Ibid., p. XLI.

23. Jenkins, 1998; and Jenkins JM. 2009. The 2012 Story. New York: Tarcher Penguin.

24. Men, p.84.

25. Ibid., p. 1.

26. Ibid., p. 5.

27. Men H. 2011. Solar Meditation. http://www.gaiamind.org/maya.html. Accessed: March 6, 2011.

28. Gilbert and Cotterell, 1995. Appendix 4. *Also*, Geryl P. 2005. The World Cataclysm in 2012. Kempton, Illinois: Adventures Unlimited Press.

29. Men H. 2010. p. 24.

30. Ibid., p. 25.

31. Paredes DM. 2010. Hunab K'u, Síntesis del pensamiento filosófico Maya. Editoria Cusamil S.A. Mexico City. 1973. *In* Hunbatez Men, 2010. p. 25–26.

32. Men H. 2010. p. 28.

33. Figures 10.12 and 10.13 are based on Men, H. 2010, figures 2.2 and 2.3.

34. Ibid., p. 68.

35. Ibid., p. 38.

36. Ibid.

37. Ibid., p. 40.

38. Ibid., p. 75–76.

39. Villela KD, Robb MH, and Miller ME. 2010. Introduction. In The Aztec Calendar Stone. Villela KD, and Miller ME, eds. p. 1. Los Angeles: The Getty Research Institute.

40. Refer to various writings in Villela KD, and Miller ME, eds. 2010. The Aztec Calendar Stone. Los Angeles: The Getty Research Institute.

41. Villela KD. and Miller ME. eds. 2010. The Aztec Calendar Stone. Los Angeles: The Getty Research Institute. p. 2.

42. Beyer H [1921]. The So-Called "Calendario Azteca": Description and Interpretation of the Cuauhxicalli of the "House of the Eagles. In Villela KD, and Miller ME, eds. 2010. The Aztec Calendar Stone. Los Angeles: The Getty Research Institute. p. 118–150.

43. Ibid., P. 144–147.

44. Navarrete C, and Heyden D. 1974. The Central Face of the Stone of the Sun: A Hypothesis. *In* Villela KD and Miller ME, eds. 2010. The Aztec Calendar Stone. Los Angeles: The Getty Research Institute. p. 185–194.

45. Khristaan D, Robb MH, and Miller ME. 2010. p. 1.

46. Gilbert A. 2008. p. xii.

47. Santayana G. 1905. Reason in Common Sense. In The Life of Reason, Vol. 1. New York: Charles Scribner's Sons.

48. Gilbert A. 2008. p. 95.

49. Ibid., p. 98.

50. Däniken E von. 1999. Chariots of the Gods. New York: The Berkley Publishing Group.

51. Gilbert A. 2008. p. 139.

52. Ibid., p. 121.

53. Ibid., p. 119–121.

Chapter 11

a. Bowman T. 1970. Hebrew Thought Compared with Greek. New York. W.W. Norton.

1. Dedopulos T. 2005. Kabbalah: An Illustrated Introduction to the Esoteric Heart of Jewish Mysticism. New York: Gramercy Books.

2. Quote by R. Buckminster Fuller. In Edmondson, Amy C. 2007.

3. http:/disctionary.babylon.com/ Merkabah/. Accessed June 17, 2011.

4. Theosophy. Vol. 10, No. 7, May, 1922. p. 218.

5. Wolf, p. 23.

6. Ibid., p. 34.

7. Online Etymology Dictionary. 2010. Douglas Harper. http://www. etymonline.com/. Viewed on: April 7, 2011.

8. Easton, MG. 1897. Illustrated Bible Dictionary, Third Edition. London, Edinburgh and New York: Thomas Nelson. http://www.ccel.org/e/easton/ ebd/ebd3.html. 9. Scripturetext.com.

9. Huller S. 2009. The Real Messiah: The Throne of St. Mark and the True Origins of Christianity. London: Watkins Publishing. Stephan Huller provides a detailed description and purpose of the throne based his twenty years of research.

10. Dedopulos. 2005. p. 10–11.

11. King James Bible. Revelation 1:10–16; 4:6–8.

12. Ibid., Revelation 22:1–2; 2:7.

13. Wolf, p. 31.

15. Ibid., p. 30.

16. Ibid., p. 31.

17. Sephir Yetzirah. *In* Dedopulos. 2005. p. 14.

18. Dedopulos. 2005. p. 15–16.

19. Weidner, J, and Bridges, V. 2003. The Mysteries of the Great Cross of Hendaye. Rochester, Vermont: Destiny Books. p. 88.

20. Wolf, p. 5–7.

21. Ibid., p. 7.

22. Ibid., p. 8.

23. Ibid., p. 13.

24. Refer to Wolf, p. 218.

25. Leet L. 1997. The Secret Doctrine of the Kabbalah: Recovering the Key to Hebraic Sacred Science. Rochester, Vermont: Inner Traditions. p. 2.

26. Leet, p. 8.

27. Leet. p. 8–9.

28. Ibid., p. 17.

29. Ibid., p. 20.

30. Ibid., p. 219.

31. Ibid., p. 232–233.

32. Ibid.

33. Ibid., p. 243–244.

34. Ibid., p. 325.

35. Santillana, G de, and von Dechend H. p. 340.

36. Leet. p. 93.

37. Ibid., p. 202.

38. Ibid.

39. Ibid., p. 92.

40. Ibid., p. 204.

41. Zohar, 1:6 (1:1b-2a).

42. Leet, p. 249.

43. Zohar, 1:20–21 (1:5a).

44. Leet, p. 258.

45. Ibid., p. 285–286.

46. This is a major theme presented throughout Leet (1997), and is the product of the two components her theory of the Hebraic priesthood's hidden knowledge as described earlier in this chapter.

47. Leet devotes chapter 7 to geometrical and philosophical construction of this concept.

48. Zohar, 1:65 (1:15b).

49. Leet, p. 9.

50. Ibid., p. 10.

51. Weidner J, and Bridges V. p. 88.

52. Ibid., p. 91.

53. Kaplan A, trans. 1990. Sefer Yetzirah: The Book of Creation. York Beach, Maine: Samuel Weisner. 6:1.

54. Refer to descriptions of the three Leviathan dragons in Weidner J, and Bridges V. p. 94–96.

Chapter 12

1. Regardie I. 2007. The Tree of Life: An illustrated Study in Magic. Woodbury, Minnesota: Llewellyn Publications. Glossary, p. 463.

2. Ibid.

3. Dedopulos. p. 83.

4. Ibid.

5. Ibid., p. 82.

6. Caubet. 1893. Souvenirs (1860-1889). Paris, cerf. p. 4. The statement by Leví was recalled by a friend of his, by the name of Caubet, who attended the reception speech.

7. Ibid.

8. Pinon. Annuire Maconnique de tous les rites 1861-1842. p. 121.

9. Symonds J. 1997. The Beast 666: The Life of Aleister Crowley. London: Pindar Press.

10. Dedopulos. p. 26.

11. Crowley, A. 1987. The Book of the Law. York Beach Maine: Samuel Weiser, Inc.

12. Regardie. 2007. Introduction to the Third Edition by Chic Cicero, p. xxix. The reference notes that the original quote is found in Regardie's book *My Rosicrucian Adventure* written in 1936.

13. Ibid., p. 1.

14. Ibid. Introduction, p. 14-19.

15. Ibid., p. 4.

16 Ibid., p. 7.

17. Ibd., p. 43.

18. Ibid., p. 41-43.

19. Ibid., p. 60.

20. Ibid., p. 72.

21. Ibid., p. 91, 23n.

22. Ibid., p. 95.

23. Ibid., p. 264. Regardie references a quote in Waite, AE. 1906. Studies in

Mysticism and Certain Aspects of the Secret Tradition. London: Hodder & Stoughton.

24. Regardie, p. 120–123.

25. Ibid., p. 157–159.

26. Ibid., p. 140.

27. Ibid., p. 157.

28. Regardie. p. 308. The Goetia is one part of the *Lemegeton* or *Complete Lesser Key of Solomon*, author unknown.

29. Dedopulos, p. 26.

30. Ibid.

31. Ibid., p. 28.

32. Dumas, A. 2006. The Three Musketeers. New York: Viking Adult.

33. Dedopulos, p. 73.

34. Bloom, H. 1997. Omens of the Millennium: The Gnosis of Angels, Dreams, and Resurrection. New York: Riverhead Books.

35. Karr, D. 2010. Approaching the Kabbalah of Maat: Altered Trees and the Procession of the Æons. http://www. digital-brilliance.com/kab/karr/maat/AKM.pdf. Viewed on: June 12, 2010.

36. Ibid., p. 2.

37. Ibid., p. 5.

38. Ibid.

Chapter 13

1. Brown. 1989, p. 36.

2. Ibid., p. 89.

3. Walker. 2006, p. 87.

4. Atiyah, M, and Sutcliffe, P. 2003. Polyhedra in Physics, Chemistry and Geometry. Milan J. Math Vol. 71: 33–58.

5. Quote from Plato's *Timaeus*.

6. Leet. p. 249. Quote from the *Zohar* 1:6(1:1b-2a).

7. Ibid., p. 258.

8. Ibid,. p. 257. Quote from Kaplan A. 1979. The Bahir. New York: Samuel Weiser. p. 34, v. 95.

9. Ibid., p. 258.

10. Ibid. Quote from Zohar, 1:13 (1:3b Mantua text).

11. Ibid., p. 258–259.

12. Ibid., p. 268.

13. Ibid.

14. Ibid.

15. Ibid., p. 270.

16. Ibid., p. 249. Quote from the Zohar 1:6 (1:1b-2a).

Chapter 14

1. Neihardt. 1985, p.

2. Refer to Burley (2007b) for further description of this process.

Chapter 15

1. Edmondson, Amy C. 1986. A Fuller Explanation: The Synergetic Geometry of R. Buckminster Fuller. Boston: Birkhäuser. Online edition: http://www. angelfire.com/mt/marksomers/40.html

2. Ibid. Preface.

3. Ibid.

4. Thurston, William P. and Jeffrey R. Weeks. 1984. The Mathematics of Three-dimensional Manifolds. Scientific American. Vol. 251, p. 108–120.

5. Edmondson. Chapter 1. p. 1–3.

6. Ibid.

7. Ibid. Chapter 1, p. 6–9.

8. Ibid.

9. Ibid. Chapter 2, p. 15–18.

10. Ibid. Chapter 2, p. 18–21.

11. Ibid. Chapter 10, p. 157–158.

12. Ibid. Chapter 3, p. 25–28.

13. Ibid. Chapter 3, p. 28–30.

14. Ibid.

15. Ibid.

16. Ibid. Chapter 3, p. 30–33.

17. Ibid.

18. Ibid.

19. Ibid.

20. Ibid. Chapter 1, p. 6–9.

21. Ibid. Chapter 3, p. 33–35.

22. Ibid. Chapter 4, p. 40–43.

23. Ibid. Chapter 4, p. 51–53.

24. Ibid. Chapter 5, p. 56–60.

25. Ibid.

26. Ibid.

27. Ibid.

28. Ibid.

29. Ibid.

30. Ibid.

31. Ibid. Chapter 6, p. 65–67.

32. Ibid.

33. Ibid. Chapter 14, p. 206–209.
34. The process outlined here more fully described in Edmondson, Chapter 14.
35. Edmondson. Chapter 4, p. 51–53.
36. Ibid. Chapter 14, p. 209–213.
37. Ibid. Chapter 7, p. 82–85.
38. Ibid. Chapter 14, p. 209–213.
39. Ibid. Preface.
40. Ibid. Chapter 3, p. 28–30.
41. Ibid. Chapter 4, p. 51–53.
42. Ibid. Chapter 5, p. 56–60.
43. Ibid. Chapter 6, p. 65–67.

Chapter 16

1. Cajete, p. 47.
2. Frank. 1981. *In* Cajete, p. 46.
3. The fact that ancient societies were capable and successful in constructing numerous buildings, walls, and other features using megaliths is attested in numerous archeological records and sacred sites around the world. These figures provided *in* Knight, Christopher, and Alan Butler. 2010. Civilization One. London. Watkins Publishing. p. 13.
4. Thom, A. 1967. Megalithic Sites in Britain. London: Clarendon Press.
5. Knight and Butler. 2010. p. 2 and p. 23.
6. Franklin, H. February 2004. Megalithic Yard Unearthed. <http://hew_frank. tripod.com/index.htm>. Viewed April 12, 2011.
7. Data provided in: Knight and Butler. 2010. Chapter 1 and Chapter 2.
8. Franklin. 2004.
9. Brown. 1989. p. 5.
10. Brown, Vinson (2004), p. 107; *also* Kaltreider (2004), p. 86–87.
11. Neihardt. 1985. p. 33.
12. Brown. 1989. p. 108.

Chapter 17

a. Susskind L. 2008. The Black Hole War. New York: Back Bay Books. p. 55.
1. Brown. 1989. p. 89.
2. Eyer, Shawn. 2011. The Vesica Piscis & Freemasonry. http://academialodge.org/article_vesica_piscis.php

3. Flower of Life Research. 2011. http://www.floweroflife.org/faqp.htm "What makes geometry 'sacred?'" Viewed on: April 22, 2011.
4. Calter, Paul. 2011. Polygons, Tilings, & Sacred Geometry. <http://www.dartmouth.edu/~matc/math5.geometry/unit5/unit5.html>. Viewed on April 22, 2011.
5. http://en.wikipedia.org/wiki/Sacred mental states not at all unpleasant and oftentimes of startling unreality_geometry#cite_note-0. Viewed on April 22, 2011.
6. Melchizedek, D 1999. The Ancient Secret of the Flower of Life Volume 1. Light Technology Publishing, Clear Light Trust.
7. These ideas are presented in detail in Leet (1997). A summary of these ideas is provided in Chapter 11 herein.
8. Weidner, J, and Bridges, V. p. 242–247.
9. Ibid., p. 244–245.
10. Ibid., p. 245. Weidner and Bridges attribute this quote to Keith Critchlow, who they describe as a geometric philosopher and student of R. Buckminster Fuller.
11. Ibid.
12. Ibid., p. 245–246.
13. Ibid., p. 88.
14. Ibid., p. 91.
15. Ibid., p. 339–340.
16. Burley. 2007a.
17. Burley. 2007b.

Chapter 18

1. Hawking and Mlodinow. 2010. p. 5.
2. Ibid.
3. Weeks, J. The Poincaré Dodecahedral Space and the Mystery of the Missing Fluctuations. Notices of the AMS. Vol. 51. No. 6. June/July 2004.
4. Ibid.
5. Ibid.
6. Ibid.
7. Ibid.
8. Personal communication. November 2, 2010.
9. Weeks. 2004.
10. Ibid.

11. Ibid.

12. Ibid.

13. Ibid.

14. Ibid.

15. Ibid.

16. Ibid.

17. Ibid.

18. O'Shea, D. 2007. The Poincaré Conjecture. New York: Walker & Company. p. 122.

19. Ibid.

20. Ibid., p. 134.

21. Weeks, 2004.

22. Ibid.

23. Mota, Bruno, Marcelo J. Reboucas, and Reza Tavakol. Observable Circles-In-The Sky In Flat Universes. WSPC Proceedings. July 21, 2010. arXiv:1007.3466v1 [astro-ph.CO] . http://arxiv.org/abs/1007.3466v1.

24. Poincaré Dodecahedral Space Model Gains Support To Explain The Shape Of Space. 2008 Feb 11th. <http://www.science20.com/news_releases/poincare_dodecahedral_space_model_gains_support_to_explain_the_shape_of_space>. Accessed April 14, 2010.

25. Roukema, B. F. 2009. Does Gravity Prefer the Poincare Dodecahedral Space? International Journal of Modern Physics. D18:2237-2241. Retrieved from http://arxiv.org/abs/0905.2543.

26. Ibid.

27. Personal communication. September 3, 2010.

Chapter 19

1. For detailed discussion see: Knight and Butler. 2010.

2. Hawking and Mlodinow. 2010. p. 5.

3. Cajete. 2000. p. 48.

4. Ibid., p. 49.

5. Garcia, F. 1990. The Ceremony of Art. Unpublished Manuscript. Santa Fe, New Mexico. Quoted in: Cajete, p. 49.

6. Cajete. 2000. p. 49.

7. Ibid., p. 49-50.

8. Marshall III, JM. 2001. The Lakota Way: stories and lessons for living. New York: Penguin Compass. p. 225.

9. Ibid., p. 225-226.

10. Ibid., p. 225.

Chapter 20

1. Rosenberg, p. 178.

2. Graves and Patai, p. 78–79.

3. Bible quotation from New International Version, 2010.

4. Ibid.

5. Arendzen, J. 1908. Cherubim. The Catholic Encyclopedia. Vol. 3. New York: Robert Appleton. <http://www.newadvent.org/cathen/03646c.htm>. Viewed 26 Nov. 2010.

6. Sandberg, A. The Priestesses of Innana. <http://www.rahoorkhuit.net/goddess/ancient_priestesses/priestesses_of_inanna.html>. http://hem.bredband.net/arenamontanus/Mage/inanna.html. Viewed Dec. 10, 2010.

7. Ibid.

8. Ibid.

9. McKenna, T. 1988. Hallucinogenic Mushrooms and Evolution. ReVISION 10(4) Spring: 51–57.

10. Allegro, JM. 1970. The Sacred Mushroom and the Cross. London: Hodder and Stoughton.

11. Knight, C, and Lomas, R. 1997. p. 114.

12. Eliade. M: Shamanism: Archaic Techniques of Ecstacy. Quoted in Knight and Lomas, 1997. p. 114–115.

13. Kinght and Lomas. 1997. p. 115.

14. Ibid.

15. McKenna. 1988.

16. Ibid.

17. Schultes, R. 1963. Hallucinogenic Plants of the New World. The Harvard Review. Vol. 1, p. 18–32.

18. Ibid.

19. Ibid.

20. McKenna. 1988.

21. Jenkins, John Major. 1998. p. 112.

22. Meng S , Slatyer, TR, and Finkbeiner DP. 2010. Giant Gamma-ray Bubbles from Fermi–LAT: Active Galactic Nucleus Activity or Bipolar Galactic Wind? The *Astrophysical Journal*. 724 1044 doi: *10.1088/0004-637X/724/2/1044*. Viewed April 14, 2011.

23. Fabbro, F. March 1996. Use of
    Hallucinogenic Substances in Ancient
    Religions. http://tribes.tribe.net/
    ambrosia-society/thread/60e3b36e-
    7dd9-4253-94e5-6b1eedcd2dae. Viewed
    December 10, 2010.

24. Ibid.

25. Schultes. 1963.

26. Ibid.

27. Ibid. The reader should refer to Schultes'
    listing for a more detailed itemization of
    the hallucinogenic effects of each plant.

28. Ibid.

29. Jenkins, p. 8.

30. Hancock, G. 2007. Supernatural. New
    York: The Disinformation Company,
    Inc. p. 71.

31. Ibid., p. 40.

32. Ibid., Refer to Hancock (2007) for
    further descriptions of the geometries he
    experienced.

33. Ibid.

34. Ibid., p. 107.

35. McKenna. 1988.

36. Fabbro. 1996.

37. Ibid.

38. Ibid.

Chapter 21

1. Leet. 1999. p. 9.

2. Ibid., p. 285.

Selected Bibliography

James Abbott, William Ranney and Richard Whitten, Report of the 1982 East River Petroform Survey, report on file at South Dakota Archeological Research Center, Fort Meade, South Dakota, 1982.

Allegro, John M. *The Sacred Mushroom and the Cross.* London: Hodder and Stoughton, 1970.

Allen, James P. *Reading a Pyramid.* Hommages Ã Jean Leclant. BdE 106/11993.

Allen, James P. *Middle Egyptian: An Introduction to the Language and Culture of Hieroglyphs.* Cambridge: Cambridge University Press, 2000.

Applegate, Melissa Littlefield. *The Egyptian Book of Life: Symbolism of Ancient Egyptian Temple and Tomb Art.* Deerfield Beach: Health Communications, Inc, 2001.

Arendzen, John. "Cherubim." In *The Catholic Encyclopedia.* Vol. 3. New York: Robert Appleton Company, 1908. Accessed August 2011. http://www.newadvent.org/cathen/03646c.htm.

Atiyah, Michael, and Sutcliffe, Paul. "Polyhedra in Physics, Chemistry and Geometry." *Milan Journal of Mathematics* 71, no. 1 (2003): 33-58. doi: 10.1007/s00032-003-0014-1.

Axil/Aion Tarot, The.ca. Bellingham: Axil Press, 1985.

Baird, Ben. 2010. "Parallelism in Sioux and Sami Spiritual Traditions." Accessed August 2011. http://www.utexas.edu/courses/sami/diehtu/siida/religion/paralellism.htm.

Bauval, Robert, and Adrian Gilbert. *The Orion Mystery: Unlocking the Secrets of the Pyramids.* New York: Three Rivers Press, 1995.

Baynton-Williams, Ashley and Miles. *New Worlds-Maps from the Age of Discovery.* Sydney: Murdoch Books, 2006.

Beckman, Petr. *A History of Pi.* New York: Marboro Books, 1990.

Berlinski, David. *The Secrets of the Vaulted Sky: Astrology and the Art of Prediction.* Orlando: Harcourt Books, 2003.

Beyer, Hermann. "The So-Called 'Calendario Azteca': Description and Interpretation of the Cuauhxicalli of the House of the Eagles," *Villela.* Khristaan D. and Mary Ellen Miller, ed. Los Angeles: The Getty Research Institute, 2011.

Neihardt, John G. *Black Elk Speaks: Being the Life Story of a Holy Man of the Oglala Sioux.* New York: State University of New York Press, 1972.

Bloom, Harold. *Omens of the Millenium: The Gnosis of Angels, Dreams, and Resurrection.* New York: Riverhead Books,1997.

Boman, Thorlief. *Hebrew Thought Compared with Greek*. New York: Norton, 1970.

Boon, George C. September 1982. "A Coin with the Head of the Cernunnos." *Seaby Coin and Medal Bulletin*. 769: 276–82.

Bray, Kingsley M. *Crazy Horse: A Lakota Life*. Norman: University of Oklahoma Press, 2006.

Sturluson, Snorri. *Edda: Skáldskaparmál. Northvegr*. Translated by Aurthur G. Brodeur. 1916, 1923.

Brown, Joseph Epes. *The Sacred Pipe: Black Elk's Account of the Seven Rites of the Oglala Sioux*. Norman: University of Oklahoma Press, 1989.

Brown, Joseph Epes. *The Spiritual Legacy of the American Indian: With Letters While Living with Black Elk*. Bloomington: World Wisdom, 2007.

Pyramid of Man. "The House of Going Forth by Day." Brown, Vincent. 2002. http://www.pyramidofman.com.

Brown, Vinson. *Crazy Horse, Hoka Hey! The Story of Crazy Horse: Legendary Mystic and Warrior*. Happy Camp: Naturegraph Publishers, 2004.

Brumley, John H. "Medicine Wheels on the Northern Plains: A Summary and Appraisal. Archeological Survey of Alberta." *Manuscript Series No. 12*. Edmonton: Alberta Culture and Multiculturalism, Historical Resources Division, 1988.

Buechel, Eugene and Paul Manhart. *Lakota Dictionary*. Lincoln: University of Nebraska Press, 2002.

Budge, E.A. Wallis. *The Gods of the Egyptians*. New York: Dover, 1969.

Burgess, C. and G. Moore. *The Standard Model: A Primer*. Cambridge: Cambridge University Press, 2007.

Burley, Paul D. 2006. "Initial Site Investigation: Cloud Peak Medicine Wheel, Cloud Peak Wilderness Area, Big Horn County, Wyoming." Report on file at Wyoming State Archeologist's office, Laramie, Wyoming.

Burley, Paul D. 2007a. "Medicine Wheels and Landmarks as Boundary Markers of the Teton World." Paper presented at the 65th Annual Plains Anthropological Conference, Rapid City, South Dakota, October 2007.

Burley, Paul D. 2007b. "The Round Stone: Gift from White Buffalo Calf Woman to the Teton." Paper presented at the 65th Annual Plains Anthropological Conference, Rapid City, South Dakota, October 2007.

Cajete, Gregory. *Native Science: Natural Laws of Interdependence*. Santa Fe: Clear Light Publishers, 2000.

Calter, Paul. "Polygons, Tilings, & Sacred Geometry." 2011. http://www.dartmouth.edu/~matc/math5.geometry/unit5/unit5.html.

Campbell, Joseph, Bill Moyers. *The Power of Myth*. Betty Sue Flowers, ed. New York: MJF Books, 1988.

Campbell, Joseph. "Gods and Heroes of the Levant" in *The Masks of God: Occidental Mythology*. New York: Viking Penguin, 1991.

Clegg, Samuel. "Lectures on Architecture: Lecture IV—Pelasgic Remains in Greece, Italy, Asia Minor. Architecture of the Jews." *The Civil Engineer and Architect's Journal*. William Laxton, ed. Volume 13, 1850.

Cole, J.H. "Determination of the Exact Size and Orientation of the Great Pyramid of Giza," *Survey of Egypt*, Paper No. 39. Cairo: Government Press, 1925.

Coomaraswamy, Ananda. *The Door in the Sky*. Princeton: Princeton University Press, 1997.

Coomaraswamy, Ananda. *Guardians of the Sundoor: Late Iconographic Essays*. Louisville: Fons Vitae, 2004.

Crowley, Aleister. *The Book of the Law*. York Beach: Weiser, 1987.

Dalley, Stephanie. *Myths from Mesopotamia: Creation, the Flood, Gilgamesh and Others*. Oxford: Oxford University Press, 1991.

Däniken, Erich von. *Chariots of the Gods*. New York: The Berkley Publishing Group, 1999.

Dedopulos, Tim. *Kabbalah: An Illustrated Introduction to the Esoteric Heart of Jewish Mysticism*. New York: Gramercy Books, 2005.

De Loof, Arnold. "Definition of Life: At last 'What is Life?' can be answered, simply and logically." *SciTopics*. November 17, 2008. http://www.scitopics.com/

Definition_of_Life_At_last_What_is_
Life_can_be_anwered_simply_and_
logically.html.

De Vaux, Roland. *Ancient Israel: Its Life and Institutions.* Translated by John McHugh. New York: McGraw-Hill, 1961.

Deussen, Paul. *The Philosophy of the Upanishads.* New York: Dover Publications, 1966.

Dilke, O.A.W. *Mathematics and Measurement.* Berkeley: University of California Press, 1992.

Doerner, John. *Road to the Little Bighorn.* Garry Owen: Friends of the Little Bighorn Battlefield & Bob Reece, 2007.

Douglas, Mary, ed. *Witchcraft Confessions and Accusations.* A.S.A. Mon. 9. London: Tavistock, 1970.

London: The Pagan Federation. Association of Polytheist Traditions. "Who is Cernunnos? Pagan Dawn #157," by Alexa Duir, 2005. Accessed October 15, 2010. http://www.manygods.org.uk/articles/essays/Cernunnos.shtml.

Dumas, Alexandre. *The Three Muskateers.* New York: Viking Adult, 2006.

Illustrated Bible Dictionary. 3rd ed. Easton, M.G, 1897. London: Thomas Nelson. http://www.ccel.org/e/easton/ebd/ebd3.html.

Eddy, John A. "Astronomical Alignment of the Big Horn Medicine Wheel." *Science*, vol. 184, no. 4141 (1974): 1035-1043.

Edmondson, Amy C. *A Fuller Explanation: The Synergetic Geometry of R. Buckminster Fuller.* Boston: Birkhäuser, 1986. Online edition: http://www.angelfire.com/mt/marksomers/40.html.

Encyclopedia Brittanica. "Vedas." Encyclopedia Britannica Premium Service, 2004.

"Enki and Ninhursag: How Enki surrendered to the Earth Mother and Queen." http://www.gatewaystobabylon.com/myths/texts/retellings/enkininhur.htm.

Edwards, I.E.S. *The Pyramids of Egypt.* New York: Viking Penguin, 1986.

Eyer, Shawn. "The Vesica Piscis & Freemasonry," 2011. http://academialodge.org/article_vesica_piscis.php.

Fabbro, Franco. "Use of Hallucinogenic Substances in Ancient Religions,"

March 1996. http://tribes.tribe.net/ambrosia-society/thread/60e3b36e-7dd9-4253-94e5-6b1eedcd2dae.

Finkelstein, Israel, and Neil Asher Silberman. *The Bible Unearthed: Archaeology's New Vision of Ancient Israel and the Origin of Its Sacred Texts.* New York: Free Press, 2001.

Firth, Raymond William. *Symbols: Public and Private.* Ithaca: Cornell University Press, 1973.

Flood, Gavin D. *An Introduction to Hinduism.* Cambridge: Cambridge University Press, 1996.

Flower of Life Research. "What makes geometry 'sacred?'" http://www.floweroflife.org/faqp.htm.

Karr, D. *The Kabbalah of Maat: Book On.* 2nd ed. 1985.

Frank, Fredrick. *Art as a Way: A Return to the Spiritual Roots.* New York: Crossroad, 1981.

Franklin, Hugh. "Megalithic Yard Unearthed," February 2004. Accessed April 12, 2011. http://hew_frank.tripod.com/index.htm.

Achad, Frater. *The Egyptian Revival: The Ever-coming Son in the Light of Tarot.* Chicago: The Collegium Ad Spiritum Sanctum, 1969.

Fuller, Richard Buckminster. *Synergetics.* New York: Macmillan, 1982.

Garcia, Frank. "The Ceremony of Art." Unpublished manuscript, 1990.

George, Andrew R., trans. *The Epic of Gilgamesh: The Babylonian Epic Poem and Other Texts in Akkadian and Sumerian.* London: Penguin, 1999.

Geryl, Patrick. *The World Cataclysm in 2012.* Kempton: Adventures Unlimited Press, 2005.

Gilbert, Adrian. *2012: Mayan Year of Destiny.* Virginia Beach: ARE Press, 2008.

Gilbert, Adrian G. and Maurice M. Cotterell. *The Mayan Prophecies.* New York: Harper Element, 1996.

Grant, Kenneth. *Hecate's Fountain.* London: Skoob Books, 1992.

Graves, Robert and Raphael Patai. *Hebrew Myths: The Book of Genesis.* New York: Doubleday, 1983.

Green, Miranda. *Symbol & Image in Celtic Religious Art*. London: Routledge, 2001.

Grimm, Jacob. *Teutonic Mythology*. James Steven, trans., ed. vol. 1–4. New York: Dover, 2004.

Hancock, Graham. *Fingerprints of the Gods*. New York: Three Rivers Press, 1995.

Hancock, Graham. *Underworld: The Mysterious Origins of Civilizations*. New York: Three Rivers Press, 2002.

Hancock, Graham. *Supernatural*. New York: The Disinformation Company, 2007.

Hancock, Graham and Robert Bauval. *The Message of the Sphinx*. New York: Three Rivers Press, 1996.

Hassrick, Royal B. *The Sioux: Life and Customs of a Warrior Society*. Norman: University of Oklahoma Press, 1964.

Hawking, Stephen, and Leonard Mlodinow. *The Grand Design*. New York: Bantam Books, 2010.

Hennepin, Louis. "A New Discovery of a Vast Country in America." Chicago,1903. http://www.americanjourneys.org/aj-124a/index.asp.

Hough, Louis, ed. *The Spanish Regime in Missouri: A Collection of Papers and Documents Relating to Upper Louisiana Principally within the Present Limits of Missouri during the Dominion of Spain, from the Archives of the Indies at Seville, etc.* vol. 1—2. Chicago: R.R. Donnelly and Sons, 1971. http://www.kathleenrogers.co.uk/work/. 2 October 2008. http://geocities.ws/alma_mia/calendar/loose.html. http://dictionary.babylon.com/merkabah/.

Huller, Stephen. *The Real Messiah: The Throne of St. Mark and the True Origins of Christianity*. London: Watkins Publishing, 2009.

Hunter, Carol and Allen Fries. *The Bighorn Medicine Wheel, a Personnel Encounter*. Meeteetse: Big Horn Medicine Wheel Research,1986.

Hyde, George E. *Red Cloud's Folk: A History of the Oglala Sioux Indians*. Norman: University of Oklahoma Press, 1975.

Jenkins, John Major. *Maya Cosmogenesis 2012*. Rochester: Bear & Company, 1998.

Jenkins, John Major. *The 2012 Story*. New York: Tarcher Penguin, 2009.

Jewish Encyclopedia. "Temple of Solomon," 2011. http://www.jewishencyclopedia.com/view.jsp?artid=129&letter=T.

Jones Celtic Encyclopedia. http://www.maryjones.us/jce/dagda.html.

Kaltreider, Kurt. *Amercian Indian Prophecies, Conversations with Chasing Deer*. Carlsbad: Hay House, 2004.

Kaplan, Aryeh, trans. *Sefer Yetzirah: The Book of Creation*. York Beach: Samuel Weisner, 1990.

Karr, Don. "Approaching the Kabbalah of Maat:Altered Trees and the Procession of the Æons," 2010. http://www.digital-brilliance.com/kab/karr/maat/AKM.pdf.

Kennedy, Maev. *Relic Reveals Noah's Ark was Circular*. London: Guardian New and Media Limited, 2010.

Guardian.co.uk. "Relic Reveals Noah's Ark was Circular," by Maev Kennedy, January 2010. Accessed January 2010. http://www.guardian.co.uk/uk/2010/jan/01/noahs-ark-was-circular.

Kepler, Johannes. *Mysterium Cosmographicum: The Secret of the Universe*. Norwalk: Abaris Books, 1999.

Knight, Christopher, and Alan Butler. *Civilization One*. London: Watkins Publishing, 2010.

Knight, Christopher, Robert Lomas. *The Hiram Key: Pharaohs, Freemasons and the Discovery of the Secret Scrolls of Jesus*. Beverly: Fair Winds Press, 2001.

Kobychev, V.V., S.B. Popov. 2005. "Constraints on the photon charge from observations of extragalactic sources." *Astronomy Letters*, 31: 147–151. doi:10.1134/1.1883345.

Kramer, Samuel Noah. 1985. BM 23631: Bread for Enlil, Sex for Inanna. Orientalia 54. Pp. 117-132.

Kramrisch, Stella. *The Presence of Siva*. Princeton: Princeton University Press, 1994.

Kuehn, David D. "The Swenson Site, 32DU627: A Medicine Wheel in eastern Dunn County, North Dakota." *Journal of the North Dakota Archeological Association*, Vol. 3. Grand Forks: University of North Dakota, 1988.

Lambert, W.G., and A. R. Millard. *Atrahasis: The Babylonian Story of the Flood*. Oxford: Clarendon Press, 1969.

Lame Deer, John (Fire), Richard Erdoes. *Lame Deer, Seeker of Visions*. New York: Simon and Schuster, 1994.

Landa, Diego de*Yucatán Before and After the Conquest*. Traslated by William E. Gates, 1937.

Leeming, David. *Jealous Gods and Chosen People: The Mythology of the Middle East*. New York: Oxford University Press, 2004.

Leet, Leonora. *The Secret Doctrine of the Kabbalah: Recovering the Key to Hebraic Sacred Science*. Rochester: Inner Traditions, 1997.

Lehner, Mark. *The Complete Pyramid*. London: Thames & Hudson, 1997.

Leick, Gwendolyn. *A Dictionary of Ancient Near Eastern Mythology*. London: Routledge, 1991.

Levi, Eliphas. *Transcendental Magic*. York Beach: Samuel Weiser, 1995.

Levy, Janey. *The Great Pyramid of Giza: Measuring Length, Area, Volume, and Angles*. New York: Rosen Publishing Group, 2005.

Lindow, John. *Norse Mythology: A Guide to the Gods, Heroes, Rituals, and Beliefs*. Oxford: Oxford University Press, 2002.

Mails, Thomas E. and Dallas Chief Eagle. *Fools Crow*. New York: Double Day, 1990.

Mallory, Garrick. "Picture-writing of the American Indians." *Tenth Annual Report of the Bureau of [American] Ethnology [for] 1882–'83*. Washington: Smithsonian Institution, U.S. Government Printing Office, 1893.

Marshall III, Joseph M. *The Lakota Way: Stories and Lessons for Living*. New York: Penguin Compass, 2001.

Marshall III, Joseph M. *The Day the World Ended at Little Bighorn*. New York: Viking Penguin, 2006.

McKenna, Terrence. "Hallucinogenic Mushrooms and Evolution," *ReVISION*, vol. 10, no. 4. 1988. Pp. 51–57.

McKechnie, Jean L., ed. *Webster's New Twentieth Century Dictionary of the English Language, Unabridged*. 2nd ed. New York: Collins World, 1977.

Melchizedek, Drunvalo. *The Ancient Secret of the Flower of Life* vol. 1. Flagstaff: Light Technology Publishing, 1999.

Men, Hunbatz. *The 8 Calendars of the Maya: The Pleiadian Cycle and the Key to Destiny*. Rochester: Bear & Company, 2010.

Men, Hunbatz. "Solar Meditation," 2011. http://www.gaiamind.org/maya.html.

Meng Su , Tracy R. Slatyer , Douglas P. Finkbeiner. "Giant Gamma-ray Bubbles from Fermi-LAT: Active Galactic Nucleus Activity or Bipolar Galactic Wind?" *The Astrophysical Journal*. 2010. Accessed April 14, 2011.724 1044 doi: 10.1088/0004-637X/724/2/1044.

Milstein, Mati. "King Solomon's Wall Found: Proof of Bible Tale?" *National Geographic Daily News*, 2010. Accessed February 2010. http://news.nationalgeographic.com/news/2010/02/100226-king-solomon-wall-jerusalem-bible/.

Miskin, Maayana. "Dig Supports King Solomon's Temple," *The Jerusalem Connection Report*, 2010. Accessed February 2010. http://www.thejerusalemconnection.us/columns/2010/02/24/dig-supports-king-solomons-temple.html. Viewed on: 24 February 2010.

Mota, Bruno, Marcelo J. Reboucas, Reza Tavakol. "Observable Circles-In-The Sky In Flat Universes," *WSPC Proceedings*, 2010. arXiv:1007.3466v1 [astro-ph.CO] . http://arxiv.org/abs/1007.3466v1.

Murray, Margaret. *The God of the Witches*. London: Sampson, Low, Marston, 1931.

Navarrete, Carlos and Doris Heyden. "The Central Face of the Stone of the Sun: A Hypothesis," in *Villela*, Khristaan D. and Mary Ellen Miller, ed. Los Angeles: The Getty Research Institute, 2010.

Neihardt, John G. *The Sixth Grandfather, Black Elk's Teachings Given to John G. Neihardt*. Lincoln: University of Nebraska Press, 1985.

Online Etymology Dictionary. 2010. Dougla s Harper. Accessed January 2010. http://www.etymonline.com/.

O'Shea, Donal. *The Poincaré Conjecture*. New York: Walker & Company, 2007.

OriginsNet. "Researching the Origins of Art, Religion, & Mind," by James B. Harrod, 2010. Accessed January 4, 2011.www.originsnet.org.

OriginsNet. "Hominid Cognitive, Artistic and Spiritual Diasporas Out-of-Africa," by James B. Harrod, 2010. Accessed January 4, 2011. http://www.originsnet.org/abouteras.html.

OriginsNet. "Upper Paleolithic Art, Religion, Symbols, Mind (circa 90,000 to 10,000 years ago)," by James B. Harrod, 2010. Accessed January 4, 2011. http://www.originsnet.org/mindup.html.

OriginsNet. "Oldowan Era," by James B. Harrod, 2010. Accessed January 4, 2011. http://www.originsnet.org/eraold.html.

OriginsNet. "Early Paleolithic Art, Religion, Symbols, Mind (circa 1.4 million to 100,000 years ago)," by James B. Harrod, 2010. Accessed January 4, 2011. http://www.originsnet.org/mindep.html.

Paredez, Domingo Martínez. *Hunab K'u, Síntesis del pensamiento filosófico Maya.* Mexico City: Editorial Orion, 1973.

Petrie, Sir William Matthew Flinders. *The Pyramids and Temples of Gizeh.* London: Field & Tuer, 1883.

Pinch, Geraldine. *Handbook of Egyptian Mythology.* New York: Oxford University Press, 2002.

News Staff. "Poincaré Dodecahedral Space Model Gains Support To Explain The Shape Of Space." *Science 2.0.* (February 2008). http://www.science20.com/news_releases/poincare_dodecahedral_space_model_gains_support_to_explain_the_shape_of_space.

Ordo Adeptorum Invisiblum. *Liber Magnus Conjunctions Workings, Sub-Figura MC.* Chicago: Stellium Press, 1982.

Ordo Adeptorum Invisiblum. *The Processes of Initiation.* Chicago: Stellium Press, 1982.

Powers, Thomas. *The Killing of Crazy Horse.* New York: Alfred A. Knopf, 2010.

Powers, William K. *Oglala Religion.* Lincoln: University of Nebraska Press, 1977.

Raman, C.V. and Bhagavantam, S. "Experimental Proof of the Spin of the Photon." *Indian Journal of Physics* 6 (1931): 353-366.

Regardie, Israel. *The Tree of Life: An Illustrated Study in Magic.* St. Paul: Llewellyn Publications, 2007.

Robins, Gay. *Proportion and Style in Ancient Egyptian Art.* Austin: University of Texas Press, 1994.

Rood, Ronald J. and Vicki OverholserRood, Report on the Class I and Class II Cultural Resource Investigations of a Portion of the Cendak Water Project Area, report on file at South Dakota Archeological Research Center, Fort Meade, South Dakota, 1983.

Rosenberg, Donna. *World Mythology: Anthology the Great Myths and Epics.* New York: Mcgraw Hill Companies, 1994.

Rosenroth, Knorr von. *Kabbala denudate.* 1684.

Roukema, B. F. "Does Gravity Prefer the Poincaré Dodecahedral Space?" *International Journal Modern Physics* (2009):2237-2241.

Royal Alberta Museum. "What is a Medicine Wheel?" Last modified December 5, 2005. http://www.royalalbertamuseum.ca/human/archaeo/faq/medwhls.htm.

Rundle Clark, R.T. *Myth and Symbol in Ancient Egypt.* London: Thames & Hudson, 1978.

Sandberg, Anders. "The Priestesses of Innana." Accessed January 2010. http://hem.bredband net/arenamontanus/Mage/inanna.html.

Sandars, N.K., trans. The *Epic of Gilgamesh.* London: Penguin Books, 1972.

Santillana, Giorgio de, and Hertha von Dechend. *Hamlet's Mill.* Boston: David R. Godine, 1992

Santayana, George. *The Life of Reason.* Amherst: Prometheus Books, 1998.

Schejter, A., Agassi, J. "On the Definition of Life." *Journal for General Philosophy of Science* 25 (1994): 97-106.

Schmidt, Emanuel. "Solomon's Temple," in *The Biblical World Volume 14.*William Rainey, ed. Harper. Chicago: The University of Chicago Press, 1899.

Schultes, Richard. "Hallucinogenic Plants of the New World." *The Harvard Review* (1963): 18-32.

Schwaller de Lubicz, René Adolphe. *Temple of Man.* Rochester: Inner Traditions, 1998.

Scripturetext.com. Accessed January 2011. http://multilingualbible.com/ezekiel/1-22.htm.

Shrestha, Romio. *Celestial Gallery.* New York: Fall River Press, 2009.

Speiser, E.A. "The Epic of Gilgamesh." In *The Ancient Near East, An Anthology of Texts and Pictures,* James B. Prtichard, ed. New Jersey: Princeton University Press, 1958.

Stanford SOLAR Center; "Bighorn Medicine Wheel" 2008. http://solar-center.stanford.edu/AO/bighorn.html.

Steinmetz, Paul B. *Pipe, Bible, and Peyote Among the Oglala Lakota: A Study in Religious Identity.* Knoxville: University of Tennessee Press, 1990.

Symonds, John. *The Beast 666: The Life of Aleister Crowley*. London: Pindar Press, 1997.

Tellinger, Michael, and Heine, Johan. *Temples of the African Gods*. Waterval Boven, South Africa: Zulu Planet Publishers, 2009.

Thom, A. *Megalithic Sites in Britain*. London: Clarendon Press, 1967.

Thompkins, Peter. *Secrets of the Pyramid*. New York: Harper & Row, 1995.

Thurston, William P., and Jeffrey R. Weeks. "The Mathematics of Three-dimensional Manifolds." *Scientific American* 251 (1984).

Utley, Robert M. *The Lance and the Shield: The Life and Times of Sitting Bull*. New York: Ballantine Books, 1993.

Valkeapää, Nils-Aslak. *The Sun, My Father*. Seattle: University of Washington Press, 1997.

Van Duzer, Chet. "The Mythic Geography of the Northern Polar Regions: Inventio fortunata and Buddhist Cosmology." *Culturas Populares. Revista Electrónica 2* (2006). http://www.culturaspopulares. org/textos2/articulos/duzer.pdf.

Verner, Miroslav. *The Pyramids: Their Archaeology and History*. London: Atlantic Books, 2003.

Villela, Khristaan D. and Mary Ellen Miller. *The Aztec Calendar Stone*. Los Angeles: The Getty Research Institute, 2010.

Waite, Arthur Edward. *Studies in Mysticism and Certain Aspects of the Secret Tradition*. London: Hodder & Stoughton, 1906.

Walker, James R. *Lakota Society*. United States: University of Nebraska Press, 1982.

Walker, James R. *Lakota Belief and Ritual*. United States: University of Nebraska Press, 1991.

Walker, James R. *Lakota Myth*. United States: University of Nebraska Press, 2006.

Weeks, Jeffrey. "The Poincaré Dodecahedral Space and the Mystery of the Missing Fluctuations." *Notices of the AMS* 51, no. 6 (2004).

Weidner, Jay and Vincent Bridges. *The Mysteries of the Great Cross of Hendaye*. Rochester: Destiny Books, 2003.

West, Delno C. "Inventio fortunate and Polar Cartography 1360-1700." Paper presented at the conference "De-Centereing the Renaissance: Canada and Europe in Multi-Disciplinary Perspective 1350-1700," Victoria University in the University of Toronto, March 7-10, 1996.

Wikipedia; Wikipedia's "Sacred Geometry" entry; Last modified August 15, 2011. http://en.wikipedia.org/wiki/ Sacred_geometry#cite_note-0.

Wilford, John Noble. "On Crete, New Evidence of Very Ancient Mariners." *New York Times* (February 15, 2010).

Wilson, Michael. "In the Lap of the Gods: Archeology in the Big Horn Mountains, Wyoming." *Archeology in Montana* 17 (1981): 33-34.

Wilson, Michael. "Sun Dances, Thirst Dances, and Medicine Wheels: A Search for Alternative Hypotheses," in Megaliths to Medicine Wheels: Boulder Structures in Archeology. Wilson, Road and Hardy, eds. Proceeding of the Eleventh Annual Chacmool Conference. University of Alberta, Calgary, Alberta.

Wolf, Rabbi Laibl. *Practical Kabbalah: A Guide to Jewish Wisdom for Everyday Life*. New York: Three Rivers Press, 1999.

MDidea Extracts Professional. "Pine Nuts or Pinus edulis, Nutritional Supplement Source and Good Nut for Art of Love!" Last modified October 21, 2010. www.mdidea.com/products/proper/ proper02306.html.

Wyoming State Historic Preservation Office. Medicine Wheel National Historic Landmark, Big Horn County, Wyoming. 2011. http://wyoshpo.state.wy.us/ NationalRegister/Site.aspx?ID=60.

Wyse, Elizabeth, ed. *Mesopotamia: Towards Civilization. In Past Worlds, Atlas of Archaeology*. Ann Arbor: Borders Press, 1999.

Index

About the Author

Paul D. Burley is an engineering consultant and environmental geologist. He has investigated natural landscapes and the human environment at thousands of sites across North America and studied the history of architecture, engineering, science, and technology, as well as Indigenous and ancient cultural traditions. Burley has presented professional papers addressing the engineering and environmental concerns associated with property management and development, with specific attention to archeological and ethnographic analyses of prehistoric America and support and protection for Native American cultural traditions. He is the author of The Business Owner's Guide to Environmental Site Assessment. He lives in Minnesota with his wife, Nancy.

Please visit:
www.PaulDBurley.com